Rethinking Warfare in the 21st Century

This book seeks to critically review and evaluate the changes and consistencies in how warfare is interpreted and represented by academics, mass media outlets and political actors in the twenty-first century. The authors suggest that it is essential to understand the evolution and transformation of contemporary warfare's conceptualisation and practice in order to make sense of the current global geopolitical transformations that are in process, from a unipolar to a multipolar global order. They therefore examine the various key actors in international relations from conceptual, theoretical and empirical perspectives through thematic chapters that demonstrate the increasingly central role played by intangible factors in the representation and management of contemporary armed conflict. The book stresses the need to reflect upon and rethink the potentially highly problematic trajectory of the global community within the framework of twenty-first-century warfare's political and informational influence and effects.

Iulian Chifu is the State Counsellor of the Romanian Prime Minister for Foreign Relations, Security and Strategic Affairs. He was the founder and President of the Conflict Prevention and Early Warning Centre, Bucharest and is a professor at the National Defence University, Bucharest.

Greg Simons is based at the Department of International Relations at the National Research University Higher School of Economics, a Lecturer at the Department of Communication Sciences at Turiba University and formerly a researcher at the Institute for Russian and Eurasian Studies at Uppsala University.

Rethinking Warfare in the 21st Century

The Influence and Effects of the Politics, Information and Communication Mix

---

Iulian Chifu
National Defence University, Bucharest

Greg Simons
National Research University Higher School of Economics

CAMBRIDGE
UNIVERSITY PRESS

CAMBRIDGE
UNIVERSITY PRESS

Shaftesbury Road, Cambridge CB2 8EA, United Kingdom

One Liberty Plaza, 20th Floor, New York, NY 10006, USA

477 Williamstown Road, Port Melbourne, VIC 3207, Australia

314–321, 3rd Floor, Plot 3, Splendor Forum, Jasola District Centre, New Delhi – 110025, India

103 Penang Road, #05–06/07, Visioncrest Commercial, Singapore 238467

Cambridge University Press is part of Cambridge University Press & Assessment, a department of the University of Cambridge.

We share the University's mission to contribute to society through the pursuit of education, learning and research at the highest international levels of excellence.

www.cambridge.org
Information on this title: www.cambridge.org/9781009355230

DOI: 10.1017/9781009355247

First published 2023

A catalogue record for this publication is available from the British Library.

A Cataloging-in-Publication data record for this book is available from the Library of Congress

ISBN 978-1-009-35523-0 Hardback

Iulian and I would like to express our deep gratitude and profound thanks to our wives and families for not only tolerating but supporting our intellectual 'obsessions' in researching and writing this book. It has taken valuable time away from them, but we remain optimistic and hopeful that we may all live in less 'interesting' times through the study and understanding of the nature of warfare in the 21st century.

Contents

1 Introduction: Motivation, Purpose and Structure of the Book

Iulian Chifu and Greg Simons

Reasoning and Motivation for the Book

The present authors have been collaborating in various research and writing projects since 2005. Most of these have involved various issues at the national (Romania), regional (Black Sea Region) or global level in the practitioner and academic fields of crisis management, politics, international relations, geopolitics and security studies. This collaboration has resulted in a book on crisis management in Romania in 2007 (Chifu & Ramberg, 2007) and followed by a book on the nature of warfare in the twenty-first century (Simons & Chifu, 2017). Different patterns and processes in international politics, international relations and geopolitics were noted in these publications and various research projects.

This has led the authors to conclude that the time is ripe to update the findings of the previous publications and consider the various observable global trends and patterns. There is an especially acute need for consideration of the evolving transformation of the global order and the possible effects of Coronavirus as the nature of risk and threats evolve and multiply in hard and soft forms. There are increasing calls for a reform to the orthodoxy of knowledge and practice in Western foreign and security policy, such as an open call for Joe Biden to reform and not simply restore US foreign policy, away from the cynical and (self-) destructive patterns of the twenty-first century (and even before).[1] This is even more essential when one considers the increased fracturing of our societies coupled with the decreasing level of civility in critical public debates concerning the direction of Western civilisation.

There is a clear and present need to rethink where we are and where we are going with the current state and trajectory of warfare as foreign policy. The authors hold, at times, diverging opinions and understandings of

[1] D. Davis, *Biden Must Not Restore Foreign Policy, but Reform It*, Washington Examiner, Biden must not restore foreign policy, but reform it (washingtonexaminer.com), 11 January 2021 (accessed 11 January 2021).

these processes that are described above. However, instead of considering this lack of consensus as a weakness and a problem, we embrace the pluralism as a strength. A lively and divergent debate on critical issues has the effect of sharpening one's critical thinking and intellectual senses. Given the timeliness of the current debates, together with the need to not present any form of group think, filter bubble or echo chamber, the qualitative assessments and evaluations of the research can vary between the chapters to present a wide range of considered and weighed views and opinions.

The title of the present book, *Rethinking Warfare in the 21st Century: The Influence and Effect of the Politics, Information and Communication Mix*, may seem to have a rather elusive meaning. But we have a clear reasoned logic and motive for this particular title. The contemporary approach to the operationalisation of warfare, is the subjective (geo)politicisation of armed conflicts through the projection of increasingly relativised and selective information through its interpretation and representation in descriptive and not analytical contexts (before, during and after) of warfare. These relationships and interactions are important to highlight, as information and communication support politics, and it is politics that drives the motivation for, and the strategy and goals of, contemporary warfare.

The real added value of the research is not to be found in the individual parts that make up the chapters of this book, even though these present compelling and illustrative snapshots of various practical and theoretical problems, but rather in the knowledge that is derived from a sum of these parts in the conclusion. The intention is to locate and highlight the red thread that unites these very diverse individual parts in order to highlight and to understand the common lessons, opportunities and threats that transcend the tactical and operational level debates and presented information. It is the intention of this book to engage in understanding the strategic picture of trends and processes of the politics and communication of conflict and warfare in the twenty-first century, rather than concentrating on specific individual tactical or operational-level events in international relations.

Questions for the Book

Each chapter in this book approaches the issue of presenting, analysing and drawing conclusions from the study of twenty-first-century warfare in rather different qualitative ways. Readers may ask, and rightly so, how is it possible to draw general conclusions from the content of this book? The topic is certainly a critical one, but it is also a diversely understood and operationalised subject. A way around this is to pose a set of theoretically and conceptually informed questions in the Introduction of the book, which cut across all divides. There are three such questions, which are to be answered in the Conclusion, which weigh and analyse the sum of the parts:

(1) Is warfare in the twenty-first century qualitatively different from earlier periods in human history?
(2) How do politics and information 'inform' and influence twenty-first-century warfare?
(3) What are the possible trends and transformations of twenty-first-century warfare from this point in time?

In answering these questions in the Conclusion, it is hoped to pull together the common knowledge and lessons that can be drawn from the diverse topics and subjects in the individual chapters. Creating this red thread will unite the key points that are hinted at in the title of this book – the influence and effect of the politics, information and communication mix. Following this thread, one will be able to critically assess, reconsider and rethink the nature of warfare across the globe in the twenty-first century, and to understand where the current trajectory is taking global civilisations and humankind.

The following sections concern various key elements and aspects that will appear across the different chapters of this book and should be taken into consideration when making sense of when, why and how conflicts and warfare begin.

Global Transformations and Changes: Away from a Western-Centric World Order?

The world of international relations and global orders are in a constant state of motion throughout human history; some hegemonies rising, others stagnating and declining, and others that maintain their power and influence. There are even observable variations within a particular branded period of geopolitical history, where a brand name signifies the implied relations between various actors and the balance of power. For example, the Cold War era that signified relations between the US-led West and the Soviet Union-led Eastern Bloc. In this seemingly homogeneous period, until the final collapse of the Soviet Union in late 1991 that led to the end of the Cold War, there were attempts to find and label variations, such as the discussions concerning the possibility of the existence of a New Cold War within the 'Old' Cold War (Kubalkova &

Cruickshank, 1986). The 'victory' of the US-led West over the Soviet-led Eastern Bloc created a sense of euphoria and self-righteous belief that has created a messianic aura of attempts to spread liberalism and US influence, without any immediate and effective checks or balances by other actors.

As such, it has created a relative complacency in the West whereby the likes of Kissinger (2015) and others have tried to warn against allowing this to become deeply embedded into policymaking and practitioner mindsets. Globalisation was accelerated in the wake of the Cold War, becoming a vehicle for both entrenching and expanding Western global influence and power. However, it is a theoretical construct that is contested intellectually and practically (Germain, 2013). One can also argue that globalisation has presented an opportunity for powers that challenge the hegemonic system to do so.

In 2018 John Mearsheimer published the book *The Great Delusion: Liberal Dreams and International Realities*, which caused a great deal of debate. He concluded that the liberal hegemony and foreign policy course of the United States was doomed to fail, where liberal hegemony was facilitated by the country's unipolarity. He argued that the emergence of China and Russia has ended unipolarity and ushered in an era of multipolarity, where realism has replaced liberalism as the objective of an ideologically based grand strategy (of spreading liberalism) with the focus on a balance of power politics.[2] This is manifested in an increasing level of competition and global instability as the strength and capacity of the current global hegemon is becoming visibly weakened in both a tangible and intangible manner.

Richard Haas has argued that stable world orders are a rare thing, that global orders have come and gone throughout history and that this is an inevitable fate. He also points out the basic fact that, even if one wished to, there is no ability to turn back the clock to an idealised period in the history of the order and freeze time and circumstances. A further point made is that new orders emerge from the ashes of an old one. There are notable signs of decay in the US-led global order, which Haas suggests requires a careful management of its deterioration rather than trying to resurrect what it was historically, and this requires a mixture of compromise, incentives and pushback.[3] However, the evolving global order

[2] R. Vivaldelli, *Interview With Mearsheimer*, https://lobelog.com/interview-with-mearsheimer/, 22 March 2019 (accessed 11 April 2019).

[3] R. Haas, *How a World Order Ends, and What Comes in Its Wake*, Foreign Affairs, www.foreignaffairs.com/articles/2018-12-11/how-world-order-ends, 11 December 2018 (accessed 18 December 2018).

involves much more than merely the influence and effect of politics and geopolitics on the system; intangibles such as culture and perception effectively shape the cognitive realm's interpretation and evaluation of the physical realm.

There is a return, if in fact it had ever departed, to popular geopolitics in influencing people's perception of other countries and parts of the world. This in turn influences their perception and opinion of actors and acts in international relations by creating 'idealised' perceptions through audience exposure to ideas and images (re)produced in popular culture (Szostek, 2017). One of the problems of contemporary Western academia is the absence of culture as a variable affecting international relations scholarship, or at least a sufficiently nuanced manner to understand its effects. Criticism of this basic fact has been noted.

> In today's world politics, culture is everywhere. The rise of non-Western great powers, the return of ethnonationalism, violent extremism justified in the name of religion, and so-called white resistance – the list goes on. Yet those who should be best placed to explain it – international relations scholars – are ill-equipped to do so.[4]

As noted by the author above, the main schools of thought retain an obsolete understanding of how civilisations function, which is perhaps a hangover from Fukuyama's triumphant and premature declaration of the 'end of history' in reference to the perceived decisive and all-encompassing victory of Western civilisation (defined as a community of nations) as a unified global culture. This is visualised through how a country views and defines itself as a nation or as a civilisation or something else. As there is a gradual move away from accepting the Western ideology of Liberalism and Globalisation by countries such as China and Russia that define themselves as civilisations (with unique cultural values and political institutions) rather than nations, there is a resulting transformation of geopolitics away from liberal universalism towards cultural exceptionalism.[5] This is already having a profound impact upon the physical realm of international relations and the informational representations of it.

The population growth and economic rise of some Asian powers, such as China and India, have already culminated in the predictions of a rise in

[4] Reus-Smit, C., *International Relations Theory Doesn't Understand Culture*, Foreign Policy, https://foreignpolicy.com/2019/03/21/international-relations-theory-doesnt-understand-culture/, 21 March 2019 (accessed 23 March 2019).

[5] Pabst, A., *China, Russia and the Return of the Civilisational State*, New Statesman, https://www.newstatesman.com/2019/05/china-russia-and-return-civilisational-state, 8 May 2019 (accessed 18 June 2019).

Asia (an Asian century) surpassing the US and Europe in terms of power and influence.[6] This fits with the overall strategic picture of the transformation in the global order away from a Western-centric construction. Diesen (2018) notes that China's continued rise through the geoeconomics of the 'One Belt One Road' strategy has the US on the defensive and seeking to push back via expanding the New Cold War to target China as a means of attempting to arrest the rise of China's global power and influence. It has also meant that Russia, as the earlier target of containment attempts in the New Cold War, has transformed the Sino-Russian partnership from a marriage of convenience to a strategic alliance. In doing so, this has broken the long-adhered-to US geopolitical Cold War rule of dividing China and Russia.

The result of these actions and reactions has been to increase the global level of tension and instability that involve various symbolic and actual demonstrations of power and resolve towards the opponent/competitor that has resulted in the narrative of a New Cold War. Some observers have referred to the situation as creating a new global tinderbox.[7] This situation has in turn created several alarmist political calls to action and mass media headlines. One such example is the US Secretary of State Mike Pompeo's claim that the US was developing a strategy for China based on the notion that for the first time in US history it was fighting 'with a really different civilisation'; the difference being that China is not a product of Western philosophy and history.[8] The clear message here is that China is profoundly different in terms of culture and approach and is a grave threat to the continued US global hegemony. A scenario of a Chinese military threat (which runs counter to the narrative of Western military technological superiority) to the US was broached in the run up to the 2020 US presidential elections. Different hypothetical scenarios were posed where US forces in the Pacific region were vulnerable to a sudden Chinese attack; a contest that the US could potentially lose.[9] These are attempts to prime and rally the public by

[6] V. Romei& J. Reed, *The Asian Century Is Set to Begin*, The Financial Times, www.ft.com/content/520cb6f6-2958-11e9-a5ab-ff8ef2b976c7, 26 March 2019 (accessed 28 March 2019).

[7] M. T. Klare, *The New Global Tinderbox*, Le Monde Diplomatique, https://mondediplo.com/openpage/tinderbox-cold-war, 30 October 2018 (accessed 31 October 2018).

[8] J. Gehrke, *State Department Preparing for Clash of Civilisations with China*, Washington Examiner, www.washingtonexaminer.com/policy/defense-national-security/state-department-preparing-for-clash-of-civilizations-with-china, 30 April 2019 (1 May 2019).

[9] J. Gehrke, *Top Republican Senator Fears China Could Defeat US in the West Pacific*, Washington Examiner, www.washingtonexaminer.com/policy/defense-national-

inducing fear, which is becoming more commonplace in the practice of international relations in an unstable era.

In 1998, in reaction to the Eastward expansion of NATO, the architect of the US containment of the Soviet Union, George Kennan, declared, 'I think it is the beginning of a new cold war. I think the Russians will gradually react quite adversely and it will affect their policies. I think it is a tragic mistake. There was no reason for this whatsoever. No one was threatening anybody else.'[10] George Kennan's analysis and foresight in 1998 seems to have been realised in the opening years of the twenty-first century. In the wake of the Colour Revolutions and the Arab Spring, Russia has been reassessing its position and role in the world and has taken a more assertive stance in articulating and defending its perceived national interests.[11] This has resulted in a mixture of cooperation and competition between Russia and other external powers seeking influence in post-Soviet space, where Russia pursues a strongly pragmatic rather than ideological foreign policy line (Wlodkowska-Bagan, 2012; Markedonov & Suchkov, 2020). The leading Western powers also demonstrate an evolving strategy in these times of global transformation.

It is evident that the US-led West seeks to retain its declining position as the global hegemon and the benefits that come with this lead position. However, the tactics employed to achieve this are likely, as noted by Waldman (2021), to have a detrimental strategic effect and consequences that potentially accelerate this erosion. Actions seem to be taken without regard to the longer-term consequences, even if they are supported by eloquent rhetoric. France's actions in the regime change of Gaddafi during the cover provided by the Arab Spring, and its subsequent actions to gain influence, serve as a reminder that there are consequences in the use of short-term tactics to achieve long-term strategic interests.[12] The US has been suffering the effects of ill-conceived strategies and hastily executed militarised foreign policy operations.

During the period of the Cold War, the US often resorted to the use of covert regime change operations. The reasons for this were: that it

security/top-republican-senator-fears-china-could-defeat-us-in-the-west-pacific, 29 July 2020 (accessed 30 July 2020).

[10] T. L. Friedman, *Foreign Affairs; Now a Word From X*, The New York Times, www.nytimes.com/1998/05/02/opinion/foreign-affairs-now-a-word-from-x.html, 2 May 1998 (accessed 24 March 2019).

[11] NAMEA Group, *Uncoiled Spring: Russia's New View of Its Role in the World*, NAMEA Geopolitical Update, www.namea-advisors.com/blog/namea-geopolitical-update, 30 October 2015 (accessed 9 November 2015).

[12] P. Taylor, *France's Double Game in Libya*, Politico, www.politico.eu/article/frances-double-game-in-libya-nato-un-khalifa-haftar/, 17 April 2019 (accessed 26 December 2019).

exposed fewer US service personnel lives to risk; and overt operations may well have triggered a Soviet counter-response that ran the risk of rapid escalation. However, with the end of the Cold War as a result of the Soviet collapse, several overt military regime change operations came into effect (1999 in Kosovo, 2003 in Iraq and 2011 in Libya). However, the outcomes are unpredictable, the lives of citizens in the target country are adversely affected and US interests and reputation can be compromised. Given the costs to political capital from overuse, a return to covert regime change has resumed; such as in Syria, Venezuela or Iran.[13] This has created a growing backlash against those that champion foreign regime change, which has resulted in the US fighting wars that cannot be won militarily ('Forever Wars') and come with political and financial losses.[14] The open assassination in January 2020 of Iranian General Qassim Soleimani in Iraq, and the subsequent touting of it, was a display of rapid and reckless foreign policy escalation. Such an act would have previously been undertaken by more covert means in order to maintain diplomatic appearances. By breaking this 'rule', the US exposes itself to the risk of the same tactics being used against it, as it has with the use of drones and cyber-attacks.[15] This act of war against another country, carried out under false pretences has not halted the continuing decline of US influence in the Middle East region or the feeling of unease and lack of support by key allies who are uncomfortable with this sort of crude and risky act.

The current global transformation is creating an environment of increasing instability and tensions. This is owing to the breaking up, or at least the weakening, of the old agreed upon sets of 'rules' and guidelines that used to regulate international relations in the era of US unipolarity, and the absence of new 'rules' or guidelines that regulate the new and emerging relations between powers of various sizes. It is a period where international actors are trying to increase or retain their power and influence on the international stage. There are also actors attempting to position or reposition themselves in the marketplace of actors (hegemon, defender, challenger, nicher and so forth) in the global order, as new powers begin to rise and old hegemons seem to be failing.

[13] S. Kinzer, *America's Legacy of Regime Change*, The Future of Freedom Foundation, America's Legacy of Regime Change – The Future of Freedom Foundation (fff.org), 10 June 2019 (accessed 18 June 2019).

[14] T. Parsi & S. Toossi, *Beware the Foreign Regime Change Charlatans*, The American Conservative, www.theamericanconservative.com/articles/beware-the-regime-change-cottage-industry, 24 January 2019 (accessed 29 January 2019).

[15] G. E. Fuller, *US Foreign Policy by Assassination*, Graham E. Fuller, US Foreign Policy by Assassination (grahamefuller.com), 4 January 2020 (accessed 7 January 2020); K. Gilsinan, *America's Self-Sabotage in the Middle* East, The Atlantic, U.S. Derails Own Middle East Goals – The Atlantic, 6 January 2020 (accessed 7 January 2020).

Not All Types of Warfare Are Equal

The nature and state of warfare in the twentieth century is a picture of a rapidly evolving and diverse way of engaging an opponent, from clear conflicts such as World Wars I and II, to those much less clear and far more ambiguous, such as the Vietnam War. There were varied elements of national interest, geography, ideology, religion, ethnicity, blood and other reasons underlying those conflicts. Sometimes these were wars of attrition, at other times they were wars of manoeuvre (Harkavy & Neuman, 2001; Maoz & Gat, 2001). The conflicts themselves were motivated and initiated sometimes by reasoned logic and often by emotional logic. Tactics and strategies used by the players were intended to try and utilise one's own strengths against the enemy's weaknesses, and to offset the vice-versa situation.

In 1991, the mostly peaceful collapse of the Eastern Bloc heralded the end of the Cold War and the relative stability of the bipolar global order. It was stable and relatively predictable owing to the balance of power, mutually assured destruction, and gradually established rules of the geopolitical game designed to prevent inadvertent conflict between the superpowers. This collapse of the Cold War predictability ushered in a reassessment of the future of conflict in the context of a unipolar global order, with the United States as the sole superpower remaining (Maoz & Gat, 2001). Therefore, there are less constraints and restraints on the US capability and willingness to engage covertly or overtly in forced regime change operations in the post-Cold War era (Walker, 2019). Wars and warfare moved from the often proxy and insurgency style of the Cold War to the façade of humanitarian intervention and 'preventative' style of the post-Cold War era.

Wars and warfare require political consensus and a certain level of public approval, or at least a lack of resistance, to be effectively imitated and waged; hence the political and social environment need to be carefully cultivated in order that war aims are seen as 'legitimate' at the elite and public level (Western, 2005). There are various and multiple political and military constraints on the enthusiasm and the ability to wage war, especially in nominally democratic countries, where different actors lobby behind the scenes at the national and international level either against, or in order to force, a course of military action (Dixon, 2019). In addition to political opportunity and expediency, there are other environmental elements that need to be considered to persuade and influence an audience into accepting, or at least not offering effective resistance against, a policy leading to war.

One of those elements is a perception of normality versus the abnormality of the represented and interpreted wider political and social human environment. As noted by Brecher (2018), the declaration or the actual eventuality of a crisis can potentially create sufficient instability, motive and opportunity for increasing the likelihood of a conflict. This is something that has been observed throughout human history, extraordinary times (whether true or fake), precipitated by the invocation of a crisis, prompt the political rhetorical call and mobilisation for action to nominally resolve the problematic situation; even if, in fact, such calls may exacerbate the crisis event in question, for example the proposal to send arms to one side in a proxy war in order to alleviate a humanitarian disaster. These are misleading interpretations communicated in the information realm that are intended to shape and influence the conclusions created in the cognitive realm of the target audience.

Another element is the role and influence of technology on the physical and cognitive realms. As technology develops increasingly sophisticated and more powerful means of communication to mass audiences, and at the same time weapons systems developed are much more destructive in their capability, Betz (2015) argues that contemporary conflicts are much less a contest of arms and are increasingly a contest of hearts and minds. This situation also allows non-state actors to increasingly challenge the authority and status quo imposed by tangibly powerful state actors. It is also noted by other observers who conclude 'the practice of military conflicts during the past decade demonstrates that the strategic advantage goes to the actor who first understands and implements new technologies, who can use them as a force multiplier and therefore overcome superior conventional forces – often without provoking a sustained response' (Danyk, Maliarchuk & Briggs, 2017: 23). Conflict and warfare are gradually moving in a direction that is much less transparent and accountable, and are more for the purposes of short-term political or geopolitical gain.

For example, Waldman (2019, 2021) has noted a gradual transformation of the manner of waging war by the United States in the post–World War II era. This has seen an evolution from conventional deployments and engagements to more covert and secretive military operations, which he labels as being vicarious warfare (please see Chapter seven of this volume for more details). This form of warfare tends to expose US combat troops less to the risks of combat operations, is financially less burdensome, accumulates short-term political capital and is done more covertly in terms of public accountability, and is politically expedient for potential rewards while trying to escape the consequences. All these aspects tend to be tactically advantageous to the political elite to fulfil their ambitions. However, it can be damaging in terms of the hidden costs and the potential for strategic harm that is caused to the national interest.

There are a number of themes that emerge from popular publications (non-academic) on the different aspects of contemporary warfare, which are not only about the questions on how and why wars are fought, but also about the manner in which they are waged. One of the perennial questions in the age of the Global War On Terrorism (GWOT) and the Forever Wars is why do Western political leaders (and especially the US) persist in creating and fighting armed conflicts that cannot be won?[16] Another pertinent set of issues pose the question, how many wars is the US currently really fighting?[17] This then raises questions about the hidden or secret wars such as Syria,[18] which in turn leads to the call for truth and accountability from political leaders.[19] In turn, when a war begins people generally expect to see an end at some stage, therefore a natural question that stems from this is what does victory look like in an Endless War and how is the victor defined?[20] As a consequence, there is a growing Western public aversion to overseas military interventions.[21] Various political narratives are communicated through mass media outlets in an attempt to support or give the false impression of answering the numerous questions concerning contemporary warfare and its practice, in which there are also evident contradictions.

Political language and embedded symbolism are important means of trying to persuade or influence an audience using rational logic (including false logics) to convey the desired meaning to resonate and pacify public concerns. When pressed to explain why the US still had troops in Afghanistan after almost twenty years in one of the Forever Wars, President Trump responded that the country was a 'lab for terrorists'

[16] W. J. Astore, *Why American Leaders Persist in Waging Losing Wars*, Le Monde Diplomatique, Why American leaders persist in waging losing wars, by William J. Astore (Le Monde diplomatique – English edition, October 2018), 26 October 2018 (accessed 28 October 2018).

[17] N. Turse, *How Many Wars Is the US Really Fighting?*, The Nation, How Many Wars Is the US Really Fighting? | The Nation, 24 September 2015 (accessed 30 September 2015).

[18] L. Sly, *America's Hidden War in Syria*, The Washington Post, www.washingtonpost.com/graphics/2018/world/syria/us-troops-in-syria/, 14 December 2018 (accessed 16 December 2018).

[19] Editorial, *Trump Owes Us the Truth About the Afghanistan War*, Washington Examiner, Trump owes us the truth about the Afghanistan War (washingtonexaminer.com), 15 May 2019 (accessed 17 May 2019).

[20] J. Gehrke, *Trump's Syria Envoy Spells Out 'What Victory Looks Like in This Awful War'*, Washington Examiner, Trump's Syria envoy spells out 'what victory looks like in this awful war' (washingtonexaminer.com), 22 May 2019 (accessed 25 May 2019).

[21] C. Sagir, *More Than Half of Brits Oppose Military Intervention Overseas, New Poll Reveals*, Morning Star, More than half of Brits oppose military intervention overseas, new poll reveals | Morning Star (morningstaronline.co.uk), 11 January 2019 (accessed 17 January 2019).

that he tried to brand as the 'Harvard of terrorists.'[22] Sometimes other actors are blamed by the political elite for the lack of progress of policy implementation. For example, the successful Taliban momentum on the battlefield in Afghanistan was blamed for the delay in withdrawing US troops.[23] There are also issues of trying to explain or justify contentious policy, such as: sending more troops to the Middle East while promising to end Forever Wars,[24] poorly thought through policy, such as using one Islamist group to fight another Islamist group,[25] or policy with high-risk potential to create strategic harm.[26] These examples illustrate the role and influence of short-term politics on contemporary warfare.

There is also the evident use of misleading or outright false claims in order to create an impression of legitimacy for undertaking questionable and risky steps towards warfare, such as the claims by the Trump administration of Iran having ties with the terrorist group Al Qaeda. This is in spite of the obvious conflict between Shia Iran and Sunni Al Qaeda and the various Saudi Arabian links. As noted by some observers, the parallel between this false claim and that of the Bush Administration concerning the non-existent links (at the time) between Iraq's Saddam Hussein and terrorism was one of the excuses for the disastrous Iraq War.[27] There were also examples of questionable narratives, where one narrative coincided with a potentially damaging one. Bearing in mind the stubborn, but poorly evidenced Russia Gate narrative that was intended to compromise the Trump administration, Secretary of State Mike Pompeo announced that the US and Russia shared common interests in the

[22] J. Rosas, *'Harvard of Terrorists': Trump Explains Why US Has Yet to Leave Afghanistan*, Washington Examiner, www.washingtonexaminer.com/news/harvard-of-terrorists-trump-explains-why-us-has-yet-to-leave-afghanistan, 1 July 2019 (accessed 5 July 2019).

[23] J. Gehrke, Senator: Taliban 'Momentum' Complicating US Peace Talks, Washington Examiner, www.washingtonexaminer.com/policy/defense-national-security/senator-taliban-momentum-complicating-u-s-peace-talks, 22 May 2019 (accessed 25 May 2019).

[24] Z. Halaschak, US Sending 1000 More Troops to Middle East, Washington Examiner, www.washingtonexaminer.com/news/us-sending-1-000-more-troops-to-middle-east, 17 June 2019 (accessed 18 June 2019).

[25] M. Rubin, *Using the Taliban to Fight the Islamic State Is Khalilzad's Worst Idea Yet*, Washington Examiner, www.washingtonexaminer.com/opinion/using-the-taliban-to-fight-the-islamic-state-is-zalmay-khalilzads-worst-idea-yet, 22 May 2020 (accessed 26 May 2020).

[26] T. Rogan, *To Respond to Russian Jets, Give the Libyan Government Patriot Missile Systems*, Washington Examiner, www.washingtonexaminer.com/opinion/responding-to-russian-jets-give-patriot-missile-systems-to-the-libyan-government, 26 May 2020 (accessed 28 May 2020).

[27] E. Wong & C. Edmondson, *Iran Has Ties to Al Qaeda, Trump Officials Tell Skeptical Congress*, The New York Times, www.nytimes.com/2019/06/19/us/politics/us-iran.html, 19 June 2019 (accessed 21 June 2019).

Syrian War and he was 'very excited' about talks with Putin.[28] Given that Pompeo's statement supported a narrative that was intended to harm Trump and his administration, shows how easily that one statement can be so self-damaging when the complete context of the information realm is not considered. There are, on occasion, public clashes between different stakeholders on the same side of a conflict. The most senior British military officer in the war against the Islamic State, Major General Christopher Ghika, contradicted the American 'assessments' of a heightened military threat from Iran to American forces in the Middle East.[29] Warfare is political by its nature, and in the contemporary age requires effective communication in order to start or continue an armed conflict.

Role of Information and Politics in Twenty-First Century Warfare

Content oriented information strategy has been in use by organised societies since the ancient times (Arquilla & Borer, 2007). Information is found in different forms and is constituted for different purposes. The basic fact that information and news is not always intended to be a public good that is objective and informative to the audience is not something that is new to the twenty-first century; it has a long history. Bernays (1947) noted that news is not coincidental in either its timing or content as it is used as an engineer of public consent. The development and acceleration of a systematic media-centred approach to campaigning and policymaking creates a logical demand and approach increasing the motivation for engaging in news management (Esser & Spanier, 2005). The use of the powerful emotion of fear is becoming more prevalent whereby political appeals target audiences in order to try to prime and mobilise them for political causes (Scheller, 2019). The situation runs the risk of constructing a black and white 'reality' that is not based on the essential element of truth that is needed as a basis for effectively understanding and responding to events in the physical realm. This has been one of the criticisms of the contemporary US approach to its foreign policy goals and agenda.

[28] J. Gehrke, *Pompeo 'Very Excited' About Syria Talks with Putin*, Washington Examiner, www.washingtonexaminer.com/policy/defense-national-security/pompeo-very-excited-about-syria-talks-with-putin, 14 May 2019 (accessed 17 May 2019).

[29] J. Fitsanakis, *British Commander in Iraq Contradicts US Assessment of Increased Iranian Threat*, Intel News, https://intelnews.org/2019/05/15/01-2548, 15 May 2019 (accessed 17 May 2019).

> The good guys and the bad guys. Friends and enemies. Black and white. Since at least World War II, American foreign policy has been viewed through an all-or-nothing lens in which actors are routinely considered either all good or all bad. It's not hard to understand why — it's a lot easier to categorize states or groups as a monolith. But doing so unnecessarily harms our interests.[30]

This national narrative and myth define and attempt to legitimise political acts of the country's elite through the engineering of a public consensus by simplistic use of persuasive mass communications. It is intended to order the perception of the categorisation of various social, political and other elements in the human environment and experience. The art of propaganda, influence and persuasion is most effective on the target audience in its simplest and suggestive forms, rather than in more complex and overt ways (Taylor, 2003). Arquilla and Borer (2007: 14) note that 'whereas forces on the battlefield can readily be commanded and controlled with ever increasing precision – thanks to the rise of advanced information technologies – influence, deception and persuasion campaigns will only be undermined by overcontrol.' In the current information age, the temptation to engage in information warfare is tempting, and it is possible given the increasing capabilities and capacities of new information communication technologies. There is, however, the human cognitive element too that is ignored at one's peril (in the ability to persuade and influence).

The increasing mistrust of mainstream media as a source of propaganda and powerful interest influence, and the migration to social media that has been trumpeted as a tool of liberation and democracy, is one of those popular myths. It creates the narrative of powerful social media in the hands of an energetic youth as a means for bringing about social and political change in regions, such as the Middle East and North Africa (MENA), where hope for change by other means is presented as being limited. A picture is painted of a clash between old media, supporting the political and social status quo, and new media struggling for change and opposed to the status quo. However, the experience of the Arab Spring has shown other patterns and more complicated relationships and interactions in the zones of conflict than these Western ideas that perpetuate the ideologically based concepts and theories of democratic change (Moreno-Almeida & Banaji, 2019). As Arquilla and Borer (2007) point out, like power, information should not be understood as an absolute concept, but rather as a complex relational one with many parts.

[30] D. Davis, *Biden Must Not Restore Foreign Policy, but Reform It*, Washington Examiner, Biden must not restore foreign policy, but reform it (washingtonexaminer.com), 11 January 2021 (accessed 11 January 2021).

In the contemporary global information environment, there are many different actors and interests that seek to compete for attention and influence with different audiences. Even though the global information sphere is radically different in terms of the dynamics of relationships and speed, some old habits remain within the mindset of governments and the military. In 2019, a US Army document was being developed that emphasized the need to dominate in the information realm. General Funk, the Commander of the US Army Training and Doctrine Command, characterised the US information dominance at the time as being too narrow, under-developed and lacking co-ordination.[31] However, this desire ignores the basic fact that the contemporary information environment is a crowded and competitive marketplace of actors and agendas. For example, social media enables diverse groups to form alliances in order to challenge the state's legitimacy (Zaharna & Uysal, 2016). The environment is much more complex and interactive, where new technology and adaptable organisational structures and operational approaches can outmanoeuvre much larger and more established organisations that lack flexibility and environmental awareness.

All of the points in the sections above provide a most compelling reason to rethink the contemporary approach to warfare, how and why it is fought, and those elements that influence and persuade audiences to act against their interests. To be aware of the intellectual and operational aspects of creating and waging warfare, and to understand the greater costs and consequences of these steps and courses of action, is a means to hopefully enable us to be brought back from the brink. The next section gives an overview of the book's structure and a brief review of the main points of each chapter's content.

Chapter Overview

This co-written book comprises twelve chapters, which include this Introduction and the Conclusion. The Introduction and Conclusion are the work of both Iulian Chifu and Greg Simons; four of the chapters are solely the work of Iulian Chifu, and six are written solely by Greg Simons. Chapters that concern subjects of a more conceptual or theoretical nature are grouped to appear in the front of the book, and the more empirical case study-like chapters are grouped together and occur towards the later part of the book. The intention is to provide the reader

[31] G. I. Seffers, *Army Adding Information Dominance to Campaign Plan*, Signal Magazine, www.afcea.org/signal-media/information-operations-sweep-across-milieu-peace-and-war, 20 August 2019 (accessed 22 August 2019).

with the authors' conceptual and theoretical approach (bordering on the philosophical) as a lens and a means with which to understand and interpret the empirical materials that involve issues related to policy-making and practitioner concerns. The authors hope to be able to balance conceptual and theoretical review with the lessons from practitioners and practice, and an analysis of the policymaker and policy-making approach to (mis)managing wars and warfare in the twenty-first century; this is in order to understand and appreciate the interactions between politics, information and communication within the current context of international relations. As a caveat, please note that the chapters of this book were written before the outbreak of the Russia–Ukraine War that began in February 2022.

Chapter two, *Informational Warfare: A Theoretical Approach* by Iulian Chifu, is an exploratory analysis of the fundamental concepts within the understanding and the practice of information warfare in the contemporary age. This includes aspects of the politicisation of the 'truth' and the relativisation of 'truth', which has been and is an evolving process and approach to influencing target publics for a political or social agenda of the initiator of the communications (or their client). It is an example of a misleading or false interpretation of the physical realm in order to influence the cognitive realm of the intended audience.

Chapter three, Greg Simons on *Understanding Political and Intangible Elements in Modern Wars*, engages in the research and explanation of the impact and effects of tangible and intangible elements in modern warfare. This is broached from a historical and philosophical perspective as well as an applied operational perspective. The chapter also notes that the intangible elements of war can influence the level of effectiveness of the tangible elements.

Chapter four written by Iulian Chifu on *The Fourth Generation of Information Warfare*, details the different generations of information warfare. The different generations of information warfare are presented and described. The connection of information warfare to hybrid operations is also presented as there is a specific purpose and agenda to engaging in these kinds of operations. Different means and avenues of information warfare, including the use of defined tools such as as character assassination, are explained to the reader.

In Chapter five, *The Culture and Language of Contemporary Armed Conflict*, Greg Simons notes the dualistic nature of war, its contradictions and dilemmas. These are conveyed and transmitted to the public through the specifics of culture and language that prevail at a given time and place, hence the national variations in attitudes and approaches to war around the globe. Language and culture are not neutral and objective aspects, but rather have embedded and subjective meanings and interpretations that influence how warfare is perceived.

Greg Simons in Chapter six, *Different Shades of Information and Communication in Armed Conflict: White, Grey and Black*, illustrates the point that information as conflict is not a homogeneous entity. It is not only a question as to how and why conflicts are fought, but how they are fought is equally as important. This chapter analyses the distinctions of the various levels of 'truth' and openness shown by politics and the mass media in using information and communication for going to war and how such wars are waged.

Chapter seven on *The Effects of Technological Evolution and Social Media on the Individual, Society and Politics* by Iulian Chifu, raises the issue of the effect and impact of technological evolution and social media on the information and cognitive realms. In addition to creating opportunities and strengths, they have created or strengthened existing weaknesses and threats. This chapter details and illustrates a number of those next-generation weaknesses and threats faced by society and amplified by social media, which create uncertainty and risk.

Chapter eight by Greg Simons on *Western Hybrid Warfare: Crisis and Subversion in Regime Change*, examines and analyses the concept of hybrid warfare and its operationalisation by Western countries in bringing about regime change in a targeted country. This chapter demonstrates how the circumstances of a declared 'crisis' are used to create the perception of an extraordinary situation in order to permit greater operational freedom; this would be deemed 'legitimate' or permissible in more ordinary situations to subvert and collapse a targeted government via a series of seemingly benevolent values and norms.

The *Geopolitics in the Age of Social Media: The Struggle for Influence on Ukraine* is the subject of Chapter nine by Greg Simons. Social media has better enabled the ability for actors to prime and mobilise support for their international political and geopolitical agenda. It is a format that permits remote mobilisation of foreign target audiences to become engaged as activists or supporters, or the contrary for a particular proposed or actual course of action. This chapter examines and analyses the role of social media in communicating the 'realities' of the regime change in Ukraine and the consequences stemming from this moment in the opening phase of the conflict in the country.

Chapter ten is the final case study chapter, *The Ukrainian and Syrian Conflicts: Civil Wars or Shatterbelts?* by Greg Simons. How a conflict is labelled or described in mainstream literature and public discourse reveals a lot about the specific associated subjective politics and geopolitics of the environment that drives it. This chapter looks at the Ukrainian and Syrian conflicts and how they are described, thereby potentially creating an orthodoxy of knowledge; this becomes harder to refute or offer alternative explanations, as it becomes an accepted discourse in a given society that is intolerant of diverging views. War is first and foremost about politics, which is fuelled by information and communication relating to it.

Iulian Chifu discusses *The Impact of the COVID-19 Pandemic on Humankind, Society, Politics and International Relations: The Role of Social Media and Information Warfare* in Chapter eleven. In 2020, the Coronavirus pandemic brought massive changes to all aspects of human life and activity, affecting politics, international relations, social life and the economy, as well as eroding aspects of democratic freedoms and liberties. The health and safety dimension was the most immediately obvious crisis to emerge and which needed to be managed. However, it soon became apparent that a parallel crisis emerged with what has been labelled an ‘infodemic’ (information epidemic) that has magnified the complexity of the situation.

References

Arquilla, J. & Borer, D. A. (Eds.) (2007), *Information Strategy and Warfare: A Guide to Theory and Practice*, New York: Routledge.

Bernays, E. (1947), The Engineering of Consent, *The Annals of the American Academy of Political and Social Science*, 250(1), pp. 113–120.

Betz, D. (2015), *Carnage and Connectivity: Landmarks in the Decline of Conventional Military Power*, Oxford: Oxford University Press.

Brecher, M. (2018), *A Century of Crisis and Conflict in the International System, Theory and Evidence: Intellectual Odyssey III*, Cham (Switzerland): Palgrave Macmillan.

Chifu, I. & Ramberg, B. (Eds.) (2007), *Crisis Management in Transitional Societies: The Romanian Experience*, CRISMART, Volume 33, Stockholm: Elanders Gotab.

Danyk, Y., Maliarchuk, T. & Briggs, C. (2017), Hybrid War: High Tech, Information and Cyber Conflicts, *Connections: The Quarterly Journal*, 16(2), pp. 5–24.

Diesen, G. (November 2018), China’s Geoeconomics and the ‘New Cold War’, Valdai Papers #96, Valdai Discussion Club.

Dixon, P. (September 2019), The British Military and the Iraq and Afghan Wars: Defeat, Popularity and Power, *Political Insight*, 10(3), pp. 22–24.

Esser, F. & Spanier, B. (2005), News Management as News, *The Journal of Political Marketing*, 4(4), pp. 27–57.

Fukuyama, F. (1992), *The End of History and the Last Man*, New York: Free Press.

Germain, R. (Ed.) (2013), *Globalisation and Its Critics: Perspectives from Political Economy*, Basingstoke: Palgrave Macmillan.

Harkavy, R. E. & Neuman, S. G. (2001), *Warfare in the Third World*, Basingstoke: Palgrave Macmillan.

Kissinger, H. (2015), *World Order*, New York: Penguin Books.

Kubalkova, V. & Cruickshank, A. A. (1986), The 'New Cold War' in 'Critical International Relations Studies', *Review of International Relations Studies*, 12 (3), pp. 163–185.

Maoz, Z. & Gat, A. (Eds.) (2001), *War in a Changing World*, Ann Arbor: The University of Michigan Press.

Markedonov, S. M. & Suchkov, M. A. (2020), Russia and the United States in the Caucasus: Cooperation and Competition, *Caucasus Survey*, DOI: 10.1080/23761199.2020.1732101.

Mearsheimer, J. J. (2018), *The Great Delusion: Liberal Dreams and International Realities*, New Haven: Yale University Press.

Moreno-Almeida, C. & Banaji, S. (2019), Digital Use and Mistrust in the Aftermath of the Arab Spring: Beyond Narratives of Liberation and Disillusionment, *Media, Culture & Society*, 41(8), pp. 1125–1141.

Scheller, S. (2019), The Strategic Use of Fear Appeals in Political Communication, *Political Communication*, DOI: 10.1080/10584609.2019.1631918.

Simons, G. & Chifu, I. (2017), *The Changing Face of Warfare in the 21st Century*, London: Routledge.

Szostek, J. (2017), Popular Geopolitics in Russia and Post-Soviet Eastern Europe, *Europe-Asia Studies*, 69(2), pp. 195–201.

Taylor, P. M. (2003), *Munitions of the Mind: A History of Propaganda from the Ancient World to the Present Day*, 3rd Edition, Manchester: Manchester University Press.

Waldman, T. (2021), *Vicarious Warfare: American Strategy and the Illusion of War on the Cheap*, Bristol: Bristol University Press.

(2019), Buy Now, Pay Later: American Military Intervention and Strategic Cost Paradox, *Defence Studies*, 19(1), pp. 85–105.

Walker, S. (2019), *American Foreign Policy and Forced Regime Change Since World War II: Forcing Freedom*, Cham (Switzerland): Palgrave Macmillan.

Western, J. (2005), *Selling Intervention and War: The Presidency, the Media, and the American Public*, Baltimore (MD): The Johns Hopkins University Press.

Wlodkowska-Bagan, A. (2012), Power Rivalry in the Post-Soviet Space in the Political and Military Domain, *The Copernicus Journal of Political Sciences*, 2 (2), pp. 51–62.

Zaharna, R. S. & Uysal, N. (2016), Going for the Jugular in Public Diplomacy: How Adversarial Publics Using Social Media Are Challenging State Legitimacy, *Public Relations Review*, 42(1), pp. 109–119.

2 Informational Warfare: A Theoretical Approach

Iulian Chifu

Introduction

This chapter is confined to the fundamental concepts used in this area. First, it is about information warfare. What is information war/warfare? The concept has, as is usual, several elements and different approaches. We examine the most important ones. We then present our take on the matter and our concept of informational warfare, as already developed (Simons & Chifu, 2017: 35–47, 118–168). However, the concept has evolved and developed along different avenues and areas of research, which are far more in depth.

We then examine the very present concept of "fake news". This has proved to be more of a label than a concept through over-use, or at most an umbrella concept that encompasses much of what we are trying to explore and to chart. The most important aspects run in parallel with, or are equivalent to, disinformation and propaganda; we do not consider areas where politics plays a role and where everything a political figure dislikes is labelled "fake news".

Another part of our theoretical endeavour refers to "post truth", that is relativizing the truth. No criteria of recognition, no means of clear identification, an extreme politicization of the term – as is the case of fake news – and falsely quoting a personal, individually perceived truth are harming the absolute truth coming from facts and direct perception via the human senses. Truth becomes subject to relativization, and it is worth looking at those developments when we are speaking about the ways and means to forge an informational warfare.

In this preliminary first theoretical chapter, we are also concerned with looking at the whys. Why do we have this problem? Why now? What happened to the society that prompted this evolution? How can we fix it? Even though we are not formulating all the answers and we are not offering absolute solutions to deal with this phenomenon, posing these real questions and making sense of this noise and relative chaos is essential in covering our subject.

Last but not least, we are looking at the avenues for future research on informational warfare. Some directions are already on the table and assumed by a number of researchers; in some cases, new ideas are being introduced, expecting a thorough inter-disciplinary approach to the new concepts and realities of the world of tomorrow. The most important part of the endeavour is to look at the evolution of technology and social media and their impact on the individual, society, politics, and democracy; the failure to manage and understand the consequences and impact of the extensive use of technology has not been at the forefront of previous studies, despite the fact that it has evolved rapidly and has become indispensable in our everyday lives.

Information Warfare: Theoretical and Institutional Approaches

The attempts to define the concepts of information/informational warfare/war are directed in several areas. Some arise from more practical experiences or empirical findings, others are already more refined. Some are broader in content and means, others are more focused on informational operations *per se*. In some cases, the reference is the target or the objective, in some others the means and acts. In other cases, interdisciplinary instruments and means are implied, in some other definitions, there are rationalizations of the matters related to informational warfare and more focused definitions. In some cases, the military approach prevails, in others conflict and destruction are present, in some others the informational warfare is thought outside of the boundaries of the conflict, destruction, rupture, but more on societal results through influencing and shaping the informational or public space. At its origins, information warfare was linked to cyber warfare. It includes hacking, cyber-attacks, and not only informational warfare *per se*. It is historically important because the first approach to information warfare was linked to the use of computers; this greatly concerned the Pentagon, which led to the appointment of the RAND Corporation (RAND) to examine the initial consequences of the extensive use of the Internet and computers, and following the evidence presented by the Head of the CIA, George Tenet, before the intelligence select committee in 1998.

This doesn't mean that there was no propaganda, disinformation, or deception long before the information age; these have been factors since the beginning of the humankind. Intelligence is considered to be the second (if not the first) oldest occupation, and just as immoral as the first. But the difference is the fact that anybody can reach everyone online, at every moment, directly, without any possibility of the authorities knowing about this communication. The level of multiplication of the communication is enormous and the impact is, as described in the hearing of George Tenet, hard to contain.

Computers and telecommunications have given rise to the Information Age, which is as different from prior Industrial Ages, since information technology has demolished time and distance (Winston, 1997). Sovereignty is rapidly eroding. No one person and no nation can block (in real terms, completely) the flow of information across national borders. The avalanche of real-time data has also transformed the international economy. Information technology has also produced a new source of non-material wealth, another important achievement of the informational era, as well as a substantial structural difference from its predecessors.

But, on another point, excessive reliance on information technology also has proven to have dangerous drawbacks (as was the case on 9/11 with excessive reliance on TECHINT (Technological Intelligence) and machines versus a limited and under- financed HUMINT (Human Intelligence)). America's information infrastructure, according to the recent Defense Science Board Working Group Report on Information Warfare, is "vulnerable to attack" and "creates a tunnel of vulnerability that has never been reached before in the history of conflict" (Department of Defense, 2012).

On another point, the link between informational warfare and cyber operations has been proved from the earliest of operations. Cyber activities offer the bases and much-needed information to document any informational operation, by knowing the target of the operation as well as the target group that needs to be convinced or derailed in its convictions, where the alternative reality should be projected and established. Today, even though an informational operation should be built without the use of computers, cyberspace offers a communication line with important multiplication skills, and social media an environment where convictions could be formed by mimetic procedures and groupthink through the "friends" that we never know of.

Information power is simply that form of strategic power operating in or through the infosphere. The main characteristics of informational power are its flexibility and accessibility. The combination of these two characteristics gives information power great potential in the strategic world (Longsdale, 1999). The accessibility of information power is mainly the result of the very low entry costs required, costs that allow small actors to operate reasonably and effectively in the fifth dimension. This is not a unique feature. Smaller actors can also operate meaningfully in the other dimensions of the strategy (Longsdale, 1999).

Information warfare did not acquire, from the first moments, a definition, but rather an attempt to define a content. It is considered both a rapidly evolving and imprecisely defined field of growing interest for defence planners and policymakers, linked to the information revolution – led by the ongoing rapid evolution of cyberspace, microcomputers, and associated information technologies (Molander, Riddile & Wilson, 1996).

Current and potential U.S. adversaries (and allies) are also looking to exploit the evolving global information infrastructure and associated technologies for military purposes. It becomes a new "but subordinate facet of warfare" in which the United States and its allies readily overcome their own potential cyberspace vulnerabilities and gain and sustain whatever tactical and strategic military advantages they can. RAND also mentions that "the ongoing information revolution be so rapid and profound that the net result is a new and grave threat to traditional military operations and U.S. society that fundamentally changes the future character of warfare" (Molander, Riddile & Wilson, 1996).

In January 1995, the Secretary of Defence formed the IWEB (Information War Executive Board) to facilitate "the development and achievement of national information warfare goals". RAND was asked to provide and exercise an analytic framework for "identifying key information war issues, exploring their consequences and highlighting starting points for informational war-related policy development" – looking to help develop a sustainable national consensus on an overall U.S. information war strategy. The most important findings of RAND are already common knowledge for the informational war (even though a definition of the concept was not formulated at that moment):

(1) *Low entry cost:* Unlike traditional weapon technologies, development of information-based techniques does not require sizable financial resources or state sponsorship. Information systems expertise and access to important networks may be the only prerequisites.
(2) *Blurred traditional boundaries:* Traditional distinctions – public versus private interests, warlike versus criminal behaviour – and geographic boundaries, such as those between nations as historically defined, are complicated by the growing interaction within the information infrastructure.
(3) *Expanded role for perception management:* New information-based techniques may substantially increase the power of deception and of image-manipulation activities, dramatically complicating government efforts to build political support for security-related initiatives.
(4) *A new strategic intelligence challenge:* Poorly understood strategic information war (IW) vulnerabilities and targets diminish the effectiveness

of classical intelligence collection and analysis methods. A new field of analysis focused on strategic IW may have to be developed.

(5) *Formidable tactical warning and attack assessment problems:* There is currently no adequate tactical warning system for distinguishing between strategic information war attacks and other kinds of cyberspace activities, including espionage or accidents.

(6) *Difficulty of building and sustaining coalitions:* Reliance on coalitions is likely to increase the vulnerabilities of the security postures of all the partners to strategic information war attacks, giving opponents a disproportionate strategic advantage.

(7) *Vulnerability of the U.S. homeland:* Information-based techniques render geographical distance irrelevant; targets in the continental United States are just as vulnerable as in-theatre targets. Given the increased reliance of the U.S. economy and society on a high-performance networked information infrastructure, a new set of lucrative strategic targets presents itself to potential information war-armed opponents.

George Tenet was making the link between the cyber-attacks and cyber warfare and information warfare, defined for the first time during his committee testimony in 1998. As was the case in those times, all was linked to terrorism and less to state rivals. "Cyber-attacks offer terrorists the possibility of greater security and operational flexibility. Theoretically they can launch a computer assault from almost anywhere in the world, without directly exposing the attacker to physical harm" stated George Tenet, the Director of Central Intelligence of the United States (Tenet, 1998).

On another point, linking cyber and information security, he noted that when we talk about cyber warfare, we cannot detach ourselves from discussing one pertinent issue in cyberspace; that is, the information security tenets. Information is the most valuable asset to an organization and information is the number one critical success factor. Therefore, it is mandatory to ensure information confidentiality, integrity, availability, authentication, and non-repudiation (that information can be trusted) (SANS Institute, 2004).

Informational Warfare: Orientation of the Concept on Means and Instruments

Another category of definitions is linking the concept closer to informational warfare *per se*, focusing on disinformation as a target of informational warfare. They identified the instruments of Behavioural Data

Collection and targeted advertising and the instruments used in political propaganda and campaigns, including disinformation operations, as sources that could be used in informational warfare (Ghosh & Scott, 2018).

> When disinformation operators leverage this system for precision propaganda, the harm to the public interest, the political culture, and the integrity of democracy is substantial and distinct from any other type of advertiser. (...) The more behavioural data they are able to collect on users, the better they are able to serve them targeted ads that cater to their unique interests. (Ghosh & Scott, 2018)

The European Union (EU) looks at information warfare strictly from the point of view of disinformation, insisting on strict defensive means; the EU has adopted a defensive strategy to deal with disinformation. It has delivered several strategic documents, including an Action Plan in December 2018 that provides a promising basis for action (Scheidt, 2019). The East StratCom Task Force (an EU institution that is a fact and information checker), "detects and debunks Russian narratives". The major online platforms are currently trying to implement a Code of Practice that the European Commission has set up with the aim of curbing the spread of disinformation on social networks (Communication of the Commission, 2018).

Informational Warfare as a Concept for the EU

The direct definition of information warfare in the EU is "The offensive and defensive use of information and information systems to deny, exploit, corrupt or destroy an adversary's information, information-based processes, information systems and computer-based networks while protecting one's own" (EU Vocabulary, Concept.[1])

The European Parliament is even closer to the informational warfare by naming the objectives of an information war: distrust and societal tensions. Mass online disinformation campaigns are being widely used by a range of domestic and foreign actors to sow distrust and create societal tensions, with serious potential consequences for our security (European Parliament, 2018). Furthermore, disinformation campaigns of third countries can be part of hybrid threats to internal security, including election processes, particularly in combination with cyberattacks. For example, Russian military doctrine explicitly recognizes

[1] See https://op.europa.eu/en/web/eu-vocabularies/concept/-/resource?uri=http://eurovoc.europa.eu/c_4e75f061, accessed 13 January 2020.

information warfare as one of its domains (The Military Doctrine of the Russian Federation, 2014). We see below that this link between informational war and Russia very often appears when it is about Russian documents in this field.

Last but not least in the debate, it is worth looking at the concept used by the NATO Centre for Excellence in Riga on information warfare. It also links information warfare to Russia, noting that the demand for such projects (information warfare) has been rising steadily following Russia's interference in Ukraine in 2014, which placed the notion of "information warfare" at the heart of the public debate (CoE NATO Riga, 2018: 60).

Panarin defines information warfare as "a form of confrontation, in which parties use special (political, economic, diplomatic, military and other) methods, ways and means to influence the information environment of the opponent and protect one's own interests and achieve set objectives" (Panarin & Panarina, 2003: 20).

Information warfare is often considered a link between information and communication, more concretely linked to infrastructure destruction. For the Federation of American Scientists (Lewis, 1997), information warfare in its broadest sense is "a struggle over the information and communications process, a struggle that began with the advent of human communication and conflict". Information and communication technologies and their increasing prevalence in our society have revolutionized the communications process and, with it, the significance and implications of information warfare.

In a far narrower sense, information warfare is "the application of destructive force on a large scale against information assets and systems, against the computers and networks that support the four critical infrastructures (the power grid, communications, financial, and transportation)" (Lewis, 1997). The same source highlights the difference between offensive and defensive information warfare, noting that information warfare is a proper option for the U.S. to employ to advance its foreign policy interests. As the pre-eminent information society, the U.S. possesses the technological knowledge to wage an effective information war. At the same time, the Department of Defense and the intelligence community are charged to identify and adjust to new national security threats posed by information warfare.

Rationalizing the Concept: Interdisciplinary Take, Interdisciplinary Means

Another school of thought concerns itself more with rationalizing the concept of information warfare and moving it into an inter-disciplinary sphere; that of understanding as well as the means used. It implies multiple components and there is a separation between the broader and narrower senses NATO has chosen to rationalize and integrate the knowledge on informational warfare, focusing, at the same time, on informational space, on the cyber, information and media links, moving closer to our view.

"Information warfare is an operation conducted in order to gain an information advantage over the opponent. It consists in controlling one's own information space, protecting access to one's own information, while acquiring and using the opponent's information, destroying their information systems and disrupting the information flow". Information warfare is not a new phenomenon, yet it contains innovative elements of the effect of technological development, which results in information being disseminated faster and on a larger scale (DEEP NATO).

In defining the concept, NATO relies on a number of theoretical sources quoted as such. Therefore, it suggests that cyberspace and the related area of new technologies provide an important field for information warfare. Cyberwar activities may consist of cyber-attacks, destroying opponents' information systems, but these may also involve so-called social cyber-attacks, creating in people's minds a specific image of the world, consistent with the goals of the information warfare conducted by a given country (Aro, 2016: 121–132).

Information war is also employed by use of the Internet and social media. The Internet enhances and expands the possibilities of data acquisition, information defence and information disruption, and makes it easy to reach both the citizens of a given country and the international community. Given the speed of communication, wide coverage and low cost of (dis)information, the Internet uses, among other campaigns, social media, which plays a crucial role. Social networking sites are also a valuable source of information on the target groups to which (dis) information activities are to be addressed (Macdonald, 2006). The approach also identifies some of the instruments to conduct information warfare over the Internet, among them:

(1) Troll factories – entities employing people who post comments on the Internet in line with the goal of the ordering party, using fake profiles in social media.
(2) Bots – programs sending out messages automatically, for example in response to the appearance of a keyword.
(3) Fake news – messages intended to mislead media users (Ohlin, Govern & Finkelstein, 2015).

Information warfare also has a relationship with mass media and journalism. Media not only report on war and armed conflicts, but they also become the targets of attacks involving, for example, disinformation through the dissemination of fake news. Journalists must be extremely careful in verifying information related to international relations, as the messages they receive may be part of disinformation activities. Another problem is often related to the hacking of websites whose profile is in opposition to the activities of the state conducting information warfare (Thomas, 2014: 101–130).

Media users become victims of information warfare conducted using both the so-called traditional media and the Internet. The signs of propaganda and disinformation are present in numerous media messages, including traditional media as well as social media. Media users are becoming increasingly aware that they are the objects of (dis)information activities aimed at affecting their perception of reality. With growing distrust towards information appearing in official circulation, Internet users are turning to alternative sources of information, including civil media (Thornton, 2015: 40–48).

An important element of individual resistance to propaganda and disinformation is to escape the "information bubble" ("echo chamber"), the situation of restricted access to information other than that provided by algorithms based on the user's previous activity, by diversifying the sources of information and acquiring it in other ways than those suggested by algorithms regulating social media (DEEP NATO).

Information warfare could also be linked to conflict, the normative-legal side of the matter, security studies, technical – IT, communication – or all the above, in an interdisciplinary approach. Information security professionals should refrain from focusing only on the technical aspects of this area, since it has been shown that legal frameworks, national as well as international, also have to be considered. The prevailing challenge for countries around the globe is to foster collaboration among lawyers, information security professionals, and technical IT professionals. They should continue striving to at least keep the registry of the IW arsenal and remedies updated, which may, in turn, incite adversaries to provide more material for research (Eloff & Granova, 2009).

Information is probably worth much more than any other commodity. Globalization, the other important phenomenon of the times we live in, has taken the value of information to new heights. That's why informational warfare is not only possible, but it has also already taken place and is growing internationally as a preferred way of warfare. It also demonstrates that successful strategies, offensive or defensive, are dependent on taking a holistic view of the matter.

The World Economic Forum also has a much-extended approach, linking warfare to non-lethal attacks. Its definition is conflict oriented. It introduces the information environment; to find a proper mix of means of offensive action between hacking and direct informational warfare. For the use of information warfare, it includes in the common approach, electronic warfare, cyber-attacks, and psychological operations (PSYOPs). The definition is comprehensive, in a broader approach. The means and tools are dependant on the field targeted in information warfare.

The familiar form of warfare in which physical damage is meted out against the opponent's military forces and infrastructure has become only one form of attack. Instead, states are increasingly launching non-lethal attacks against an enemy's information systems – this is the rise of information warfare (Stupples, 2015). The World Economic Forum use Dan Kuehl's (of the National Defence University) concept of information warfare as the "conflict or struggle between two or more groups in the information environment". Political leaders are establishing military commands for attacking, defending and exploiting the vulnerabilities of electronic communications networks. Information warfare combines electronic warfare, cyberwarfare and PSYOPs into a single fighting organization, and this will be central to all warfare in the future.

When it concerns the anatomy of information warfare, the World Economic Forum uses information as a major tool essential to business, international relations, and social cohesion, as it is essential to a military force's ability to fight. Communications today lean heavily on the Internet, or on communications using various parts of the electromagnetic spectrum (such as radio or microwaves) through terrestrial communications networks or satellite networks in space. We live in a highly connected world, but it doesn't take much to tip it into instability or even chaos. Therefore, the concept and tools that combine electronic warfare – used to disrupt or neutralize electromagnetic transmissions, to cripple military communications or weapons guidance systems, or harm air traffic control systems – with cyber-attacks launched through the Internet against digital networks, can make it impossible for businesses to operate. Enormous damage can follow, in both cost and reputation, causing economic chaos, such as: bringing down a stock exchange; taking over industrial control systems used in manufacturing plants or in power, water, and gas utilities, putting lives at risk; and PSYOPs aimed more at degrading the morale and well-being of a nation's citizens by the spreading of false information, rumour, and fear through social media and news outlets. Misinformation and fear can spread rapidly, resulting in panic. Information warfare, then, is the integration of electronic warfare, cyberwarfare, and psychological operations, for both attack and defence (Stupples, 2015).

The Military Approach to Information Warfare and Information Operations

In the military field, the developments are as substantial as those in the research on security and communication. A full catalogue on information warfare and information operations was developed in 2008, covering the early stages of the concept, beginning in 1995 (Marlatt, 2008).

Under the operational definition of information warfare, the focus moves to the situation where information is not collected, stored, moved, or used to reduce uncertainty (Borden, 1999). Information warfare is more of an information-based warfare, centered on the idea that information is generated in the course of reducing uncertainty so that decisions can be made (Campen, 1995: 67–69). In fact, "information is the reduction in uncertainty ... measured in bits". This is consistent with the definitions of information and uncertainty given by Claude Shannon of the Bell Laboratories in the late 1940s, definitions which are now standard in the mathematical theory of information and communications (Shannon, 1948: 379–423, 623–656):

> Information Warfare is any action to Deny, Exploit, Corrupt or Destroy the enemy's information and its functions; protecting ourselves against those actions and exploiting our own military information functions. (Widnall & Fogelman, 1997)

In another source, we have a list of not less than six subareas (highlighted below) where information warfare is developed. Information warfare is often seen as a new threat involving break-ins to Pentagon computers, disabled satellites, and downed phone networks, therefore requiring information warfare defence (Nichiporuk, 1999: 179–215). The possibility for these new information-warfare tools to threaten the ability to project power or to realize national interests is real and deserves analytical attention and public awareness.

The paper recognizes the "potpourri of different definitions for the term information warfare". In its own approach, information warfare is the process of protecting one's own sources of battlefield information and, at the same time, seeking to deny, degrade, corrupt, or destroy the enemy's sources of battlefield information (Nichiporuk, 1999: 180). This includes six pre-existing subareas: operational security, electronic warfare (EW), psychological operations (PSYOPs), deception, physical attack on information processes, and information attack on information processes

(Hutcherson, 1994; Joint Staff, Joint Doctrine for Command and Control Warfare, 1996). Operational security is all about defensive information warfare; offensive information warfare consists of the aggregation of EW, PSYOPs, deception, physical attack, and information attack.

Disinformation and Informational Warfare: Russia and the Militarization of Information

Another approach to the concept of information warfare is linked to disinformation, as well as to Russia and its documents in the field. When we refer to informational warfare, we are talking about the "militarisation of information". This is the case with the Canadian Intelligence Service, which sees information warfare linked to disinformation (Canadian Security Intelligence Service, 2018).

The reach and speed of the Internet and social media have escalated the potential impact of disinformation. Increases in data transmission capacity coupled with a shift towards programmatic advertising (involving targeted individual consumers via cookies, device IDs, and algorithmic software, automating the sale of advertising using real-time-bidding) have resulted in a precipitous decrease in the ability of traditional journalism to mediate the quality of public information. Conventional journalism has been partially displaced by a torrent of data from an infinite number of originators. Within that torrent is a current of lies and distortions that threatens the integrity of public discourse, debate, and democracy (p. 3).

The apparent physiological component of disinformation is likely enhanced by the effects of computer-mediated communications and experience. According to this source, the history of the Soviet use of the term disinformation is itself an example of disinformation. First coined in Russia, the intelligence services of the Soviet Union and their allies were ordered in the early 1950s to spread a story indicating that the term was actually French, and described a weapon of information warfare deployed by the capitalist West against the USSR and people's democracies throughout the world (Pacepa & Rychlak, 2013: 33).

Among other crucial developments, the year 2013 witnessed Russia openly declaring an information war on the West. The first wave of onslaught was directed against states placed between the Baltic and the Black Sea (so-called Intermarium): Estonia, Latvia, Lithuania, and Ukraine – countries that have remained the prime target of Russian intimidation and aggression since the dissolution of the USSR in 1991 (Pacepa & Rychlak, 2013: 39). It was also Russia which declared and applied a militarization of information. Russian military strategists consider disinformation as an organic part of future conflict (Chifu & Țuțuianu, 2017).

Theoretical research and practical steps resulted in the creation of "research units" and "cyber troops". According to Russian Minister of Defence Sergey Shoigu, those "will be much more efficient than the 'counterpropaganda' department of the Soviet period" (Pacepa & Rychlak, 2013: 40).

How Russia or Putin understand information warfare is of primary importance in defining the concept of informational warfare. In the Russian construct, information warfare is not an activity limited to wartime. It is not even limited to the "initial phase of conflict" before hostilities begin, which includes information preparation of the battle space (Čižik, 2017: 9–10).

Western understanding of information warfare can be defined as tactical information operations carried out during hostilities to deceive an adversary and indirectly influence its decision making and the actions that follow based on this decision-making process. Based on the previous definition, Western understanding of information warfare is focused mainly on its military use and only in time of war.

Informational warfare is also linked to Open Source Intelligence (OSINT) (Williams & Blum, 2018). Rethinking OSINT as an intelligence discipline and linking OSINT with information warfare is important since OSINT is used in discovering and defending traces of information warfare; its aims, targets, and means used in order to alter Western democratic societies and their cohesion.

In the same family of weaponization as social media, the NATO STRATCOM (Strategic Communications) Centre of Excellence in Riga offers another linkage between the cyberwar debate and information warfare. Even if it is strictly restrained at the level of understanding of the conflict in question, the definition is useful for linking all the tools to a common purpose. Therefore, cyber-attacks give to a whole series of actors the means and opportunities for influencing key strategic audiences' understanding of the conflict. As these conflicts are about perception, legitimacy, and credibility, those about "information warfare" are more than just about initially controlling the territory; policy objectives are achieved through diplomatic, political, legal, informational/psychological and cyber means – supported by armed action or activities – which is why social network media can play an important role in this kind of warfare (Nissen, 2015). In particular, that aspect is explored in the social media battle for Syria, pointing out "that it is hard to overstate how heavily both sides depend on cyber tools to articulate their information warfare tactics in an unprecedented way" (Farwell & Arakelian, 2013).

Information Warfare: Our Take on the Concept

A component of the hybrid war, and a type of warfare in its own right, is informational warfare (Simons & Chifu, 2017: 35–47). Nicholas O'Shaughnessy reproduces an epitomized history on how informational war has been engaged since ancient times, from Napoleon and Frederick the Great, to the twentieth century, which may be termed the "propaganda century". The process of expanding and developing the informational type of warfare was concomitant with the technical evolution of communication. Hitler's rise and sophisticated propaganda, a subdivision of informational war, was completed by Russian informational warfare, mainly focused on *spetspropaganda* – firstly taught as a subject at the Russian Military Institute of Foreign Languages in 1942 and meaning a combination between agitation and propaganda – and on *dezinformatsiya* (disinformation). These pillars became the features of the Cold War (Lucas & Pomeranzev, 2016: 2–6).

A summary of the evolution of modern Russian informational war includes elements of fake elements and forgeries, reflexive control, and active measures. The active measures became essential in influencing the policies of other governments, undermining confidence in their own leaders and institutions, disrupting a country's relations with other nations, and discrediting and weakening governmental and non-governmental opponents (Lucas & Pomeranzev, 2016: 7).

Valery Gerasimov, the Chief of the General Staff of the Armed Forces of Russia, brought a fresh approach to informational war in 2013, when he described it as incremental for achieving political and military goals, through "indirect and asymmetric methods" outside conventional military intervention (Polyakova et al., 2016: 3). Concerning informational warfare, the debate is far more complicated because the definition has to be nuanced and adapted through a family of actions; legitimate and legal, as well as illegitimate and profoundly illegal. On one hand, there is the expression of the fundamental human rights of opinion, free thinking, and free circulation of ideas, as well as the democratic principles of the freedom of speech. Then, we have the legal use of communication and PR, but also immoral and illegitimate polit-technology and agit-propaganda (Simons, 2010). Last but not least, in the same family, we can find the illegal use of components of psychological operations and informational war.

Informational war, according to our own definition of the concept, signifies creating alternative realities by perverting the truth based on data, facts, and concrete arguments, and twisting it by using a combination of facts, syllogisms, sophisms, propaganda, interpretation, and a lot of lies. The alternative reality perverts the perception of the targeted population employing a combination of PSYOPs – psychological operations, misinformation, and propaganda which use basic beliefs, feelings, and created images in order to lead the public to a predesigned perception. At the end of the day, since the public already has an opinion, its perception has taken the place of the reality (Stern, 1999), and whatever valid argument and proof of the truth exists, the false perception has already been established.

This strategy perfectly coalesced with Russia's strategy on Ukraine. The NATO STRATCOM Centre of Excellence offers extensive research on how asymmetric and information activities were instrumental to achieve political and military goals in Ukraine; by deception through the use of information and psychological operations, social media, English- and Russian-language satellite TV-based propaganda, and older Soviet-style techniques such as "active measures" and "reflexive control", especially in Crimea. The main component has been based on the Russian perception of the world, meaning Russia, Ukraine, and Belarus as the three pillars of the Slavic Orthodox civilization jointly share history, values, culture, and the recognition of Russian supremacy (Bērziņš et. al., 2014: 6–7).

The Russian informational strategy galvanized a network-flow model, representing the "unvirtuous circle" of Russian influence; this aimed to gain influence over state institutions, bodies, and the economy, achieving power to shape others' national decisions, and using methods in which corruption became a "lubricant on which the system operates" (Conley et al., 2016). Russia's influence in Central and Eastern Europe has mainly followed two pathways during Vladimir Putin's third mandate. Firstly, manipulating one country by dominating or abusing strategic sectors of the economy translates into economic capture. The second refers to cultivating political relationships with autocrats, nationalists, populists, Eurosceptics, and Russian sympathizers, thus achieving the political capture of another state (Conley et al., 2016: 1–2).

As Edward Lucas and Ben Nimmo demonstrate with concrete examples, the Russian Doctrine of informational warfare is monolithic, while its implementation is segmented. In the Baltic States, the Russian-owned media outlets continue to exploit fears of U.S. abandonment of their own country, while stimulating Soviet nostalgia. In Romania, the informational type of war seeks to erode the public faith in democratic institutions and consolidate the impression that EU accession was a failure. Concomitantly, in the Czech Republic and Slovakia, the Russian strategy focuses on local environmental and anti-war themes to bring a negative perspective to Western energy politics and the U.S./NATO security reassurances (Lucas & Nimmo, 2015: 3).

In a world where information quickly reaches a large number of people in real time – via television, the Internet, and social media – the perception of a specific event is easily formed, derailed, twisted, and imposed. Presenting the truth later would have a limited, large-scale impact on public opinion due to the lack of critical thinking within the majority of the general population, as well as the conservative approach in the wider acceptance of its own mistakes and the easy association with the use of the explanation already internalized by the average person, especially in a community that has developed a conformism and has its own viewpoints, perceptions and "truth".

The Informational War: Concept, Instrumentation, Functionality

For the most part, informational war represents a "contact-less" conflict that is fought in the realms of perception and the human mind. It continues through both official peace and wartime (Lucas & Nimmo, 2015: 8). The battle of alternative narratives, of "alternative truths", has become the most insidious way of constructing beliefs (Simons & Chifu, 2017: 118–168). But it relies on the groups and audiences targeted for each individual operation, on a vast labyrinth of knowledge preparation, inclinations and expectations of the target audience, its propensities, and frustrations. Obviously these aspects do not concern only the information as such, but refer to other subtler components as well, related to basic emotions, context, and the capitalization of opportunities provided by the events in progress, and the state of mind created in a target population, in order to inculcate a certain reference to the subject.

The alternative reality distorts the perception of the target population, with a combination of psychological operations: PSYOPs, together with disinformation and propaganda; using fundamental beliefs, feelings, and images with impact; and aiming to lead the target audience towards a pre-defined perception. And, finally, as opinions have already been instilled, public, perception has replaced reality (Stern, 1999), and, no matter what arguments or proof of the truth are presented, these cannot challenge the already formed perception.

The overall objectives of the informational war are: to determine, to control, or at least to alter, the strategic decision of foreign policy, security, and defence; to pervert or hinder the instruments intended for the military component of a state; and to obstruct the functioning of, if not to block, some elements related to the security of a state. The instruments and mechanisms to achieve this goal are to determine how to influence the public, the citizens, and the pressure groups that are already prepared and conditioned, organized and directed, to put pressure on the authorities in order to alienate the state from its identified objectives due to the lack of public support, or even opposition.

The characteristic elements, principles, and values of an open and democratic society are being exploited and used against the states and their institutions. An undermining from the inside of a community is accomplished through building within groups of enemies. Moreover, the undertaken operational approach is unitary, integrated, and not rarely a fact or a component of the informational war, taken separately, and appears only as an oddity, a singularity, an accident, in no circumstances an act aggressively planned together with a suite of other elements. The insidious manner of action and this integrated approach creates the favourable conditions for a credible denial of such action of informational aggression on a target population by the attacker.

The informational war uses three distinct levels of action – tactical, operational and strategic – with different relevance and degree of legal and moral significance. An integrated and sequential approach, dependent on the target audience, increases the efficiency of the informational war. The first component is the visible one, what is evident, and regards the engagement in altering the public space of a subjected target by involving the media, which nowadays is considered an instrument of war. Winning modern wars is as much dependent on winning over domestic and international public opinion as it is on defeating the enemy on the battlefield (Payne, 2005: 81); the Internet and social media operate at more subtle levels, injecting ideas and information that valorize the principles and fundamental values of democracy and human rights, including freedom of expression, of speech, and the free circulation of ideas, elements that are not only legal and moral, but are among the fundamental values that democratic societies defend.

At a lower level, the open injection of propaganda can be seen as a form of coercion without the appearance of coercion (O'Shaughnessy, 2012: 29); manipulation and disinformation are difficult to prove. Moreover, this type of activity requires constant monitoring, analysis, and investigation; public exposure of deceptive activity, deliberate misinterpretation of facts and of alternative narratives behind these ideas, are using the misinterpretation to support an alternative package of truth proposed by the author. Also, selective news coverage can be a powerful propaganda tool by tending to promote the kind of reporting that is favourable to an actor and its cause, and thereby restrict information that could be damaging (Zhukov & Baum, 2016: 8).

Not infrequently, activities may capitalize on the vulnerabilities of the system: failures to comply with, bypassing, or avoiding legal sanctions; weaknesses of the control institutions, their legal limitations, or the absence of professional bodies capable of morally sanctioning excesses in promoting one approach or another on any particular subject or narrative; the absence of responses to the same news story due to the lack of expertise to validate the reported information; the poor training of employees working in the media; and failures in information dissemination and effective marketing. False or selective information coming from obscure sources and penetrating the mainstream media leads to a forced interpretation and the piloting of judgements to a default conclusion. There is an absence of alternative independent media, which would keep to strict rules of narrating events; providing balanced news reporting and alternative approaches, polemic or at the same time 'pros and cons' arguments, all help to build the informational war element.

In developed societies, with an established and mature mass media, the ability to take over, influence and control the public space is greatly reduced. Levels of education also matter as well as the democratic culture of the target population. But there are societies where the public space is controlled by third-party actors; for example, the Republic of Moldova, where 80 per cent of the audience consumed news that was produced in Moscow. When this is packaged in extremely well-made and attractive entertainment programmes, these 'genuine' news 'injections', situated in the space of a preconceived narrative, makes the impact of the projected alternative truths greater. Propaganda has its limits however, and the use of a set of integrated instruments is required; these are developed and utilized simultaneously with a distinct target audience in mind. At this point we consider another element, that of lobbying or public relations, whether we are referring to a company or a personality or public diplomacy, if concerning a state. This is a perfectly legitimate approach in those countries where a law on lobbying or its equivalent exists.

The method works as follows: decision makers must be influenced by ideas launched in the public space by legitimate and credible sources that control the impact on the general population as well as the state's decision makers in power at the time. Sums of money are paid to credible persons, such as analysts, experts, former and current politicians, who can impact a target population or foreign media, but in a credible and legitimate manner to defend the views of those who are paying them.

The paid media or expert, for example an analyst expressing an opinion, are paid to create a certain image, to defend a certain actor, to carry a certain message, a vision, or a previously prepared narrative. Therefore, a credible character, usually a highly regarded impactive individual, suddenly voices a viewpoint in the public domain, spreading ideas that are not the result of his analysis, of his thinking, but are the broadcast material paid for by the silent partner. The public is unaware that the person before them is not speaking freely, but is expressing a set of ordered ideas. Therefore, an assessment is made in the context of the credibility and legitimacy of the person concerned; what the analyst is actually selling is the message from the lobbying party. Moreover, after fulfilling the mission, this individual returns to their own personal analyses; legitimate, correct, and a result of their own thinking, experience, and ideas. The person becomes once again themself, preserving their future credibility and may then be used subsequently on another project involving the informational war or promoting a different theme.

While there are laws on lobbying in the US and public exposure of certain positions are paid for, with a transparent formula of the amount of money received to convey certain themes, this instrument is missing in Europe, but not in respect of the lobbying agencies. It is a legal instrument, sometimes immoral, but which is used in combination with propaganda and taking control of or altering the media space in the informational war framework.

The third level is the most insidious, deeply illegal, which targets subtler components of the informational war. It is linked to PSYOPs – elaborate psychological operations geared towards influencing the behaviour of the target, while the target can be military or civilian depending on the situation (Naef, 2003). The level of access of this component is much more profound, in the fibre of the social organism, of the common beliefs, of the psychological state, and of the deeper fears. Here the disseminated information matters, and especially the effect created by information in the target audience, namely the shaping and creating or deepening of fears, creating collective emotions, preparing the public to react to future events in a directed formula, previously prepared and pondered.

The objectives are achieved using false news, personalities and experts with known moral references within the society, directed to certain positions, but the target is not the immediate information but the creation of the context for an emotional public reaction to a future event. Dirigisme in relation to the subsequent reaction is created by accessing subtle levels of the subconscious and creating patterns of thought by repeating some sequences of this kind at predetermined intervals, teaching the brain and the subconscious to react in a certain manner to this type of information and emotional pre-set stimuli. The goal is to determine a specific and collective public response, when needed, in order to put pressure on and alter the actions of a decision maker in times of crisis.

The instruments used are a combination of narratives, alternative ideas, planting doubt, promoting and validating lies in as credible a way as possible, that will enter the subconscious as perceptions of truth; exaggerated and directed interpretations, manipulation, misinformation (with all its components), the effects of trolls' actions (directed commentators in cyberspace). Here as well, the components for conditioning the target audience are built, actions at the subliminal and subconscious level, the inoculation of the constructed perception, accessing the basic emotions and their orientation – fear, humiliation and hope.

The functioning of the informational war is achieved in an integrated manner, on all three dimensions, with deliberate dosages and a vast instrumentation built over time. Important resources are used for such activities and the components are most frequently moved towards the military space. It is an instrument whose relevance is just being discovered and the impact can be extremely powerful.

Fake News: An Umbrella Concept – An Oxymoron for a Star Concept of Information Warfare

"Fake News" is the concept/reference most used in relation to informational warfare. The language in itself is highly disputed since it is about an oxymoronic format; news usually means that it is fact-checked and involves clear data obtained by professional journalists from at least three sources, with solid evidence, and judgements based upon clear factual data and a transparent logic. So, "fake news" is, in fact, no news at all; on the contrary, it is never news, by any means. It is about a false content, either by direct lack of facts, or by the interpretation or false perception of somebody using 'wishful thinking' and distorting the material as if it was a real fact. It is an evident imposture used in order to distort the truth by adding false arguments or untrue facts presented in order to reach a conclusion that harms the person receiving it, the individuals it is about, or the institutions and democracy. It is certainly not information, real information based on facts.

But "fake news" is also a label covering a plethora of concepts and definitions, more commonly seen as equivalent to disinformation and propaganda. References to "fake news" have nothing to do with the real concept, with the fake, but are only an action of labelling/naming a fact that somebody, usually a political figure, disagrees with. Politicization of the concept of "fake news" itself is misleading and, as we have seen in the introduction, could also harm the truth.

There are institutions and organizations that deal with "fake news", trying to eliminate or expose it by cleaning up the informational space for users. Sometimes these institutions linked to "Stop Fake" activities are even part of the strategy to distort information; by making a lot of noise, accumulating unimportant information, in order to cover and hide the obvious fake news used by those organizations which resort to such techniques. Quite often, it is even about states themselves that are involved in the cover up of fake news, through confusion, sheer quantity of information and use of third-party sources.

Moreover, UNESCO is tackling fake news through journalism and the media approach. It focuses mainly on all those who practice or teach journalism in this digital age. The changes in the communication landscape are linked to the 'new journalism' but without the professional fact-checking on the scale of the entire Internet and social media platforms. Therefore, the main linkage of fake news is with the quality of the information spread via those instruments when compared to old-style professional journalism (UNESCO, 2018). It can be seen that the lack of credibility of the Internet and social media impacts upon the generalization of the credibility of journalism itself.

In the US, for instance, the law prevents any responsibility for the message that is transmitted. Social media and the Internet are associated with telecoms, and the interpretation of the law is that a communication platform is not responsible for the messages that the users are transmitting. Therefore, social media and the Internet are not considered to be public space but informational space, placed under the rules of journalism, but are still responsible for spreading that fake news to large numbers of consumers.

In the European Union (EU), it is considered that the spreading of fake news is disinformation and a breach of the Charter of Fundamental Rights of the European Union. Therefore, the EU has introduced a Code of Practice on disinformation. In both the EU and US, social media platforms are requested, but not forced, to take action against fake news, hate speech, and so on. In Romania, for example, social media is considered to be a public space, accessible to everybody, so it is under the scrutiny and responsibility of the law as with any other electronic media.

Moreover, UNESCO is coming down hard on orchestrated campaigns that are spreading untruths by disinformation, mal-information and misinformation that are often unwittingly shared on social media. These are defined as follows:

(1) Disinformation: Information that is false and deliberately created to harm a person, social group, organization or country.
(2) Misinformation: Information that is false, but not created with the intention of causing harm.

(3) Mal-information: Information that is based on reality, but used to inflict harm on a person, social group, organization or country.

So, the handbook (UNESCO, 2018) explores the very nature of journalism with modules on why trust matters; thinking critically about how digital technology and social media platforms are conduits of the information disorder; fighting back against disinformation and misinformation through media and information literacy; fact-checking; social media verification; and combating online abuse. It avoids the assumption that the term "fake news" has a straightforward or commonly understood meaning (Tandoc, Lim & Ling, 2018). This underlines the oxymoron, because 'news' means verifiable information in the public interest, and information that does not meet these standards does not deserve the label of news. In this sense then, it notes that "fake news" lends itself to undermining the credibility of information which does indeed meet the threshold of verifiability and public interest – that is, real news.

Moreover, NATO has at least two complementary approaches to fake news. The first belongs to the NATO StratCom Centre of Excellence in Riga, that notes that the term "fake news" has become the default catchphrase for truth-seekers wishing to label inaccurate reporting, truth-obscurers spreading malevolent assertions, or the unprepared who simply want to shut down uncomfortable discussion. The shorthand expression "fake news" may fit neatly into tweeted messages, but willing amplifiers have spread it across all media, traditional and social, without necessarily giving it a meaningful definition (CoE NATO Riga, 2018).

The StratCom Centre has vast sources of reference on post-truth and populism, as well as on the impact of social media rules on the political scene, together with popular social media comment versus verified facts. It provides a good overview of the truth, proof, evidence, arguments, and also lies. The decline of the role of truth is just touched upon, since the Centre is less inclined to investigate philosophical approaches, but reflects those realities with concrete examples – most frequently, extracts from Russian informational warfare – from traditional mass media and social media, underlining the lack of checks and any kind of verification of social media posts.

Systems that oversee media platforms are large enough to be out of reach: users flagging cases of fake news or breaches of the rules to the administrators of the network, who take action on request, are not related to a system which monitors the whole platform. A lot of fake news does not end up being reported, checked, and erased. The same happens with hate speech, quarrels between group members, or harsh language and psychological violence.

Another NATO approach to countering disinformation is related to the COVID-19 pandemic. Moreover, NATO views disinformation as the deliberate creation and dissemination of false and/or manipulated information with the intent to deceive and/or mislead (NATO 2020). Disinformation seeks to deepen divisions within and between the Allied nations, and to undermine people's confidence in elected governments. The NATO Alliance has been dealing with these challenges since its inception and has been actively countering a significant increase in disinformation and propaganda since Russia illegally annexed Crimea, part of Ukraine, in 2014.

The EU also refers to fake news in its framework involving disinformation and propaganda. The exposure of citizens to large-scale disinformation, including misleading or outright false information, is a major challenge for Europe. The European Commission is working to implement a clear, comprehensive, and broad set of actions to tackle the spread and impact of online disinformation in Europe, and to ensure the protection of European values and democratic systems (European Commission, 2020: 1).

For the EU, disinformation is "verifiably false or misleading information created, presented and disseminated for economic gain or to intentionally deceive the public" (European Commission, 2020: 1). It may have far-reaching consequences, cause public harm, be a threat to democratic political and policy-making processes, and may even put the protection of EU citizens' health, security, and their environment at risk. Disinformation erodes trust in institutions and in digital and traditional media and harms democracies by hampering the ability of citizens to take informed decisions. It can polarize debates, create or deepen tensions in society and undermine electoral systems, and have a wider impact on European security.

That's how, in October 2018, the EU launched the Code of Practice on disinformation – a self-regulatory set of standards signed by platforms, the leading social networks, advertisers and the advertising industry, including Facebook, Twitter, Mozilla, Google, and associations and members of the advertising industry (European Commission, 2018: 1).

Signatories of the Code pledged to take action in five areas:

(1). disrupting advertising revenues of certain accounts and websites that spread disinformation;
(2). making political advertising and issue-based advertising more transparent;
(3). addressing the issue of fake accounts and online bots;
(4). empowering consumers to report disinformation and access different news sources, while improving the visibility and traceability of authoritative content;

(5). empowering the research community to monitor online disinformation through privacy-compliant access to the platforms' data. (European Commission, 2018: 2)

At the same time, the EU funded and launched the European Digital Media Observatory, a project aimed at creating a European hub for fact-checkers, academics, and other relevant stakeholders to link with media organizations and media literacy experts, and to provide support to policymakers. Its activities are based on the following pillars:

(1) mapping fact-checking organizations;
(2) supporting and coordinating research activities;
(3) building a public portal;
(4) ensuring secure and privacy-protected access to platforms' data;
(5) support to public authorities. (European Commission, 2020: 2)

Moreover, the EU adopted a Commission Communication called "Tackling online disinformation: a European approach", which aimed at tackling the problem of fake news in an extensive and comprehensive manner (European Commission, 2018: 3). Here we can see the direct link that the EU makes between fake news and disinformation.

At the same time, the Council of Europe finds itself in the forefront of the responsibility to deal with fake news once it realized through its own polls that two thirds of EU citizens have reported coming across fake news at least once a week. Over 80 per cent of EU citizens say they see fake news both as an issue for their country and for democracy in general (EU Open Data Portal, 2018: 1). Half of the EU citizens aged 15–30 say they need critical thinking and information skills to help them combat fake news and extremism in society (EU Open Data Portal, 2018: 2).

For the Council of Europe, it is clear that propaganda, misinformation, and fake news are one and the same threat. The terms "propaganda", "misinformation", and "fake news" often overlap in meaning. They are used to refer to a range of ways in which sharing information causes harm, intentionally or unintentionally – usually in relation to the promotion of a particular moral or political cause or point of view (Council of Europe).

It repeats the distinction used by UNESCO between mis-information – false information shared with no intention of causing harm (false connection, misleading content) –, disinformation – false information shared intentionally to cause harm (false content, imposter content, manipulated content, fabricated content) –, and mal-information – true information shared intentionally to cause harm (leaks, harassment, hate speech) (Wardle & Derakhshan, 2017).

The news that the Council of Europe is bringing in the debate about fake news is testament to the type of forensic information study taking place of the 'elements' of a piece of information – the agent/source, the messages and their contents, and the interpreters of the information – each adding to the information disorder. The study also underlines that, although none of these phenomena are new, they have taken on new significance recently with the widespread availability of sophisticated forms of information and communication technology.

The sharing of text, images, videos, or links online, for example, allows information to go viral within hours. This closes down the debate whether we have an old or a new instrument in the informational warfare, and establishes that we have an old tool but with new technological instruments, the Internet and social media, that dramatically change the substance, speed, and context of the threat (Chifu & Nantoi, 2016).

The Whys

There's an abundance of questions, I named them the "Whys", which remind us how little we still know about the impact of technology, informational war, and the fake news occupying our minds, behaviour, and day-to-day life. There are no strict and complete answers, but the Whys are opening up avenues for research. Some knowledge we already possess, and there are other studies already on the way. This is examined in the two chapters dealing with the impact of technology and social media, as well as the impact of the pandemic; although the studies refer just to the first wave – on the human being, society, political life, and democracy. This is not to excuse previous evolution and pre-existing rifts inside democratic societies, which cannot be blamed only on the crises of the past few years, but on the entire situation that led us here.

There are several Whys on the table, and some are more important to our initial responses. We are going to focus on the most obvious ones: Why do we accept fake news (altered information)? Why don't we identify fake news? Why the explosion of disinformation right now? What are the mechanisms in the human mind that make us ignore even the obvious, when we know it is not the truth nor factual reality, to fit our interests and our will?

Feelings, Emotions, Beliefs: Shading Rational Thinking

The reality of this acceptance comes from a lot of processes, but the first and most important one is linked to emotions. Any human being is defined by emotions, and those emotions are like drugs; able to shade away rationality, to put our critical thinking under anesthetic, and to make us believe what we know is not true. On another hand, yes, there is a lot of information that we don't know if it is fake, but we take it as such and consider it true, even disseminating it with the credibility that we have in our own circles of "friends" on the Internet.

Dominique Moisi was someone who made a chart of our fundamental emotions, using a psychological postulate that any emotion is a composition, in due quantities, of three fundamental ones – Fear, Humiliation, and Hope (Moisi, 2010). Attributing a certain dominant fundamental emotion to a category of people in the world allowed him to solve the most important criticism of Samuel Huntington's book on the clash of civilizations (Huntington, 2002), about the borders and artificial lines of demarcation of civilizations. For Moisi, all those emotions could be at the same place, defining different people.

Emotions can change our rational behaviour. This fact is well known, and there is a lot of literature to support it. Dan Gardner has provided an excellent overview on feelings and their relationship with emotions; the sense of "Danger to the Herd" that could export emotions to a group, especially Fear; attempts to describe the "Chemistry of Fear; and the extreme fear or terror, when analyzing the context and perspectives of those "Terrified of Terrorism" (Gardner, 2009). Playing with feelings, creating reactions and using the human mind in order to distort the (assumed) reality – or at least put an emphasis on a needed aspect – could alter dramatically the rational block of the human mind, even at the group level.

Social media plays and important role when it involves echo chambers and informational bubbles. Emotions are exacerbated through a lack of public space and the gathering of like-minded people together. Opposing arguments are missing, there is no debate, and we are no longer on a public space, so even democracy appears faked. We just have one and the same idea presented at full speed, aimed at prompting the required emotion enabling us to create a general rejection of any argument that doesn't fit into our thesis and beliefs. Indeed, these emotions and feelings move closer to religious beliefs; no proof, evidence, or argument are required, we just *believe* in the narrative of the group because we *feel* it is right, and it gives sense to our emotions.

The mechanism has been analyzed thoroughly by psychologists, socio-psychologists, and sociologists, as well as communication experts who realize how it is possible to act, to obstruct and to transform rational thinking. We have already developed some ideas. A major error is the one related to the perspective and dimensions of each issue that concerns the individual society, or even the agenda brought to the attention of the public (by the media, officialdom, social media). If you look at only one thing, it seems to fill the entire landscape and the view captures your full interest.

This error of perspective can also be influenced by the conspiracy theories that take over the public agenda; labelled as "Official", they try to drag you into an alternative world, an alternative reality, within those social media groups that are offering a different agenda. But there is not always only one "objective" or the "independent" view, as all the needs and interests of different groups are catered for, including those of aggressive foreign countries. One must adopt a realistic perspective and approach it in a rational manner, using critical thinking and fact-checking, in order to realize who are in fact profiting from your time and attention on a particular issue.

Sometimes, the subject at stake fits into your area of interests or expectations or, moreover, the subject touches upon some very intimate emotions and feelings of your own, dragging you into the group and keeping you away, eventually, from opposing arguments and real pragmatic thinking on the issue. And it is not only the subject matter that needs to be "sold" convincingly; it also needs to be assessed and addressed in the right way, through a professional approach, in order to attract new followers, new believers.

Can We Do Everything with the Informational War (Fake News, Disinformation, Propaganda)?

The big debate is what came first, the egg or the chicken? The problems within our societies are speculated upon and amplified by the use of informational warfare, or the informational warfare projects alternative realities within our societies, giving a voice to those outside of the mainstream – societies, groups, and followers of "new religions" – which are based on emotions, feelings, and fake news. A society may be exposed and "conditioned" by the existence of such groups (for example the isolation during a pandemic that pushed more people to use social media and for longer), after social media itself splits the society into pieces and divides the public space, it is easier to forge an informational warfare against a democratic society, taking advantage on the principles and values that it defends, including freedom of speech and the free flow of ideas.

Critics are coming from both sides. Presuming that propaganda can do everything in any society is a false axiom. Moreover, if we put the blame for everything that goes wrong in our societies on the informational warfare, propaganda, disinformation, and fake news, we miss the point.

We risk ignoring the real social tensions, divisions, and rifts inside our society that we need to address and to solve for the health of our societies. Having the information war and fake news under its control could lead to an over-confident government and increase the effectiveness of its propaganda unless we correctly monitor and have an overview of the real problems and concerns of our society.

We know for a fact that perception can take the place of reality. Once established via a general perception, a "perceived reality" cannot be shaken by any argument. This happens when communication is missing or is badly conducted, and the public lacks trust in a leadership on a specific badly promoted issue. It happens in times of crises, when solving the crisis but not communicating this with the public can cause the decision maker to lose their job. On another point, the so-called Thomas Theorem tells us that a falsehood perceived as real, could become real through its consequences. It produces real consequences, even if there is fake news at its origin (Dungaciu, 2017: 11–17). This is the case of logic, where a false proposition can be manipulated to give an impression of a true one through the application of rational thinking.

Moreover, the context can create the effectiveness of the falsehoods. In the instance of a pre-existing trust crisis of a state and its institutions, among the political leadership or a professional strata in a specific discipline, this may lead to an exacerbation of the real crisis from a fake news or false premise. The lack of trust and credibility of the official decision makers provides the basis for a larger share of the population to believe and to trust fake news. Critical thinking and a rational approach are put aside. On the other hand, when many individuals trust in institutions and the state, fake news and informational warfare often only has a marginal and insignificant effect on an irrelevant or marginal number of people. But what we have discovered is that any lie, even the most unbelievable, always has a public; a small, irrelevant, or marginal one, but still a public (Chifu & Nantoi, 2016).

How we can build resilience in society against fake news, propaganda, and disinformation depends on the following: good communication, which is both timely and credible; and a high level of trust in societal representatives, its leaders and elected or appointed officials. A leadership must be interested in public trust, and concerned about policies that really reflect people's hopes and expectations, as well as the true capacity of the society to solve those issues facing it. This leads to matching the leadership and political elite with the natural and professional elite of a society. At least, it could not be enough, especially when the level of public expectation is higher than the society in question could provide/offer them.

Fake news cannot change the reality *per se*, but it could definitely influence it, and amplify some trends. Fake news may benefit from pre-existing realities – difficulties, rifts, and divisions in a society may be amplified – but it cannot change reality. People and individuals need to trust somebody, and put their trust somewhere. If one does not believe and trust in their society, your political leaders, then one could trust in the source of fake news (including "my friend" from Facebook). The lack of trust in the official narrative creates the avenue for an informational warfare.

Trust can also come indirectly from the lack of hope that a government or a leadership is offering to the society. People need hope so if none is being offered by the political class of their country, they are inclined to find this hope elsewhere. It is also about emotions, fundamental emotions. People will go where somebody offers hope, because they are being offered a different outcome than the gloomy or dark prospects predicted by the leadership in a crisis situation. This was the case during the pandemic. The need for hope makes individuals look for alternative narratives, even if these are only fake news stories or outright lies.

Vulnerability and lack of resilience in the face of informational warfare arises from a low level of trust in institutions, elites, or national founding myths. This makes a whole society vulnerable to those acts of informational warfare, or to disinformation or to foreign propaganda, that accentuate disbelief. Therefore, we cannot ignore the sociological approach; it is only by knowing the real issues of a society that need to be addressed first, that we can then deal with the fake news, disinformation, and propaganda that the informational war unleashes upon it. Deal with the sociological driven issues first, before problems associated with communication techniques.

The obsession over propaganda can become propaganda in itself. It fills the space of a rational explanation for a crisis, blaming external factors other than the leadership of a country. Transforming everything into propaganda, by labelling everything as propaganda means not identifying the actual propaganda, meaning the defender risks missing the real informational aggression. But this also means not tackling the real social problems within a society. Nuances are always necessary. Therefore, sociology is back on the agenda and is needed to support the political leadership. By this is meant qualitative sociological research, not only superficial questionably qualitative polls. A study about trust, the Whys of the public's disbelief in the national leadership, especially the political one, is necessary before beginning the crusade against propaganda and informational war.

The Nuances and Sophistication of Disinformation 2.0

If we are used to discussing fake news, disinformation, and propaganda, all elements of the informational warfare, in a very black and white format – it is either true or false – "Disinformation 2.0" comes with more nuances and a higher degree of sophistication. It is no longer that easy to prove each of the attributes – true or false – so it is difficult to deal with the new generation of fake news. Nuances are just as important, because they need far more steps to prove each value. Fake news is neither true nor false. From completely false to untrue, there are 50 shades of gray. The reality is no longer only black and white.

Disinformation 2.0 is a mixture of true and false, but in different doses. As much as the false part is less perceivable, the story is better constructed, and the fake news/disinformation/propaganda (equivalent terms somehow) are more difficult to expose. However, a story does not just involve this, but is also about other ingredients of a more subjective nature like observation, impressions, feelings, perceptions, and opinions of witnesses. That situation leads directly to the post-truth era.

Surely, however, informational war is neither immaculate in scope, nor impeccable in logic, truth, and presentation of the facts. It not only about the vulnerabilities within a society. It is not innocent, with some of the activities influencing the environment itself, preparing it to become receptive to future actions of the informational warfare, with a greater impact. It is an aggression on our societies, it takes advantage, like all other components of a hybrid war, on the characteristics of a liberal democracy, on the principles and values that we are cherishing and defending, including freedom of expression and of the media.

And this situation comes from creating a complete mess about the truth, as was already underlined before. Who owns the truth? Who tells what is true? What is the value of knowing what is true and false? Therefore, we are living in times where there is a complete relativization of the truth. The real question now is: Do you believe me or not? There is no need for any argument or any evidence. Actually, disinformation 2.0 destroys evidence, criteria, and arguments. We are caught somewhere between "the truth and false are equal as importance and moral relevance" (Nietzsche) in terms of the logic, and "the truth is what I'm telling you is the truth" (Gobbels).

Fake news becomes completely different to false news. News could be counterfeit, credible, plausible, not only false, in order to be fake. There are nuances. Disinformation 2.0 is a plethora of nuances of gray inside the truth, not just nakedly false. Propaganda, disinformation, communication errors, moral panic, inuendo, collective hysterias, intoxications, diversions, conspiracies, partisanship, all are parts of the informational warfare, both Old and New. But today's means are different. Information, disinformation, persuasion, and propaganda are parts of the story, with different instruments attached and different values of the truth (Bârgăoanu, 2018: 133–139).

Fake news is a term found in nineteenth-century English vocabulary. The term exploded when it was politicized. Donald Trump played the leading role, labelling CNN and mainstream media as fake news (hiding information inside of a lot of noise) back in 2018, establishing the costs of fake news!!! Since then, the term has not gained further importance since it has become politicized and covers a wider area of use than the original concept. Fake news became an umbrella term for nearly everything. Through politicization, fake news has become the equivalent of hostility towards the media, with stories that we refuse to acknowledge, interpretation that we disagree with, and points of view that are detrimental to us.

On another point, there is an important part of society that believes that fake news is a motif for censoring the freedom of the media, an opportunity to limit the freedom of expression. And here the fight against fake news needs to pay more attention to the perceptions of the population, and to find genuine and more largely accepted ideas for limiting the freedom of the media or of expression. Such is the case with hate speech, verbal violence, and tarnishing a person's image.

As we have seen, fake news is a name/label that could be attributed to everything that we disagree with. The term is indiscriminately used and does little to identify the "real" fake news. It becomes an excess through politicization and generalization, a strategy to undermine the credibility of, to discredit everything to do with, or at least to question, the genuine truth. In the discipline of semiotics, we talk about a pair of words – for example, significant and signified – the name or label of a word and its content or substance. A way of building fake news is either to mix them, or to alter the substance of a concept. In the end, all this leads to undermining the trust in what's real, obvious, concrete and visible around us. It leads to a perfect relativization of all things.

It all begins like in a soap opera – stories inside the stories, that turn on the hero and the villain, the bad and the good. It is a work of the relativization of good and bad and the story helps to make the roles interchangeable. The bad becomes the good and vice versa. And that paves the way for huge uncertainty. That's how we begin to build conspiracy theories – stories that seem incredible but are needed to be able to shift good into bad and bad into good, or at least to add more nuances into each of the actors so that the result is that there is no difference between the hero and the villain. That's how conspiracy theories begin to be acceptable and are even welcomed within such a milieu; those who are used to consuming soap operas. Under stress, and in times of crisis, the situation becomes even worse.

Yuval Harari said that *homo sapiens* is a post-truth species (Harari, 2018). Survival is dependent on creating and believing in fictions. This is partially true, since sophistication multiplies the nuances of the truth and that of the false, and nuances and shades of gray amplify the difficulty of dealing with fake news. Old phenomena coupled with new technologies – social media, metadata, algorithms, virtual platforms, artificial intelligence, research engines – directly changes the society. How does this happen? We still don't know all of the consequences and mechanisms of the impact of social media on society and the individual. But we need to quickly realize where we go from here.

The basic approach of the EU is that disinformation means intention. I am not very sure that even in this field we can be so radical. Disinformation 2.0 can use pieces of intentional informational warfare techniques that are taking advantage of vulnerabilities already created in a society and some parts of disinformation without intention (or malinformation, as it is labelled). The side effects of political or electoral campaigns in a society can create fertile ground to accentuate its vulnerabilities or the fractures in societal cohesion; this allows an intentional Disinformation 2.0 campaign to be much more effective when needed, in an informational war that is not always linked to a conflict or physical war.

This is not without consequences. As Condoleezza Rice put it, the political risk is coming from everyone with a cell phone/photo camera/social media profile. It is a new type of media channel better fitted to exploit any story with an ideological component or political issue inside (Rice & Zegart, 2018). Since business and politics are once more closely linked following the change of geopolitics after the Cold War, and people have enough knowledge and numerous causes to align themselves to, it is easy to transform everything into politics and to shake up the government.

The above-mentioned logic is not really like that: politics has first and foremost embraced the agenda of non-governmental organizations (NGOs) and moved all issues on to the political stage, politicizing the agenda and democratizing society as a whole. But not all the pieces of this agenda are of interest to the population in their politicized form. A government cannot be shaken by practically anything, unless it requires immediate attention and as it fills a need or an expectation of the public. In other words, we need to have at least the context prepared and the trust in the leadership shaken before moving to action. If not, the impact would not be meaningful. But the fact is that megatrends in politics, business, and technology have transformed political risk-taking, making it more diverse, pervasive, and consequential (Rice & Zegart, 2018: 10).

Conclusion: Complementary Issues and Avenues of Research for Contemporary Information Warfare

Informational/information war/warfare allow us to examine other complementary concerns that need to produce suitable and comprehensive answers in order to help research and make sense of these issues. There is a lot to do in an interdisciplinary approach, and the difficulty rests with those matters. That is why we have presented some parts of the reality where there are no answers or the level of the research is still at the beginning, with no convincing results in place.

A first area of research is the one related to "Fake News", false news, or deep fake news. This relies on the capacity of our senses to determine and establish the truth, the realities, and facts. When the senses create fake news, we have a big problem, especially when it concerns our own view – which gives us more than 80 per cent of the information. We saw images that prove not to be true, to be misleading and to create fake news. When we can no longer rely on our natural senses, as humans, we have a level of relativization of the information coming from our natural senses that is no longer acceptable, and a large part of humanity can no longer cope with this level of relativization.

The Coronavirus and its sense of danger has already been evoked here. In what respects there is fear, or terror, when thinking about Coronavirus, it is difficult to realise today. Sociology could help in order to realize where distrust within our societies related to the Coronavirus pandemic comes from. If the level of impact of the illness on the human being is less important or perceived as comparable to normal mortality in human society, we could accept it and reject inconvenient measures taken by our officials in order to cope with this crisis. It is like death during wartime or by traffic accidents; even if the impact is high, nobody refrains from driving or using their car.

Another part of the research needed relates to Rhetoric. It is a science that has being marginalized or forgotten. The great speeches of our time are full of content and creative wording that cross the years and eras and are still quoted. Now, populism needs to be addressed not only from the charismatic angle of personalities, but also from the rhetorical point of view and natural abilities to make great speeches at any moment.

Communicating feelings and emotions, not only stories, is also an ability that needs to be observed and studied. We need to investigate the ways and means to fulfil the expectations of the public and realize how the acceptance of obvious and visibly untrue messages or direct lies happens in rational individuals. An interesting start could be that of Donald Trump's ability to build very vivid stories, how he creates a captivating scenography for his fantasies that makes the public trust him; people become more fascinated by the teller and the story itself than they are inclined to reject the verbal false claims and factual lies from his imagination.

We have examined some elements of StratCom, the strategic communication organization fighting fake news, and the way that our governments and the international institutions are approaching the issues. This is not enough, however. We need our reactions to be more effective and proactive; it is about penetrating and influencing the bubbles and echo chambers or trying to use our available tools to combat the populist success of communicating via social media, particularly during a pandemic.

Defence against informational warfare also needs to be achieved in conjunction with the use of more offensive instruments, tools, and techniques. If we cannot share experiences when working in that part of the informational front, we will not be capable of reacting to the art of influencing. The techniques of manipulation are generally well known, but there is less knowledge and research on the capacity of changing the shadows in a scene in any given playground: How you apply light to a scene so that it becomes a *trompe l'oeil*, falsifying the direct perception from the eye and the viewed perspective. Once those techniques are employed, we can see how this shapes fake realities to be absorbed via the senses, making them extremely credible. Rationalizing information warfare as magic tricks are not meaningful in terms of being a comparison and deriving lessons learnt, which is the intention of this research.

Finally, more detailed research should address the way democracy has evolved without a proper and genuine public space, without a real debate, and in a fragmented informational space that has become increasingly less transparent and public. We have less of the sense that we are sharing the same information that we know is true, or common knowledge about our day-to-day lives. This lack of shared information and debate, of thesis and anti-thesis, discipline of dialogue and arguments, a rational and critical-thinking approach, have all affected democracy as too many feelings and emotions are becoming involved. If we add the areas and means for relativization of the truth, we are reaching the point where our entire democracy needs to be reset, adapted, and updated to cope with the new Disinformation 2.0 world.

References

Aro, J. (2016), The Cyberspace War: Propaganda and Trolling as Warfare Tools, *European View*, 15(1), pp. 121–132.

Bârgăoanu, A. (2018), *Fakenews. Noua Cursă a înarmării*, Bucharest: Evrika Publishing.

Bērziņš, J., Jaeski, A., Laity, M., Maliukevicius, N., Navys, A., Osborne, G., Pszczel, R. & Tatham, S. (2014), *Analysis of Russia's Information Campaign against Ukraine*, Riga: NATO StratCom Centre of Excellence.

Borden, A., Colonel USAF (Ret.) (1999), What Is Information Warfare?, *Aerospace Power Chronicles*, 1999(11), 1–1, at www.airuniversity.af.edu/Portals/10/ASPJ/journals/Chronicles/borden.pdf.

Campen, A. D., Colonel USAF (Ret.) (July 1995), Rush to Information-Based Warfare Gambles with National Security, *Signal Magazine* (Signal is the official publication of the Armed Forces Communications and Electronics Association (AFCEA)).

Canadian Security Intelligence Service (February 2018), Who Said What? The Security Challenges of Modern Disinformation, Academic Outreach, World Watch: Expert Notes series publication No. 2016-12-05, Canada, www.canada.ca/content/dam/csis-scrs/documents/publications/disinformation_post-report_eng.pdf).

Chifu, I. & Nantoi, O. (2016), *Information Warfare: The Pattern of Aggression*, Bucharest: The Publishing House of the Institute of Political Sciences and International Relations "Ion I. C. Brătianu" of the Romanian Academy.

Chifu, I. & Ţuţuianu, S. (2017), *Torn between East and West: Europe's Border States*, London and New York: Routledge.

Čižik, Tomáš (Ed.) (May 2017), *Information Warfare – New Security Challenge for Europe*, NATO PDD, Marketing and Business Group, s. r. o., www.researchgate.net/publication/322695565_Information_Warfare_-_New_Security_Challenge_for_Europe.

Conley, H. A., Mina, J., Stefanov, R. & Vladimirov, M. (2016), *The Kremlin Playbook. Understanding Russian Influence in Central and Eastern Europe*, Washington: CSIS.

Council of Europe, *Dealing with Propaganda, Misinformation and Fake News*, www.coe.int/en/web/campaign-free-to-speak-safe-to-learn/dealing-with-propaganda-misinformation-and-fake-news.

DEEP NATO, www.nato.int/nato_static_fl2014/assets/pdf/2020/5/pdf/2005-deepportal4-information-warfare.pdf.

Department of Defense (2012), *Report of the Defense Science Board Task Force on Information Warfare*, Defense (IW - D) Paperback – October 25.

Dungaciu, D. (2017), Triada Gândirii Războiului informational in Dumitrescu, L. (Ed.), *Războiul informational sub lupă. Concepte, metodologie, analize*, Bucharest: The Publishing House of the Institute of Political Sciences and International Relations "Ion I. C. Brătianu".

Eloff, J. & Granova, A. (2009), Information Warfare in Vacca, J. R. (Ed.), *Computer and Information Security Handbook*, pp. 677–690. Burlington: Elsevier.

EU Commission Communication, *Tackling Online Disinformation: A European Approach*, 26 April 2018, Communication of the Commission, https://eur-lex.europa.eu/legal-content/EN/TXT/?uri=CELEX:52018DC0236.

EU Open Data Portal, Flash Eurobarometer 455, 2018: 2, https://data.europa.eu/euodp/en/data/dataset/S2163_455_ENG.

Flash Eurobarometer 464, 2018, https://data.europa.eu/euodp/en/data/dataset/S2183_464_ENG.

EU Vocabulary, Concept, https://op.europa.eu/en/web/eu-vocabularies/concept/-/resource?uri=http:///eurovoc.europa.eu/c_4e75f061.

European Commission (2018), *Code of Practice on Disinformation*, https://ec.europa.eu/digital-single-market/en/news/code-practice-disinformation.

(26 April 2018), *Communication from the Commission to the European Parliament, the Council, the European Economic and Social Committee and the Committee of the Regions*, 2018: 3, https://eur-lex.europa.eu/legal-content/EN/TXT/PDF/?uri=CELEX:52018DC0236&from=EN.

(2020), *EU and Fake News, Tackling Online Disinformation*, https://ec.europa.eu/digital-single-market/en/tackling-online-disinformation.

(2020), *The European Digital Media Observatory*, 2020: 2, https://ec.europa.eu/digital-single-market/en/european-digital-media-observatory.

(2018), *Roadmaps to Implement the Code of Practice on Disinformation*, 2018: 2, https://ec.europa.eu/digital-single-market/en/news/roadmaps-implement-code-practice-disinformation.

European Parliament (24 April 2018), *Online Disinformation and the EU's Response*, chrome-extension://efaidnbmnnnibpcajpcglclefindmkaj/https://www.europarl.europa.eu/RegData/etudes/ATAG/2018/620230/EPRS_ATA(2018)620230_EN.pdf

Farwell, J. & Arakelian, D. (December 2013), *What To Do Next About Syria*, www.huffingtonpost.com/james-p-farwell/what-to-do-next-aboutsyria_b_4494532.html.

Gardner, D. (2009), *Risk. The Science and Politics of Fear*, New York: Random House Group.

Ghosh, D. & Scott, B. (January 2018), *#DigitalDeceit. The Technologies behind Precision Propaganda on the Internet*, Public Interest Technology, New America, www.newamerica.org/pit/policy-papers/digitaldeceit/; https://d1y8sb8igg2f8e.cloudfront.net/documents/digital-deceit-final-v3.pdf.

Harari, Y. N. (2018), *21 de lecții pentru secolul XXI*, Iași: Editura Polirom.

Huntington, S. P. (2002), *Ciocnirea civilizațiilor și refacerea ordinii mondiale*, Bucharest: Editura Antet, București.

Hutcherson, N. B. (1994), *Command and Control Warfare: Putting Another Tool in the War-Fighter's Data Base*, Maxwell AFB: Air University Press.

Joint Staff, (1996), Joint Doctrine for Command and Control Warfare, Joint Publication, 3–13(1), 7 February.

Lewis, B. C. (1997), Information Warfare, Federation of American Scientists, https://fas.org/irp/eprint/snyder/infowarfare.htm.

Longsdale, D. J. (1999), Information Power: Strategy, Geopolitics and the Fifth Dimension, *Journal of Strategic Studies*, 22(2–3), pp. 137–157.

Lucas, E. & Pomeranzev, P. (2016), *Winning the Information War. Techniques and Counter-Strategies to Russian Propaganda in Central and Eastern Europe*, Washington DC: CEPA's Information Warfare Project in Partnership with Legatum Institute.

Lucas, E., & Nimmo, B. (November 2015), *Information Warfare: What Is It and How to Win It?*, CEPA Infowar Paper No 1, at chrome-extension://efaidnbmnnnibpcajpcglclefindmkaj/https://cepa.ecms.pl/files/?id_plik=1896.

Macdonald, S. (2006), *Propaganda and Information Warfare in the Twenty-First Century: Altered Images and Deception Operations*, London & New York: Routledge.

Marlatt, G. E. (2008), *Information Warfare and Information Operations (IW/IO): A Bibliography*, Dudley Knox Library, Naval Postgraduate School, Revised and updated January 2008, Homeland Security Digital Library, www.hsdl.org/?abstract&did=443229.

MEDIA – (DIS)INFORMATION – SECURITY, Defence Education Enhancement Programme (DEEP) NATO, www.nato.int/nato_static_fl2014/assets/pdf/2020/5/pdf/2005-deepportal4-information-warfare.pdf.

Moisi, D. (2010), *The Geopolitics of Emotion*, New York: Anchor Books.

Molander, R. C., Riddile, A. & Wilson, P. A. (1996), *Strategic Information Warfare: A New Face of War*. Santa Monica: RAND Corporation, www.rand.org/pubs/monograph_reports/MR661.html. (Also available in print form.)

Naef, W. E. (2003), *Psychological Operations Interview with Larry Dietz*, London: Infocon Magazine Issue One, 29 April 2003.

NATO (2020), *NATO's Approach to Countering Disinformation: A Focus on COVID-19*, www.nato.int/cps/en/natohq/177273.htm.

NATO Stratcom CoE (2018), *Fake News. A Roadmap*, Riga, www.stratcomcoe.org/fake-news-roadmap.

Nichiporuk, B. (1999), *US Military Opportunities: Information Warfare Concepts of Operation*, Chapter 7, in Khalilzad, Z. & White, J. (Eds.), *Strategic Appraisal: The Changing Role of Information in Warfare*, Santa Monica, US: Rand, www.rand.org/pubs/monograph_reports/MR1016.html.

Nissen, T. E. (2015), *#TheWeaponizationOfSocialMedia:@Characteristics_of_Contemporary_Conflicts*, Royal Danish Defence College.

O'Shaughnessy, N. (2012), The Death and Life of Propaganda, *Journal of Public Affairs*, 12(1), 29–38.

Ohlin, J. D., Govern, K. & Finkelstein, C. O. (Eds.) (2015), *Cyberwar: Law and Ethics for Virtual Conflicts*, Oxford: Oxford University Press.

Pacepa, I. M. & Rychlak, R. J. (2013), *Disinformation*, Chicago: WND Books.

Panarin, I. & Panarina, L. (2003), *Information War and Peace* [Информационная война и мир] Moscow: ОЛМА-ПРЕСС.

Payne, K. (2005), The Media as an Instrument of War, *Parameters*, 35(1), Spring, pp. 81–93.

Polyakova, A., Laruelle, M., Meister, S. & Barnett, N. (2016), *The Kremlin's Trojan Horses*, Washington DC: Atlantic Council.

Rice, C. & Zegart, A. (2018), *Political Risk. Facing the Threat of Global Insecurity in the Twenty-First Century*, London: Weidenfeld & Nicolson.

SANS Institute (2004), *Global Information Assurance Certification Paper*, GIAC Certifications, Information Warfare: Cyber Warfare is the Future Warfare, www.giac.org/paper/gsec/3873/information-warfare-cyber-warfare-future-warfare/106165

Scheidt, M. (2019), *The European Union Versus External Disinformation Campaigns in the Midst of Information Warfare: Ready for the Battle?*, College of Europe, EU Diplomacy Papers, 1/2019.

Shannon, C. E. (1948), A Mathematical Theory of Communications, *Bell Syst. Tech. J.*, 27, 379–423.

Simons, G. (2010), *Mass Media and Modern Warfare: Reporting on the Russian War on Terrorism*, Farnham: Ashgate.

Simons, G. & Chifu, I. (2017), *The Changing Face of Warfare in the 21st Century*, London and New York: Routledge.

Stern, E. (1999), *Crisis Decision Making. A Cognitive Institutional Approach*, Stockholm: Dept. of Political Science, University of Stockholm.

Stupples, D. (December 2015), What Is Information Warfare?, World Economic Forum, www.weforum.org/agenda/2015/12/what-is-information-warfare/.

Tandoc, E. & Wei, L. & Ling, R. (2018), Defining "Fake News": A Typology of Scholarly Definitions, *Digital Journalism*, 6(2): "Trust, Credibility, Fake News".

Tenet, G. J. (24 June 1998), "Testimony by Director of Central Intelligence" before the Senate Committee on Government Affairs,www.cia.gov/.

The Military Doctrine of The Russian Federation (2014), www.rusemb.org.uk/press/2029.

Thomas, T. (2014), Russia's Information Warfare Strategy: Can the Nation Cope in Future Conflicts?, *The Journal of Slavic Military Studies*, 27(1), pp. 101–130. DOI: 10.1080/13518046.2014.874845.

Thornton, R. (2015), The Changing Nature of Modern Warfare. Responding to Russian Information Warfare, *The RUSI Journal*, 160(4), pp. 40–48. DOI: 10.1080/03071847.2015.1079047

UNESCO (2018), *Journalism, "Fake News" and Disinformation: A Handbook for Journalism Education and Training*, UNESCO, https://en.unesco.org/sites/default/files/journalism_fake_news_disinformation_print_friendly_0_0.pdf.

Wardle, C. & Derakhshan, H. (2017), *Information Disorder: Toward an Interdisciplinary Framework for Research and Policy Making*, Strasbourg: Council of Europe, https://rm.coe.int/information-disorder-toward-an-inter disciplinary-framework-for-researc/168076277c.

Widnall, S. E. & Fogelman, R. R. (1997), *Cornerstones of Information Warfare*, US Air Force.

Williams, H. J. & Blum, I. (2018), *Defining Second Generation Open Source Intelligence (OSINT) for the Defence Enterprise*, Santa Monica: RAND, www.rand.org/pubs/research_reports/RR1964.html.

Winston, W. B. (1997), Bits, Bytes and Diplomacy, *Foreign Affairs*, 76(5), pp. 172–182.

Zhukov, M. Y. & Baum, M. A. (2016), Reporting Bias and Information Warfare, International Studies Association Annual Convention Atlanta, GA.

3 Understanding Political and Intangible Elements in Modern Wars

Greg Simons

Introduction

Organised warfare has been waged for many centuries; certain elements of this pursuit have remained constant despite the advancement of technology and the scale of destruction that is visited upon those societies that are affected.[1] Betz (2015; see also Hageback & Hedbloom, 2022) argues that the ways wars are fought are rapidly changing owing to increased global connectivity, which impacts upon social and political relations. It is also an era where some of the most persistent challengers of the status quo are not only state but non-state actors. There are two sides to war: the tangible and the intangible. The tangible elements are those physical components that can be seen, touched, felt and heard – soldiers, weapons, the terrain and weather, for example. The intangible elements may not be seen or heard as cognitive-based psychological factors, but they exert an impact upon the outcome of a war. Among the intangible elements, one can begin the list with politics, information, reputation, morale and perception. Intangible factors can be influenced by the nature of the tangible representations in the information domain.

Payne (2005) argues that mass media and journalism play a critical role in the life cycle of politics in modern warfare, which is evidenced by the 'Forever Wars'[2] being waged by the Western powers. Winning the media war is crucial to military planners as war is a political enterprise that makes use of hard power, but requires soft power (social and political capital) for the perception of legitimacy and righteousness of something that is by nature anything but benevolent or humanitarian.

[1] This chapter is an updated and revised version of: G. Simons, *Understanding Political and Intangible Elements in Modern Wars*, Государственное управление. Электронный вестник, Выпуск № 34. Октябрь 2012 г.

[2] See www.csis.org/analysis/current-military-operations-and-concept-forever-wars and https://foreignpolicy.com/2019/07/17/limited-wars-are-forever-wars/ for a discussion on the issue or listen to https://warontherocks.com/2020/03/are-the-forever-wars-really-forever/.

When the public reads about war in newspapers or on the Internet, hears it on the radio or watches it on TV, there is an overwhelming tendency by journalists and the mass media to be descriptive about the conflict that they are covering. It touches the surface; you see and hear the sights, sounds and horrors or war. Yet the public may actually fail to understand how and why a war occurs in the first instance as it is void of analytical context. The outcome of wars is increasingly determined by political factors and less through kinetic[3] military operations (Payne, 2005; Betz, 2015; Dimitriu, 2018). Given the rise in the usage of the term 'humanitarianism' as an excuse to justify military action, there is a critical need to understand the intangible factors that influence the political environment and lead to the manner in which wars are fought.

This chapter will examine a number of different intangible factors that shape why and how wars are currently fought. A historical understanding and conceptualisation of wars will be established through a brief introduction to a limited number of key historical treatises on the subject; namely, Sun Tzu's *The Art of War*, Niccolò Machiavelli's *The Art of War* and Carl von Clausewitz's *On War*. On occasions, Sun Tzu and von Clausewitz have been criticised in these contemporary times. These criticisms can lack an understanding of an entirely different cultural, political and strategic environment (Milevski, 2019). How do these influential writings rate in terms of their relevance to today's contemporary world?

As it will be shown, in some instances, these writings are still relevant and help to generate some further understanding of the interactions and effects of these intangible elements on war. In the age of new communications technology and instantaneous global transmissions, information can be viewed by various actors as simultaneously a threat and an opportunity. A threat owing to the potential to cause damage to the intangible assets of an actor, especially those political factors related to legitimacy and credibility. But it is also an opportunity for the same reasons; it can be wielded against an opponent's intangible assets in order to weaken and/or isolate them.

The final section deals with the actual targeting of both tangible and intangible assets. This shows the potential that can be harnessed through waging war, not only in the traditional spheres of land, sea and air, but also within the information space, which may be considered as an additional component of the battleground.

[3] Kinetic refers to military operations that use deadly force against an opponent as opposed to psychological and cognitive factors in non-kinetic operations.

Historical Framing of the Elements and Pursuit of War

From a historical point of view, there has been an age-old interest in the philosophical and theoretical aspects and components of war. The study and development of these points tend to be aimed at improving the prosecution of war. This has been carried out through the ages, ever since humankind started to wage organised forms of warfare. This section will look briefly at the views on the role of politics and intangible elements in influencing warfare of three influential authors – Sun Tzu, Niccolò Machiavelli and Carl von Clausewitz.

Sun Tzu was an influential Chinese general and tactician. Born in the sixth century BC in China, he has greatly influenced East Asia, and continues to influence Western thought on the topic of war. According to Sun Tzu, 'warfare is the greatest affair of state, the basis of life and death, the Tao to survival or extinction. It must be thoroughly pondered and analysed' (Sun Tzu, 2003: 10). He broke down the structure of warfare into five different components in order to allow for a comparative evaluation based on estimations. The five factors or components are:

(1) The Tao[4] – influences peoples' relationship with their leader, for instance their willingness to die for them (or not), and not to fear danger;
(2) Heaven – includes the yin and yang, cold and heat. Constraints based upon the seasons of the year;
(3) Earth – refers to the nature of the distances and terrain, far or near, difficult or easy;
(4) General – the possession of knowledge, discipline, credibility, benevolence and courage;
(5) Laws for military organisation and discipline – involves organisation and regulations, the Tao of command and the management of logistics. (Sun Tzu, 2003: 10–12)

What is proposed by Sun Tzu is a mix of tangible and intangible elements that need to be not only known but understood by the commander in order to be successful on the battlefield. The tangible factors are Heaven, Earth and some elements of the Laws for military organisation and discipline. Intangible elements include the Tao, General factors and parts of the Laws for military organisation and discipline. These all touch on the issues of motivation, perception, persuasion and influence, which cannot be

[4] Taoism philosophy is based on three pillars: **simplicity, patience and compassion**. Lao Tzu said that these three principles are our greatest treasures. He further explained that by being simple in actions and in thoughts, you return to the source of being.

physically felt, yet still exert an effect upon the outcome of warfare, not the least of which is the ability to motivate the military and the population to trust, follow and fight for the political and military leadership.

One of the key instruments of warfare, according to Sun Tzu, is the application of deception and disinformation. He stated that, 'Warfare is the Tao of deception' (Sun Tzu, 2003: 14). Various deceptive ploys were suggested, such as to feign incapability, 'when committed to employing your forces, feign inactivity', creating an illusion as to the location of an objective and other ploys (Sun Tzu, 2003: 14–15). Thus, the point of being able to deceive the enemy in order to gain advantage on the battlefield is recognised as being a critical asset in creating an advantage against the enemy. The key to this is the ability to fool the senses of the enemy so that they make a mistake or miscalculation based upon what they think they see or hear, rather than on what actually exists. This is therefore an 'attack' on the enemy's decision-making process and capability, but does not exclude such methods being used against a domestic civil population.

The tactics of Sun Tzu inspired and influenced the tactics of the Communist guerrillas in China (Griffith, 2000: 37) and, as Mao noted in the pamphlet 'Yu Chi Chan', 'without a political goal, guerrilla warfare must fail, as it must if its political objectives do not coincide with the aspirations of the people and their sympathy, cooperation and assistance cannot be gained' (Griffith, 2000: 43). Mao stressed the need for military forces and operations to have the concrete conception of a political goal and the political organisation to attain this objective. It was understood by Mao that political affairs and military affairs are not identical, yet they cannot be separated or isolated from each other, where military action is used to attain the political goal (Griffith, 2000: 89). This was seen as central to the political struggle of leveraging intangible elements in favour of the guerrillas and against the Nationalist government forces.

Niccolò Machiavelli was born in 1469, living in Florence during the Renaissance period, and is considered the father of modern political science. During 1519–1520 he wrote the work *The Art of War*. He stressed that war must be clearly defined, and developed the philosophy of 'limited warfare' that occurs when traditional diplomacy fails and war becomes an extension of politics (Sun Tzu also warned of the dangers of protracted warfare, from which, he said, no one profited). Several basic relationships between politics and war were enumerated by Machiavelli:

(1) Military power is the foundation of civil society.
(2) A well-ordered military establishment is an essential unifying element in civil society.

(3) A policy of military aggrandisement contributes to the stability and longevity of civil society.
(4) Military art and political art possess a common style.
(5) A military establishment tends to reflect the qualities of the civil society of which it is a part. (Machiavelli, 1965: xlvii)

According to Machiavelli, politics and the military possessed a number of similarities, and a strong statesman must also be a capable general (Machiavelli, 1965: liii). He referred to politics and the military as being creative arts, 'moulding raw human material into the desired form, as well as the necessary personal qualities for successful leadership (spirit, creative energy, personal resolve and will power … etc.) in difficult and trying environments' (Machiavelli, 1965: liv). This is in keeping with Sun Tzu's fourth factor, which relates to the personal qualities of the general and a conscious effort to develop and refine the intellectual qualities of human resources.

When it comes to the issue of deception, there is agreement among many that the use of deception and trickery against an enemy is completely justifiable. However, when it comes to fellow citizens and allies, Machiavelli does not characterise them as being a homogeneous and loyal mass, but rather a collection of different interests that can either more or less coincide with the leadership's. Therefore, he sees no moral dilemma in deceiving or tricking fellow citizens and allies in order to create a sense of unity that permits the leadership to pursue its objectives (Machiavelli, 1965: lviii). This can certainly be observed in the contemporary context, such as the use of the pretext of the presence of weapons of mass destruction and links to terrorism as a means to initiate a seemingly legitimate war against Iraq; the deception went as far as the former Secretary of State, Colin Powell, presenting 'irrefutable' evidence of weapons of mass destruction before the United Nations.

Carl von Clausewitz was born in 1780 in Prussia and was a veteran of the Napoleonic Wars. Although a soldier, he could at times be better described as a philosopher. Von Clausewitz was concise in his characterisation of the nature and purpose of the pursuit of war. He defined war as 'an act of violence intended to compel our opponent to fulfil our will' (von Clausewitz, 1982: 101). The fulfilment of 'our will' therefore seems to be the spark that initiates the process that leads to war. What is the factor that decides what 'our will' is to be? This is tied to the influence of, and issues that are guided and formed by, politics: 'Thus, therefore, the political object, as the original motive of the War, will be the standard for determining both the aim of the military force and also the amount of effort to be made' (von Clausewitz, 1982: 109). It is noted by Betz (2015:

12) that the character of politics and war is constantly evolving and follows von Clausewitz's 'Trinity of War' – hatred and animosity (passion), probability and chance, and the quality of reason (political purpose). The relationship between these elements is not fixed and tends to fluctuate in terms of relative strengths and influences. Dimitriu argues that von Clausewitz's concept of politics cannot be misunderstood or misconstrued as it is a broad understanding of power struggles: 'The political logic of war is defined here as the convergence of interrelating factors of power struggles and policy objectives within a given polity that restrains and enables these political forces' (Dimitriu, 2018: 1). The Bolshevik leader, Vladimir Lenin, an admirer of von Clausewitz, developed and tested his own dictum which the experience of the First World War seemed to confirm – warfare is an extension of politics and politics is a prolongation of warfare (Boucher, 2017: 142). This was the basis for developing the idea and practice of Marxist revolutionary warfare.

An important point and distinction made by von Clausewitz is that 'the conception that war is only part of political intercourse, is therefore by no means an independent thing in itself' (von Clausewitz, 1982: 402). This is especially relevant for Western countries in an age where the buzzwords and concepts revolve around such slogans as 'the democratic control of armed forces', which amounts to establishing a political monopoly on the control and use of state-based military force. There is little political transparency or accountability in terms of wielding military power. The obligations are from the military to the political. With these three historical authors, who have impacted upon political and military thought, there is a distinct difference in the Western approach as opposed to Chinese thinking and practice.

There are several differences noted in the Western ways of warfare as opposed to the Chinese approach, which is a product of contrasting intangible factors such as history, culture, values, politics and so forth. Hagen (1996) notes that there are two pervasive and enduring Western attitudes towards waging war: one is the romantic idealism concerning violent conflict and the other concerns the moral justification for going to war. This stands in contrast to Chinese thinking on war. The differences, according to Boylan (1982: 341), are 'an emphasis on stratagem over brute force; attacks on military as opposed to economic targets; a willingness to end a conflict once essential political goals have been met; and an emphasis on man as opposed to machinery (i.e. technology) as being the most vital element determining military strength'. Waldman (2018: 198) lays out the self-defeating approach of US-led military operations – 'vicarious warfare encourages excessively "kinetic" military approaches which serve to exacerbate complex problems, prevent progress toward more realistic negotiated solutions, or, through forms of operational tunnel vision, cause planners to miss potentially mission-fatal blind spots'. The short-term overuse of kinetic military force and the lack of viable long-term political considerations has produced a situation that places the continued military and global hegemony of the West at severe risk, which seems to be interpreted as the gradual breakdown of the US-led unipolar world order for a more non-Western-dominated multipolar constellation (Simons & Glasser, 2019). This is seen in the West's risky approach to military interventions, which are seemingly based on the tangible military superiority of the West in relation to its opponents as well as subjective political factors (domestic and international) that influence political decisions in favour of war.

The influence of political considerations can be seen in various modern international military interventions, such as the low level of interest and commitment in the marginalised conflicts of Rwanda and Darfur; and, conversely, the high level of interest and commitment in Iraq and Syria, where the objective is certainly a matter of imposing one's political will and influence over the targeted country. The use of war can be the avenue to either a means or an end, in answer to the political objective at hand. For instance, the 2003 invasion and occupation of Iraq was stated as being pursuant to the political ends of the Bush administration. That is, to rid the United States of the supposed security threat of weapons of mass destruction and links to terrorism in Saddam Hussein's Iraq. However, it seems, with hindsight, that the true purpose of the invasion and subsequent occupation of the country was to secure access to its rich natural resources, which necessitated the removal of the old regime and the installation of one that was compliant with the Bush regime's will.

There does not have to be an open and full use of military force in order to bring about violence that ultimately compels an opponent to yield to one's political will. The US policy of regime changes through orchestrated revolutions has already proven this point. Additionally, the 'limited' war against Libya did not see a full deployment of military power against the Gadaffi government. Decades of being the world's sole remaining superpower seem to have caused a sense of overconfidence that has seen the neglect of Sun Tzu's warning of involvement in a protracted war and a disregard for all the military philosophers' warnings of the need for clear political goals and concise objectives in war. This has been the case since September 2001 with no clearly defined end to it.

Information as a Threat and an Opportunity

When one speaks of *information security* there are a number of implications (see Buzan, Waever & de Wilde, 1998 for a detailed account of securitisation). Firstly, information is potentially some kind of threat and therefore must be secured. Flowing on from this basic underlying assumption that something is a threat, there is the other side of the equation: that something is being threatened. Therefore, two immediate questions need to be asked before proceeding any further in order to gain a clear understanding and perspective: 'How is information a threat?' and 'To whom is information a threat?'

Not all information is intended to be or is handled equally; information possession means power, as the absence of it potentially disempowers. With the different types of information, the basis or nature of the power possessed or sought becomes more apparent. *Free information* is created by those who are willing to create and distribute information free of charge. The benefit for the sender is if the receiver believes that information, which is related to the goals of persuasion and influence. *Commercial information* is when a sender produces information, but it comes with a price tag. Therefore, the motivation and basis for producing and distributing the information is based upon the idea of commercial gain. *Strategic information* occurs when an actor possesses certain information that their competitors do not. The power is found in the adversary not knowing that an actor possesses or has access to certain information (Keohane & Nye, 1998: 85) (such as the British possession of the German Enigma codes in World War Two). Lonsdale (1999: 144) noted that 'the primary characteristics of information power are its flexibility and accessibility.' Information is part of the fifth dimension of strategy according to Lonsdale (1999); the other four earlier dimensions of strategy include land, water, air and space. This makes the fifth dimension the only intangible dimension of strategy, as the four earlier dimensions are tangible in nature. However, belligerents are able to use the fifth dimension of strategy as a means to affect strategy and operations in the other physical dimensions by disrupting information flows and influencing conclusions in the cognitive realm of the target (Simons, 2022). Therefore, information is potentially a double-edged sword.

It should be noted that the institution that is responsible for defining and declaring a security threat is most often the state. By 'the state' I am referring to the work of a country's parliament and/or president in terms of debate, declarations and laws passed. However, determining what is to be judged a threat and what is not is somewhat subjective and tied to the interests of the ruling power, rather than to what is in the national interest or that of the people. This is a process that involves a deliberate and decisive political process that intangibly separates the friendly 'us' from the threatening 'other', and that involves a weighting of potential political and military opportunities or gains against risks or losses. A perceived threat is often framed as being something that will harm the national interest or that of the peoples, rather than as harming the interests of the incumbent political power, as this is a concept more easily 'sold' to the public. Specific interests of the United States were secured through regime change. Unstable political and economic environments in the region were taken advantage of by subversion and manipulation, with the communication of selective words and information that is intended to undermine the intangibles of the target.

> This *revolution* consisted of various forms of psychological manipulation, crowd organising, populist slogans, and publicity techniques designed not for popular empowerment but as an instrument, a template for short term, euphoric political upheaval and defeat of nationalist incumbents – the political analogue of victory on the football field. (Sussman, 2010: 140)

Thus, the façade of a grass roots and spontaneous popular uprising is in fact a highly organised event with a great deal of foreign input. Sussman also notes that although these events were characterised as 'revolutions', they were in fact 'little more than intra-elite power transfers' (Sussman, 2010), he also notes that key words are 'the artillery of propaganda' (Sussman, 2010: 157). Keywords frame and restrict how an event is communicated and perceived, and consequently the opinions and actions of people that stem from this projected belief.

One can deduce that a security threat is something that can potentially weaken or threaten the continued existence of a political entity (for the emphasis in this work is on political rather than social or economic entities), particularly a political entity that at the time holds power. Where does 'information security' fit into the process then? Its importance lies in the tenet that perception is more important than reality (Louw, 2001: 1–35). That is, people (the public) tend to react to what they perceive as being reality rather than what may actually be the real case. Therefore, one's communication potential, assuming that the requisite tangible (physical) and intangible (reputational etc.) assets are present, is geared towards information dominance. That is, in the marketplace of ideas, the aim is to crowd out potential competitors in order for one's own message and vision to prevail and so influence the target audience.

Mass media and journalism can help to shape people's opinions and values by bringing often remote events to the vast majority of a country's population. As such, the mass media sphere is a hotly contested arena that transcends a number of planes – political, social, educational and business. It is a mechanism that can unite these diverse sites and bring a common meaning. This is of even greater importance during times of political and economic instability.

It can be said, in a general sense, that there is an association between security and stability. This means that with security comes stability. The stability being sought and desired can be measured in both economic and political terms. An entity seeks a stable environment, to which it is best adapted, in order to maximize its chances of survival. When an entity is unable to adapt to a changing environment it can be replaced by another entity that is more suited to it and is able to make the necessary modifications. At times it is necessary to embark on a course of change in order to survive.

Whether it is to ensure a stable political environment or to embark upon a course of change, an incumbent political entity needs to be able to master the situation and to navigate it. This often requires that entity to control symbolism and perception, firstly in order to give meaning and understanding to an event as it unfolds, and then to guide (or at least predict) the reaction of the public to those events. The penalty for failing to master this difficult and unpredictable process is well illustrated by the events of the Appeasement Process in the mid-1930s, when Britain and France sacrificed Czechoslovakia to Nazi Germany in order to buy time. British Prime Minister Neville Chamberlain lost control of both the symbolism and the perception of the process, the result was the collapse of his reputation and political leadership, despite buying enough time for Britain to prepare against the coming German attack in the summer of 1940. He is an infamous figure in history. President George W. Bush's re-evaluation of the underlying reasons for the US defeat in the Vietnam War laid the blame with the American political and military leadership; the traditional reason of the undermining of public morale and belief in the war by the mass media and US journalists was jettisoned. He blamed them for lacking the political will in ultimate victory and not communicating a firm conviction that the war could be won, which led the public to believe that the war was lost.[5] Such a re-interpretation shaped how the Bush Administration communicated 'progress' in the 2003 Iraq War, such as the premature 'mission accomplished' banner on an air craft carrier, where Bush was giving a briefing. It provides a good demonstration of

[5] A. Zenilman & D. P. Kuhn, *Bush Chastised for Vietnam Analogy*, Politico, www.politico.com/story/2007/08/bush-chastised-for-vietnam-analogy-005507, 24 August 2007 (accessed 18 June 2020).

political thought about the role and effect of information and the nature of its communication in the context of armed conflict. Therefore, to some extent, information cannot be regarded as being strictly neutral in nature, but is somewhat ideological instead.

This situation can be contrasted with more modern contexts where a definite relationship between war and politics has been seen. In 1982 the Falklands War saved the government of Margaret Thatcher from electoral defeat; the euphoria and patriotism of the moment was harnessed for political gain. In a similar scenario, the 9/11 attacks on the United States handed George W. Bush another term as President as the American people rallied around their leader. Conversely, the end of the First Chechen War (1994–96) was hastened by the incumbent Russian President, Boris Yeltsin, owing to his extremely poor ratings in the polls and the coming Presidential election, which gave him no choice but to end the unpopular war in order to survive politically. This brings back the dictum that war is politics by another means, and the lifeblood of all politics is its ability in the art of communicating information and ideas.

Information security has become more difficult to defend and has created more opportunities for attack in the global information age of Web 2.0; in some respects it can be said that matters are more acute than in previous technological ages owing to the advances in: internet communication technologies (providing tangible means), an increasingly turbulent world order (providing the motive), and a lack of a commonly agreed-upon legal and regulatory framework to manage and regulate the problem (providing the opportunity) (Simons, Danyk & Maliarchuk, 2020). Using a von Clausewitz framework of war as compelling another to do our will and as an extension of politics, Niekerk (2018) applies this to information warfare tactics that are executed through cyber means. This is a new means of achieving an old end, reducing the capability and capacity of an opponent's political and intangible elements. It also represents the possibility where an intangible operation can lead to wider tangible consequences in terms of damage and disruption, and even potentially trigger kinetic military operations in response.

In connection with these different aspects, which have been described above, an overarching motivation for controlling the information sphere is that a belief exists that in doing so there is a greater chance of regime survivability and that political, social and economic goals are more readily realisable. Attempts to control the content and flows in the information sphere are attempted through such means as strategic communication, communication management, public relations and so forth. These assumptions are derived from the view that by being able to determine the reality of an event by massaging public perceptions there is a greater possibility of being able to predict and determine events and reactions. This thereby creates the idea that having greater control over the shaping of the process can be achieved through the exercise of a degree of information management. It amounts to the situation where the preparation and cultivation of the information space is used to effectively communicate the desired representation of the physical domain in the cognitive domain of the target audience, which is to prepare for the informational battle before the physical war has begun.

Targeting Tangible versus Intangible Assets

In a military environment, information operates at three different levels – tactical, operational and strategic (Armistead, 2004). This mirrors the physical work of a military organisation in the physical world. Information and the ability to shape an audience's perception of an event and the actors involved are seen as being crucial tasks. The ability to handle this information and effective messaging to assist physical military operations needs to be carefully considered and executed. If it is not, and is poorly managed, it has the potential to adversely affect operational aspects of a military campaign.

This chapter concerns the issue of warfare and the informational and intangible aspects that influence it, the differences that exist in the relationship and the nature of interaction between information, politics and armed conflict, and between the various kinds of armed conflict – in both regular warfare and irregular warfare. Regular warfare involves an armed conflict between various state-based military forces on a defined line of military operations. Irregular warfare involves state-based actors versus non-state actors, such as in terrorism or insurgency, where there is a lack of a clearly defined front line and a blurred distinction between combatants.

Terrorism or insurgency as a weapon and tactic of choice is often selected by the weaker side owing to the asymmetric tangible aspects of the conflict. That is, they are unable to compete owing to the lack of requisite materials, manpower and finances to wage a short and successful war against their opponent (Ganor, 2005). Moreover, there needs to be another factor to, in effect, equalise this imbalance. Information and the ability to reach and influence an audience takes on a central role as a result.

As stated above, terrorism or insurgency is the weapon of the weaker opponent. Therefore, the role of information and the campaign to influence the minds and actions of a selected audience takes on greater meaning – waging an information campaign alongside a military campaign. This is not necessarily a contest to win over hearts and minds by the different sides, although this can also be a valid objective. Another central point is the issue of security, and above all human security. If the domestic civil population does not feel safe this may translate into political demands, which may actually be of benefit to the political agenda of the terrorists. This translates into a situation whereby a sense of tangible hazard or risk is converted into action that creates intangible social or political changes in the immediate environment. Information and communication are used as the means to project certain images, opinions and beliefs, which can be used either as a deceptive ploy or to shape the environment of the battle space by influencing the information space.

One of the aims of terrorists is to create a certain sense of fear in society. Fear exists when two sets of circumstances are simultaneously present in a society under terrorist threat. Firstly, that there must be a perception that something bad/negative has a high likelihood of happening, such as a bombing or hostage taking, is an abstract concept of collective risk. Secondly, that there is a perception that the bad/negative event can happen to me (the individual) is also essential. An important point to understand is that this is based upon the perception of the audience/public and is not necessarily a reality. For all intents and purposes, perception is reality, and it is what people react to and base their decisions upon (Simons, 2010). The sense of uncertainty and risk can elevate the public sense of outrage, which may be used to pressure political decision makers.

The methods used by terrorists and exponents of other forms of irregular warfare can be explained by analysing the role of tangible and intangible assets, and how these two sets are related to each other in the greater strategic considerations of terrorism. Tangible assets are those physical objects that can be attacked by terrorists, such as buildings, people and vehicles. These are normally of symbolic or military value and are intended to weaken their opponent physically or psychologically. The US strategy in Vietnam was to inflict greater casualties on the North Vietnamese forces so that at some stage the losses would be 'unacceptable' to Hanoi, and thereby a US military victory would be achieved. Kissinger noted that this strategy had two fundamental flaws: 1) the nature of asymmetric warfare and 2) the inherent asymmetry between the sides in defining what 'unacceptable losses' were (Kissinger, 1969: 212). In other words, occupy land and kill the enemy, whereas the aim of the insurgents and North Vietnamese was not to hold land, but to undermine the sense of security and trust in the government and authorities of the population of South Vietnam. Hence, there is a clear mismatch in understanding of the notion and commitment to 'security' and an attempt to substitute short-term military tactics for any viable long-term political aims.

Intangible assets, on the other hand, are objects that cannot be touched physically, but they exert an impact upon the outcome of a battle or campaign. These assets are such things as reputation, brand, sense of security, legitimacy and the resolve or will to fight. Although these aspects cannot be physically touched, they have a profound impact upon an opponent's ability to wage or continue to wage an effective military campaign. These intangible assets are adversely affected by attacks upon tangible assets, which indicates that the primary aim of attacking tangible assets is to diminish an opponent's intangible assets.

This degradation of an actor's intangible assets shall only be effective though, if knowledge of the terrorist events is widespread and vivid. The most usual and effective means of transmitting news of terrorist acts is through the mass media, and more increasingly through social media and the Internet. It links people to events that may otherwise be remote or unnoticed. Mass media can relay events which are often descriptive rather than analytical in nature. An event is shown to happen, but not why. These events are newsworthy in themselves as they are dramatic and of interest.

There is a side effect to this process, however. Stories constructed in this manner can create a greater sense of risk and uncertainty, even if this is not the intention of the media outlets or journalists. By covering a story concerning a terrorist act there is a possibility of amplifying the operational capacity of a terrorist group. This is because after the story it is not known whether the terrorist group that was responsible has the ability to launch another attack immediately or whether they have exhausted that capacity for the time being. This then can be potentially translated into political pressure through demands to assure or guarantee various intangible assets.

There is no guarantee that the terrorist group will receive precisely the coverage that they want from an act of terrorism. They simply do not have the ability to strictly control how the message is framed and covered in the mass media. However, by committing acts of terrorism there is the possibility of influencing what hot topics are covered in the mass media and thereby what is in the news (which may not be ordinarily covered).

Regular warfare differs in several different respects to irregular warfare. One of the chief differences is the involvement of state-based actors that oppose each other on a defined area, where front lines exist, and various belligerent forces are identifiable from the uniforms, symbols and weapons used. Although there may be inequality of access to mass media and communication assets, all parties are likely to have national and/or international media in order on board to communicate their values and narrative to domestic and foreign publics. Using modern mass communication, a clear pattern leading up to a number of different wars is clearly observable, which dovetails with Sun Tzu's notion of deception, although applied more liberally, according to that proposed by Machiavelli. A great deal of effort is made in establishing the narrative of the reluctance to go to war, and that it is done as a last resort only. The wars against Iraq, Pakistan, Libya, Serbia (over Kosovo), and the worrying rhetoric and sabre rattling with Syria and Iran, demonstrate (as posited by von Clausewitz) that the policy and heavily political nature of wars in the current context is still extremely valid.

There is a distinct pattern of attempting to influence perceptions, and therefore influence the relationship between the people and their political leaders. As noted by Sun Tzu, a leader that is perceived in a positive light will accumulate a greater sense of legitimacy and following. This means that, unlike the irregular war, a clearly defined enemy is created by the political leader of the country, whilst simultaneously protecting their own image. This makes the use and conveyance of values and narratives important. For instance, trying to shift the reputation of historically and commonly understood enemies or injustices is applied to the contemporary named enemy (or 'bad guy'). Thus, nonsensical vocabularies and comparisons are made; for instance, the comparison of the Soviet Union's crushing of the 1956 Hungarian Uprising to the 2008 Georgian–Russian War, or the attempt to tar Saddam Hussein with the characteristics of Adolf Hitler. These are attempts at deception owing to their false-logic basis. These communications are directed at civil publics rather than military ones, in order to influence consensus on the 'righteous' nature of the war in question.

When it comes to the issue of the definition of contemporary regular wars, a factor that is often absent is that of a clear definition of the conflict. Certainly, there are slogans and branding applied to the various wars, and political and armed conflicts. In terms of regular wars, these are increasingly being fought under an umbrella of 'humanitarianism', which is an oxymoron when applied to compelling one state to act according to another state's will. Examples of branding are clear as well; to make something that is contrived appear to be natural, in line with this trend, the *Colour Revolutions* of the first decade of the 2000s have given way to the current *Arab Spring*. These are politically branded entities that are intended to rally the emotions and the opinions of the target audiences, through trying to project a relatable (by Western audiences' familiarity with the embedded symbolism of the values expressed) and 'contagious' sense of hope in order to mobilise mass publics to tangible (to become physically engaged in an event) and intangible action (opinion and perception to support).

However, despite this use of masking the true nature and identity of the wars and revolutions being initiated and fought, there are serious flaws in the current system. These wars are not defined, and certainly the political statesmen of today are not capable generals (a condition noted as being desirable by Machiavelli), whose tendency to interfere in military matters and yet not understanding them normally result in severe future problems. The nature of war can also transform with time and events; the March 2003 invasion of Iraq started as a regular war between the American-led coalition but transformed quickly into an irregular conflict (involving elements of insurgency and terrorism) following the occupation of the country. There was seemingly little done in terms of defining this particular war, apart from the primary goal of removing the political regime of Saddam Hussein and installing a government that would be compliant to the political and economic demands imposed upon it. The US seemed to be somewhat unprepared for the insurgency that resulted from their heavy-handed occupation of Iraq, which hints at the lack of clear definition and wider planning.

When looking at the above-mentioned problems, there is a hint of something wrong or faulty in the Western political–military mix. This is revealed by Clarke (2015: 8), who says that rather than the standard use of military power to establish conditions for the subsequent political effect, 'Western powers have persistently tried to use their forces in order to have intrinsic political effects in and of themselves.' The general lack of clearly measurable results and failures of Western kinetic military operations in the twenty-first century, in spite of a current tangible advantage over competitors and opponents, has resulted in calls for the West to rethink its strategy and application of their approach to armed conflict in order to be able to continue to shape the international order (Ucko & Egnell, 2014; Clarke, 2015). Clarke (2015) adds that this makes it difficult to clearly define the lines of victory or defeat in political terms as even a decisive military victory may not translate into a decisive political outcome.

With the endless cycle of mostly irregular conflicts of choice, in particular counter-insurgency and counter-terrorism, Western foreign and defence policy politics, are being driven by the illusion of the invincible and unparalleled hard power in the unipolar global order. Therefore, the US has willingly got itself embroiled in such a cycle since the declaration of the Global War On Terrorism has also produced noticeable intangible effects in the cognitive domain that influence the effectiveness of the capacity of using physical military force projection. The cycle of endless wars that are waged without good reason (solid strategic reasoning and interests) has had a detrimental effect on serving and former military personnel. The number of suicides among serving US personnel has increased significantly from 2005 (after four years of the Global War on Terrorism (GWOT)) (Hoge, Ivany & Schoenbaum, 2016). In 2012 more US military personnel died from suicide than combat-related deaths.[6] In 2019, there were over twenty suicides by serving and former US military personnel per day.[7] As the junior military partner in the GWOT, Britain has also been experiencing a surge in the number of suicides of serving and former military personnel,[8] and at earlier points in time the number of suicides exceeded Afghan deaths.[9] In fact, polls and surveys show an increasing level of disillusionment in political and military leadership in what are seen as being unnecessary wars by veterans.[10] This has led to increased advocacy by veterans' groups against war.[11] Even though there is a clear advantage in terms of the tangible aspects of Western military hardware in terms of budgets, the technical and technological superiority over military and geopolitical rivals (which is gradually eroding), the intangible elements in the cognitive domain are likely to weaken the capacity of these weapons.

Within the framework of the New Cold War, intangible strategies are being used increasingly as a mechanism to influence the global transformation away from the Western-centric United States unipolar order towards a non-Western-centric multipolar configuration. Operations in the information realm are intended to shape the perceptions and judgements of the cognitive realm that impact upon what does and does not happen in the physical realm. This is seen in the US-led Western intangible-based strategy to preserve its hegemony at the expense of

[6] B. Briggs, *MSRC's David Rudd consulted: 'Why Modern Soldiers Are More Susceptible to Suicide'*, Military Suicide Research Consortium, https://msrc.fsu.edu/news/msrcs-david-rudd-consulted-why-modern-soldiers-are-more-susceptible-suicide, no date given.

[7] C. Giacomo, *Suicide Has Been Deadlier than Combat for the Military*, New York Times, www.nytimes.com/2019/11/01/opinion/military-suicides.html, 1 November 2019 (accessed 5 September 2020).

[8] *What's Behind the Surge in Suicides by UK Veterans of the Afghan War?*, The Week, www.theweek.co.uk/105953/what-s-behind-the-surge-in-suicides-by-uk-veterans-of-the-afghan-war, 2 March 2020 (accessed 5 September 2020).

[9] *UK Soldier and Veteran Suicides 'Outstrip Afghan Deaths'*, BBC, www.bbc.com/news/uk-23259865, 14 July 2013 (accessed 5 September 2020).

[10] R. Igielnik & K. Parker, *Majority of US Veterans, Public Say the Wars in Iraq and Afghanistan Were Not Worth Fighting*, Pew Research, www.pewresearch.org/fact-tank/2019/07/10/majorities-of-u-s-veterans-public-say-the-wars-in-iraq-and-afghanistan-were-not-worth-fighting/, 10 July 2019 (accessed 5 September 2020).

[11] See for example, Veterans for Peace (US) – www.veteransforpeace.org/ or the UK Veterans for Peace – https://vfpuk.org/.

rising competitors and challengers, where a standardised approach to propaganda (cookie cutter propaganda) is used in an attempt to shape processes and outcomes in international relations.[12] Information flows are used to influence the production of knowledge as a means to create a consensus on perception of the human environment (orthodoxy of knowledge), which is used in turn as a means of obstructing other powers in international relations from realising their foreign policy aims and goals.[13] Hence, by default, a relative advantage for US hegemony in different global regions is delivered.

Although these tactics concern an aspect of maintaining an equilibrium (US global hegemony), Henry Kissinger warns that this can't be of value in itself. He also warned about the apparent lack of purpose and the consequences apparent in risky US foreign policy. Furthermore, Kissinger is concerned about what he terms as international relations verging on a dangerous disequilibrium: 'We are at the edge of war with Russia and China on issues which we partly created, without any concept of how this is going to end or what it's supposed to lead to.'[14] The situation of disequilibrium as characterised by Kissinger is a potential case where politics without purpose can potentially lead to war. This is coming at a time and place that ignores Sun Tzu's *The Tao*, concerning the relationship of Western political leaders to their publics in a period of increasing distrust and war wariness. There is also relevance for Machiavelli's lessons on the relationship between war and politics, where increasing tensions and hardships that can lead to war may work at dividing and not unifying society and where increasing instability is likely to develop politically, economically and socially. The threat of war, which von Clausewitz defined as a mechanism intended to compel an opponent to do one's bidding, will only work if the threat is seen as credible by the opponent. In the wake of the debacle that was the rushed exit of the US from Afghanistan, the threat of war by an increasingly disoriented and weakened hegemon is not a credible threat that shall likely compel China or Russia to carry out US demands. This is especially the case when

[12] G. Simons, *Cookie Cutter Propaganda and Herd Mentality in an Era of Geopolitical Transformation*, Propaganda in Focus, https://propagandainfocus.com/cookie-cutter-propaganda-and-herd-mentality-in-an-era-of-geopolitical-transformation/, 17 August 2022.

[13] G. Simons, *International Relations in the Age of US Decline: Orthodoxy of Knowledge and Obstructive Foreign Policy*, Opinion, Russia in Global Affairs, https://eng.globalaffairs.ru/articles/us-orthodoxy-of-knowledge/, 2 August 2021.

[14] L. Secor, *Henry Kissinger Is Worried About 'Disequilibrium'*, Wall Street Journal, www.wsj.com/articles/henry-kissinger-is-worried-about-disequilibrium-11660325251, 12 August 2022 (accessed 16 August 2022).

Chinese and Russian concerns are not understood or considered, hence the response is likely to result in a calling of the bluff.

Conclusion

Although the manner of weaponry that is used to wage war has quickly evolved into every more increasingly destructive technology, there are other factors and elements that have remained relatively constant over time. The tangible factors and means of fighting have evolved with the economic and technological changes that have occurred in society. The Heaven and Earth elements described by Sun Tzu still have an impact upon military operations, but to a lesser extent as technological innovation has been minimising this issue with some success.

Many intangible elements and factors have proved to be resilient in their continued importance in creating and fighting wars. The continued relevance of politics imposes an increasing level of influence on the outcome of wars. This is in contrast to the military influence on the outcome of individual battles within a war, where wars are used as a coercive instrument of policy whereby another country is forced to submit to the political will of the aggressor. This has been a constant through the centuries. However, some elements have been ignored, such as becoming embroiled in a lengthy and poorly defined war.

The intangible elements of legitimacy, reputation, public will and opinion are all critical factors when shaping the information environment to support launching or continuing a military conflict. Both Machiavelli and von Clausewitz noted the centrality of politics to war, which still remains. War and politics have become perhaps even more inseparable in the contemporary context; politics determines where and when armed force is used, and to what extent it is deployed against the opponent. As noted by Machiavelli, there are several similarities between the military and political spheres, such as the need to understand and shape the tangible and intangible environment around them.

As such, there is a need to shape and influence the opinions and perceptions of various publics when preparing to wage a war. The situation creates an environment where an information war runs before and then parallel to the physical fighting in regular warfare. In irregular warfare the information war runs parallel to the fighting. The narratives, vocabulary, values, frames and images need to be carefully nurtured and controlled in order to cultivate the desired effects upon the target publics. Intangible assets before or during a war can be both an opportunity and a threat to political regimes, depending on the prevailing conditions in the society concerned.

There are some interesting differences revealed in these political and military philosophies that create underlying intellectual differences in the conception of what is judged to be the 'correct' ratio in the political and military mix of armed conflict today. After several centuries of technological superiority that has aided in granting the West global political and military ascendency over other powers, there is a seeming reliance on tangible factors to decide the outcome of battles at the expense of suitable intangible factors to influence the outcome of wars. There is a focus on the holding of territory and the use of overwhelming technologically based force to physically destroy an enemy and on the short-term use of intangible factors to legitimise it. This is found, for example, in the romantic notions of armed conflict and the moral/ethical constructs that are used to frame it. Opponents of Western military operations have developed an alternative path, owing in part to the historical technological asymmetry. Much greater use of intangible factors is evident, which are designed and intended to gradually wear down the intangible factors (will to fight and belief in the fight) of a more tangibly powerful enemy. This has the knock-on effect of degrading the capacity and capability of the opposing military forces and political leadership. Added to this is the gradually narrowing technological superiority that is enjoyed by the West.

References

Armistead, L. (Ed.) (2004), *Information Operations: Warfare and the Hard Reality of Soft Power*, Washington DC: Brassey's Inc.

Betz, D. (2015), *Carnage & Connectivity: Landmarks in the Decline of Conventional Military Power*, New York: Oxford University Press.

Boucher, G. (2017), The Long Shadow of Leninist Politics: Radical Strategy and Revolutionary Warfare after a Century, in Sharpe, M., Jeffs, R. & Reynolds, J. (Eds.), *100 Years of European Philosophy since the Great War*, Cham (Switzerland): Springer International, pp. 141–159.

Boylan, E. S. (1982), The Chinese Cultural Style of Warfare, *Comparative Strategy*, 3(4), pp. 341–364.

Buzan, B., Waever, O. & de Wilde, J. (1998), *Security: A New Framework for Analysis*, Boulder: Lynne Rienner Publishers.

Clarke, M. (2015), The Ending of Wars and the Ending of Eras, *The RUSI Journal*, 160(4), pp. 4–9.

Dimitriu, G. (2018), Clausewitz and the Politics of War: A Contemporary Theory, *Journal of Strategic Studies*, DOI: 10.1080/01402390.2018.1529567.

Ganor, B. (2005), *The Counter-Terrorism Puzzle: A Guide for Decision Makers*, New Brunswick: Transaction Publishers.

Griffith II, S. B. (Trans.) (2000), *Mao Tse-Tung on Guerrilla Warfare*, 2nd Edition, Urbana: University of Illinois Press.

Hageback, N. & Hedblom, D. (2022), *AI: For Digital Warfare*, Boca Raton: CRC Press.

Hagen, K. (1996), A Chinese Critique on Western Ways of Warfare, *Asian Philosophy: An International Journal of the Philosophical Traditions of the East*, 6(3), pp. 207–217.

Hoge, C. W., Ivany, C. G. & Schoenbaum, M. (2016), Death by Suicide in US Military during the Afghanistan and Iraq Wars, *The Lancet*, 3(11), pp. 1001–1003.

Keohane, R. O. & Nye, J. S. (1998), Power and Interdependence in the Information Age, *Foreign Affairs*, 77(5), pp. 81–94.

Kissinger, H. A. (1969), The Viet Nam Negotiations, *Foreign Affairs*, 47(2), pp. 211–234.

Lonsdale, D. J. (1999), Information Power: Strategy, Geopolitics, and the Fifth Dimension, *Journal of Strategic Studies*, 22(2), pp. 137–157.

Louw, P. E. (2001), *The Media and Cultural Production*, Thousand Oaks: Sage.

Machiavelli, N. (1965), *The Art of War*, Cambridge (MA): Da Capo Press.

Milevski, L. (2019), The Idea of Genius in Clausewitz and Sun Tzu, *Comparative Strategy*, 38(2), pp. 139–149.

Payne, K. (2005), The Media As an Instrument of War, *Parameters*, Spring, 35 (1), pp. 81–93.

Simons, G. (2010), *Mass Media and Modern Warfare: Reporting on the Russian War on Terrorism*, Farnham: Ashgate.

(2022), 'Inevitable' and 'Imminent' Invasions: The Logic Behind Western Media War Stories, *Journal of International Analytics*, 13(2), 43–58.

Simons, G., Danyk, Y. & Maliarchuk, T. (2020), Hybrid War and Cyber-attacks: Creating Legal and Operational Dilemmas, *Global Change, Peace & Security*, DOI: 10.1080/14781158.2020.1732899.

Simons, G. & Glaser, M. (2019), New Cold War and the Crisis of the Liberal Global Order, *Outlines of Global Transformations: Politics, Economics, Law*, 12 (3), 61–77.

Sun Tzu (2003), *The Art of War*, Philadelphia: Running Press.

Sussman, G. (2010), *Branding Democracy: US Regime Change in Post-Soviet Eastern Europe*, New York: Peter Lang.

Ucko, D. H. & Egnell, R. C. (2014), On Military Interventions: Options for Avoiding Counterinsurgencies, *Parameters*, 44(1), pp. 11–22.

Van Niekerk, B. (2018), Information Warfare as a Continuation of Politics: An Analysis of Cyber Incidents, 2018 Conference on Information Communications Technology and Society (ITACS).

von Clausewitz, C. (1982), *On War*, London: Penguin Books.

Waldman, T. (2018), Vicarious Warfare: The Counterproductive Consequence of Modern American Military Practice, *Contemporary Security Policy*, 39(2), pp. 181–205.

4 The Fourth Generation of Informational Warfare

Iulian Chifu

Introduction

Informational warfare has developed as a component of hybrid operations and as a type of aggression. The purpose is to project an alternative reality to a designated target population in order to create a perception which could enable it to put pressure on the decision makers and to alter the well assessed, evaluated and planned strategic decision on a very narrow and clear subject or theme. It is done by using a combination of sequences of truth, deprived of their context, lies, innuendo, sophism, and predetermined reflections, secret sources and some pieces of conspiracy theories (Chifu, 2020: 67–105).

We currently have three generations of informational warfare: the first includes disinformation and propaganda operations, troll wars, lobbying, psychological operations, recruiting and conditioning; the second uses so-called sociological groups objectivized in the virtual space; the third generation is that of a micro-targeting at the group level of the entire population, a Cambridge Analytica-type of access. As a result of this, we have reached between 82 and 95 per cent probability[1] of impact using big data and a high number of targets. The fourth generation makes the jump to targeting the differences between personalities: highly educated targets: high-value targets: targets which are protected due to their jobs and level of influence, who cannot be reached or influenced at a statistic level of impact.

For these complex components, news methods are involved in obtaining the same result. Offensive and defensive weapons are put in

[1] The percentage is referring to different types of experiences: identifying the gender, race, age average, general perceptions and preferences of the target group. Cambridge Analytica has been highlighted through its experiments and applications in electoral campaigns, specifically Brexit and the 2016 US Presidential elections. This can be found in quoted references related to Cambridge Analytica, including the public presentation of Steve Bannon in Washington DC right after the 2016 elections (see Pădure, 2017).

place, tailored attacks are planned, and individual instruments used for each of the high-value, difficult-to-reach targets. Character assassination, labelling, changing or cloning identity, are individual programs that are put in place to exploit weaknesses and obtain a "genuine response" which meets the goal of projecting the alternative needed reality and "convincing" the unbeliever of a certain required interpretation, idea or reality. Some of the instruments are examined that do not use criminal approaches (such as bribery, blackmail or any other kind of coercion), and the focus is instead on informational instruments and "the genuine conviction of the subject".

The instruments do not belong only to Western countries, or to the inventors of the Internet, they were in existence way before Russia bought the capabilities of Cambridge Analytica, for instance. Russian military deception operations are used in conjunction with non-military means to compete with the West in the grey area (Morrell & Kosal, 2021). Activity in the grey zone is defined by Branding as "coercive and aggressive in nature" but "designed to remain below the threshold of conventional military conflict and open warfare between states" (Popp & Canna, 2016).

By manipulating intelligence, Russia can pursue objectives traditionally achieved through military conflict while avoiding the costs of war. According to Vladimir Slipchenko, a Russian military academic, "information has become a destructive weapon like a bayonet, a bullet or a projectile." Effective use of this weapon allows changes in the status quo to be made gradually, rather than violently (Weiner, 2020).

Three Generations of Informational Warfare

Caveat: The reference to 'Generation' is not used in a sociological sense or time frame, but refers to the jump in quality and characteristics of information warfare. This approach has been used previously and is common in discussing technology, warfare, energy sources etc. The concept of "Four Generations of War" (4GW) (Hammes, 2007) suggest that there are a number of major developments under way: a strategic shift; an organizational change; and a change in the type of participants that define different generations (Lind, 2004).[2] The wireless mobile phone industry began its creation, revolution and technological evolution in the early 1970s. In the last few decades, mobile wireless technologies

[2] See also T. Benbow, *Talking 'Bout Our Generation? Assessing the Concept of "Fourth-Generation Warfare"* (2008); W. S. Lind et al.; *The Changing Face of War: Into the Fourth Generation* (1989).

have seen five or even six generations of technological revolution and evolution (Bhalla & Bhalla, 2010).[3]

Informational warfare has been launched in a context of influencing public opinion and decision makers by applying public pressure through targeted issues debated within a democratic society. Social media and unlimited access to information has changed perspectives and instruments used. But we are far from the Cold War instruments of propaganda, *desinformatsia* (Holland, 2006) and active measures. Today, direct access to all connected citizens offers new instruments for altering their will and available options. The three generations of informational warfare have seen the development of various instruments that could be used in a war against society; the very principles and values are considered vulnerabilities of democratic societies.

In terms of what we know about the first generation of informational warfare, a component of the hybrid war and a standalone instrument is informational warfare. Informational warfare is the creation of alternative realities by perverting the objective truth – using data, facts, and concrete arguments – its misinterpretation by using a combination of selective elements; facts and fragments of the truth, interpreted and combined with reasoning altered by the use of syllogisms, sophisms, propaganda, forced interpretations, and everything mixed together with a multitude of lies (Chifu & Nantoi, 2016).

The battle of alternative narratives, of "alternative truths", has become the most insidious way of constructing beliefs. It is based on the groups and targeted audience for each operation, on a vast fabric of knowledge of the training, inclinations and expectations of the target audience, its propensities, and frustrations. Obviously, the instruments do not only concern information as such but also refer to other more subtle components of fundamental emotions (Moisi, 2009), context, and to the seizure of opportunities offered by ongoing events and the mood created inside a particular target population in order to inoculate a certain type of opinion on the subject.

Alternative reality perverts the perception of a target population using a combination of psychological operations – *PSYOPS* (Air Force Doctrine Document, 1999) – along with misinformation and propaganda (Wierzbicki, 1996; Volkoff, 2009; Ficeac, 2014), fundamental beliefs, sentiments and impactive images, which aim to lead the audience towards a pre-defined perception. And finally, as the audience has now

[3] See also Sreeramana Aithal and Shubhrajyotsna Aithal, *Conceptual Analysis on Higher Education Strategies for Various Tech-Generations* (2020); R. Sackmann & O. Winkler, *Technology Generations Revisited: The Internet Generation* (2013).

formed an opinion, its perception has replaced reality (Stern, 1999) and, no matter what argument or proof of truth has been presented, it will clash with the perception already formed.

In a world where information travels quickly and reaches large numbers of people in real time, through television, the Internet and social media, the perception of a certain event is easy to form, derail, alter, and enforce. Subsequent presentation of the truth will lead to a minimal change of opinion on a large scale due to the lack of critical thinking in the majority of the population, the conservative approach to assuming recognition of one's own mistakes within the general population, and the ease of use of the explanation already internalized by an ordinary person, especially in a community that has developed a conformism and has its own description, perception and "truth".

The main aims of the informational war are: to determine, control or even alter the strategic decisions of foreign policy, security and defence; to corrupt or hinder the instruments destined for the military component of a state; and impede the functioning, if not completely block, elements of state security. The instruments and mechanisms to achieve these goals are used to identify the audience, citizens, the organized and well-guided pressure groups, and to persuade the authorities to change policies due to either a of lack of support, or worse still, total opposition of the population (Chifu & Nantoi, 2016).

The main features, principles and values of open, democratic society are used against the states and its institutions. They are undermined from within by building up groups of enemies inside the fortress. Moreover, the approach is unitary and integrated, and, often a component of informational warfare, viewed as an isolated and opportunistic incident, not a planned aggressive act alongside a set of other elements. The insidious mode of action and this integrated approach create the advantageous elements of plausible deniability of such an act of informational aggression on a target population (Chifu, 2015).

Informational warfare uses the three distinct levels of action, with different degrees of relevance and legal and moral significance. Together, the integrated and tacit approach to the target audience increases the effectiveness of the informational warfare (Simons & Chifu, 2017). The first component is visible and evident; it involves altering the public space of an intended target, involving the media, Internet and social media, but also at other subtle levels, an injection of ideas and information that use the fundamental principles and values of democratic human rights (including freedom of expression, freedom of speech, free movement of ideas), elements that are not only legal and moral, but are among the fundamental values that democratic societies defend. Underneath, we are dealing with an open injection of propaganda, manipulation and misinformation that is hard to prove. Moreover, this type of action requires constant responses, investigations and public debunking of false voices and untrue interpretations of facts, alternative narratives behind these ideas, and finding a way of interpreting facts to support an alternative truth package.

Some activities exploit vulnerabilities within the system (Chifu, 2016). These can include: circumventing the law or the weaknesses of control institutions; limitations of legal powers or the absence of professional institutions able to sanction or promote an approach to a subject or proposed version of events, without proper context; a lack of expertise to clarify the proper release of information; or the speed and poor training of employees in the field of media and broadcasting. False information from obscure sources entering the mainstream, verified information that is forcefully argued to reach a predetermined conclusion (Chifu & Nantoi, 2016), the absence of alternative, independent media, strictly observing the narrative of events, and proper reporting that provides alternative polemic approaches, all help to build the elements of informational warfare.

In developed, mature societies, with established media and many alternative sources, the ability to seize, guide or control the public space is significantly reduced. Education also matters, and of course the democratic culture of the people. But there are societies where the public space is controlled by third-party actors, such as the case of the Republic of Moldova, where a news-impact report shows that 80 per cent of the audience consumes news produced by Moscow (CCA Report, 2016; Chifu & Nantoi, 2016). And when they are packed up in highly creative and attractive entertainment programs, these genuine "injections" of news stories in the space of a preconceived narrative have a major impact on the proposed alternative truths.

Propaganda has its limits, however. It is therefore necessary to use an integrated set of developed tools to reach the target audiences, and to adopt a distinct approach. Here we arrive at the second element; *the Lobby* or *Public Relations* as it is called, if the communicator is a person or company, but it is called public diplomacy, when carried out by a state. It is a perfectly legal approach in states where there is a law against lobbying or its equivalent.

For a tool to influence the current decision makers within a society, it must spread ideas and messages within the public space by the use of legitimate and credible message bearers who can have an impact on their own population. Money is paid to credible people who are seen as legitimate in the eyes of the target population – analysts, experts, current and well-known former politicians, foreign parties – in exchange for their support of the thesis proposed by those holding the purse strings.

The aforementioned media institution or expert, or the analyst who presents their version, are paid to create a certain image, defend a certain actor, carry a particular message, vision, or a previously prepared narrative. So a credible, otherwise rational, impactful character suddenly presents themself in the public space, conveying ideas that are not the result of their own analysis or expertise, but the result is delivered for the purpose of being disseminated for money, paid for by an interested party.

The public is not warned and does not know that the person in front of them is not speaking on their own account, but is expressing somebody else's ideas. Therefore, an assessment is made in the context of the credibility and legitimacy of the person in question (in fact, this is what the lobbyist is selling) (Lobbying Reform, 2006; Simons & Chifu, 2017). Moreover, after completing the mission, the person returns to their legitimate, correct analysis, which is the result of their own thinking, luggage of knowledge and ideas. The person re-establishes or preserves their credibility further and can later be used on another project of informational warfare, or on another subject.

While the US has a lobby law, requiring the public exposure of paid positions, and a transparency formula for the amounts of money received to convey certain themes, this tool is missing in Europe. It is a legal instrument, somewhat immoral, but used in combination with propaganda and taking control of or altering the media in informational warfare.

The third level is also the most insidious, and profoundly illegal, targeting more subtle components of the informational warfare. It is tied to PSYOPS – elaborated psychological operations. The level of access to this component is much deeper, at the heart of the social body, in common beliefs, in the psychological state, and in people's deepest fears. Here, it is not only the widespread information that matters, but above all, the effect created by information on the target audience, namely the birth or accentuation of fears, the creation of collective emotions (Moisi, 2009), preparing the public to react to future events in a previously guided, prepared and pre-formulated way.

The objectives are achieved by using fake news, personalities and experts with moral references known in society, directed to certain positions; however, the target is not immediate information but the creation of the context for a certain emotional public reaction triggered by a subsequent event. Controlling the subsequent reaction is achieved by subtly accessing levels of the subconscious and creating patterns of thought by repeating sequences of a certain type at predetermined intervals, teaching the brain and the consciousness to react in a certain way to predetermined informational stimuli and emotions (Chifu & Nantoi, 2016). The purpose is to determine a certain public response, collectively, when it is needed to put pressure on and alter the action of a decision maker in moments of crisis.

The tools used are a combination of narratives, alternative ideas, planting of doubts, the promotion and most credible validation of subconscious lies as perceptions of truth, exaggerated and guided interpretations, exaggerations, manipulation, misinformation (with all its components) and trolls (paid commentators in the virtual space). The components for conditioning the target audience are built upon: actions at the subliminal level and the subconscious; the inoculation of the constructed perception; and the access to fundamental emotions and their orientation – fear, humility, hope (Moisi, 2009).

The functioning of the informational warfare is integrated, in all three dimensions, with thought-based steps and vast instruments built over time. Major resources are used for such actions and the component is most often moved to the military space. It is a tool whose relevance is barely discovered, and the impact can be extremely powerful.

The clearest instances in the case of Romania were two impactful examples of fake news:

(1) "Romanian traffickers of Ukrainian weapons sold to terrorists attacking the West" from Sky News. Here the reaction was immediate, the DIICOT (the Romanian organized crime Directorate) intervened, and the misinformation broadcast to 1.2 million viewers was stopped abruptly, proving that the news story was false and had used "actors" appearing in the gun-trafficking scenario through the mountains of Romania, with the author of the piece standing up to these alleged smugglers.

(2) "American Nuclear Weapons in Turkey moved to Romania" (Andreescu, 2016) on EurActiv, a case of absolute falsity that destroyed the credibility of the European Union website, but also demonstrated the vulnerability of the Romanian press that worked on the principle of "why ruin a good news story with the truth", go for the sensational instead. The best article was published the next day, as the Romanian MFA vehemently denied the fact, was: "US nuclear weapons are moving to Romania. Romanian MFA denies" (EurActiv, 2016; ProTV, 2016).

The Second Generation: Objectification of the Social Construct in the Virtual Space

To address the second generation of informational warfare, another caveat is required. Not every instrument in the usable space (or used) by the informational war is wrong, illegal or bad. We must also note this evolution and the fact that an instrument comes from the organic development of society, in the face of scientific evolution, resulting in an instrument that can be used against a state.

This is an old debate; from the emergence of social networks in which every person can be a widely quoted journalist from the mainstream media, to the unlimited and free access to information. Crisis management has become more complicated for decision makers; communication in a crisis has acquired new valences (Chifu, 2016; 2017; 2018), but also the instruments of participation, persuasion and convocation in the protest actions has expanded.

The Twitter Revolution in the Republic of Moldova (9 April 2009) (Barry, 2009), which was recounted during a visit to the US in 2010 for the IVLP Conflict Resolution programme (a two-hour presentation about these developments, and the impact of the Twitter Revolution), was followed by the Arab Spring with the large-scale use of such instruments mainly in Egypt, which encountered the concept of "preparing colourful revolutions"; this new social media technique also provided strategies and doctrines for Security and Defence of the Russian Federation (Concept of the Foreign Policy of the Russian Federation, 2013; The Military Doctrine of the Russian Federation, 2014; Chifu & Țuțuianu, 2016).

Evolution in this area was furthered by the concept of "Leaderless Revolutions" where the *Occupy Wall Street* case was an example and a catalyst (Chifu & Ivan, 2013: 221–239). Additionally, the principle was developed in Romania and what we call the phenomenon of "objectivity of the social construct in the virtual space" has been theorized and analyzed (Chifu & Ivan, 2013: 221–239). It is, in essence, the phenomenon of coagulation of groups in the virtual space, from real people and unknown avatars, who beyond a certain point acquire self-consciousness independent of the members and the components of the group.

This composite collective consciousness begins to reject individual gestures or opinions, refuses to assume the responsibility or leadership of any member of the group, including the founders, and makes it difficult to communicate with the group. It would be nothing out of the ordinary if such groups did not trigger actions in the real world that fundamentally disturbed the decision makers. Such was the case in

Romania of the phenomenon 'Uniţi Salvăm Roşia Montană' (Uniţi Salvăm Roşia Montana! Prima lună de proteste, 2013; Jula, 2014), and the failure of the Parliament/Government negotiations with representatives of the group who did not want to speak in the name of the group; in addition, those who claimed to be representatives were not accepted by the group or their agenda was rejected by the collective conscience. As a result, the Gold Corporation's investment (about $US1 billion) was halted with little regard to the consequences.

Another example is the case of the Colectiv nightclub fire where more than 60 people died; street demonstrations followed it, claiming that the authorities did not correctly manage clearance of the fire regulation licence through the Mayor's office due to corrupt practices, which led to the fall of a government (ProTV, 2015; Sandru, 2015).

In a third case in January 2017, the so-called OUG 13, the newly elected government, with an impressive majority, retracted its surprise emergency ordinance (implemented in a secretive manner during the night) after less than a week of protests; it had wanted to decriminalize charges of corruption brought against the leaders of the winning coalition (Diacu, 2017; Digi24, 2017; Zachmann, 2017). The action generated massive protest movements started in the virtual space, with 600,000 people on the streets, which triggered the withdrawal of the emergency ordinance after four days, its final definitive cancellation in Parliament, the majority's drop in polls of ten per cent (although the opposition was not directly involved, it was licking its wounds after the elections) and the loss of moratorium and natural growth following a victory. The government fell less than five months later, after a rather unique no-confidence vote where the majority voted against its own prime minister, after the withdrawal of political support, in unusual circumstances.

In all three examples, which took place within a particular context, the action led to the development of a vibrant and solid civil society, and participation in the public debate when certain red lines are overcome. Their coagulation in the virtual space was achieved through social networks, and the movement quickly acquired a purpose and self-awareness (Hardt & Negri, 2005), becoming a revolution without leaders. We excluded the case of the second round of the 2014 presidential election, because here we were in an electoral process, there was a beneficiary, the future President of Romania, Klaus Iohannis, and there was a party/electoral apparatus that managed the process, even though the essence of the movement was similar.

In each of the above examples, an important part of society emerged that would normally be disinterested in participation in electoral debates – we are talking about young people, highly educated and employable, with sufficient resources, not dependent on government policies, who have no candidate able to meet their strict criteria and high aspirations (except for the Iohannis election, which was a vote to keep Victor Ponta from power). Such massive gathering is primed by the emergence of anti-moral illegality/access or massive anti-logical narrative that go beyond the red lines that lead to massive participation, the activation of groups and individuals, and the mobilization effect in the revolution without leaders (Hale & Slaughter, 2005).

Objection of social constructs in the abstract has become a tool, with negative effects because if it is general or directed from the origins on relevant topics of national interest, such a movement can cause major effects in reality without classical political approaches, dialogue, negotiation, used as tools to respond to or clarify the protesters' demands. Any objections to a group by the collective consciousness in the virtual space only cease when another's claims are fully satisfied, and participation is no longer encouraged by promises of expectation and excitement. However, disquiet can re-emerge, ready for the next phase of action. There are, of course, tactics to manage this situation through the "isolation and forgetting" of the group, through a reactive posture over time, and by infiltrating and controlling these types of groups possessing the potential for objectivity over time during the preventive phase of their construction (Chifu & Ivan, 2013: 221–239).

The Third Generation: Cambridge Analytica, Likeography and Individual Targeting of Electoral Subjects

The difficulty concerning reactions to the informational war is the fine line between legitimate, legal (but immoral) and illegal factors and their related instruments. We have seen this from the definition of the concept, where the first dimension – disinformation, propaganda and active measures – has a legal, legitimate composition that most often employs the ideals of freedom of speech, the right of association and the free circulation of ideas; these are principles and values which we fully defend (but which are open for debate in societal confrontation precisely because some consider these principles and values to be the vulnerabilities of a democratic system that are worth risking and used as opportunities to undermine this system).

The lobby is legal but immoral, because an opinion-maker is always supposed to be the bearer of independent ideas and opinions, not of prepaid messages for their own audience credibility. Finally, the third component, consists of: psychological operations, troll warfare, profound emotional approaches and catalysts for emotions. All of these are highly illegal, and part of the arsenal directly related to active military and psychological operations (Chifu, 2016).

Another issue is that part of the informational warfare instrumentation can be used by parties in electoral campaigns. This third-generation informational warfare, which initially uses a targeted and individual-based marketing tool centred on preferences, has been developed and used in campaigns, such as Brexit and the election of President Trump. One instrumentation was owned by Cambridge Analytica, whose work in this area, was banned in the UK.

The OCEAN psychometric model is the basis for the classification of personalities and behaviours and has been introduced to differentiate individuals; it is a model that seeks to evaluate people based on five personality traits, known as the "Big Five". These are: Openness (how open are you to new experiences?); Conscientiousness (how much of a perfectionist are you?); Extravert (how sociable are you?); Acceptability (how preventive and cooperative are you?); and Neuroticism (are you easily irritable?). Based on these five features – also known by the acronym OCEAN – we can make a relatively accurate assessment of the kind of person standing before us. This includes their needs and fears, and how they will likely react (Pădure, 2017).

Cambridge University has developed the second part of the research project that underpins the identification of each person's tastes and preferences. Michal Kosinski's Psychometrics[4] – with its relevance in individual and customized sales and marketing – led to like-orthography; the identification of personality and character through OCEAN, based on Facebook 'likes'. Electronic preference information at this level has led to the identification of personality and, therefore, of the approach to selling something, depending on the appetite. It was purely a matter of approach using psychological techniques.

Kosinski and his team have continuously improved their model. In 2012, they showed that, based on an average of 68 user-like styles, it was possible to guess their skin colour (95 per cent accuracy), sexual orientation (88 per cent accuracy), or sympathy for the Democratic or Republican Party (85 per cent). They did not stop there. The coefficients of intelligence, religious beliefs, as well as the consumption of alcohol, tobacco or drugs etc., could also be determined. From the data, it was possible to even deduce whether someone's parents were divorced.

The effectiveness of their simulation was illustrated by how well they could predict the answers of a subject. Kosinski continued to work on

[4] Cambridge University, 2014, at the Kosinski Psychometrics Centre.

perfecting simulations; in a short time, he was able to evaluate a person better than his colleagues, based on ten Facebook likes. Seventy likes were enough to go beyond what a friend knows about them, 150 about what the parents know, and 300, what their partner knows about a particular person. Several likes could surpass even what a person is believed to know about theirself (Pădure, 2017).

Aleksandr Kogan, involved in the Cambridge Analytica scandal, moved the whole business into the space of political communication. The level that was reached was to map all voting-age US citizens, establish the OCEAN traits, and create an application to address them based on voting options and influencing short-term behaviour. This is how it worked during Brexit. The methodology works on dichotomous choices, dividing the spectrum of choices between determined, undecided (but inclining towards the same decision) and the other two opposite categories, with the aim of influencing the behaviour of those favourable to the desired option and the spreading of doubt or to keep those undecided from voting, but choosing instead the opposite option, which avoids wasting resources on those already decided on the opposing option.

The result is a major influencing formula for the short and medium term – 10–14 days – but with a very high impact and increased likelihood of success. It was used in two cases for the US election campaign, the Cambridge Analytica director being Steve Bannon, until recently President Trump's strategy adviser. But the instrument, with its major psychological impacts, has other levels of ability to alter will and choices, and is able to identify not only OCEAN traits and psychological character, but also behaviour at a given time. The likes have been abandoned since the invention of smart phones, and the method identifies these features with a high probability (85–95 percent) from merely analysing the mobile phone and the actions performed on it.

In order to understand the sophistication level of this tool, imagine going to a commercial negotiation, where the user of such a tool determines how you will approach that negotiation based only on the activities performed on the phone that morning until arrival at the business meeting. The effect of using such an instrument is obvious, it is suggested, extending from business decisions and choices to electoral and many other processes. It is true that the instrument has its limitations and may produce errors, but it has proven to be extremely effective as an element of influence in the short term.

This is where the Third Generation, third-person informational warfare toolkit has been developed, within a large adult population. These tools need to undergo vital research in how to block their harmful effects and unwanted influences, alongside access to such subtle data and interpretations of this nature. Until the full effects of these instruments are known, and the necessary regulations implemented, informational warfare (although not a panacea) is the most impactive. The introduction of these elements into the commercial and business space, where they can be virtually purchased by anyone, creates fear and negative responses, just like any extremely powerful and impactive techniques that can be widely applied to the population, through easy-to-use distortive scenarios.

Character Assassination: An Offensive Weapon of Informational Warfare

Informational warfare comes with multiple instruments, tools and techniques complementary to the general attitudes and mechanisms specific to the field of alternative narratives, to take advantage of the perception of people on the one hand, and to take advantage of the vulnerabilities of a targeted society on the other (Chifu & Nantoi, 2016). This arises from the fact that conflicts and wars have moved from a situation of confrontation between two armies, to that between two states/two actors, or even to a confrontation between two entire societies (Simons & Chifu, 2017).

When discussing offensive and defensive weapons in informational warfare, the direction of where the draft or plans for each operation begin evolves by identifying the vulnerabilities of the society that represents the target in an informational warfare. In democratic societies, its principles and values, human rights and the rule of law, including freedom of speech, free flow of ideas, debates in the public space, mediating between opposite narratives, and even freely gathering to protest, are instrumentalised as vulnerabilities of a society (Chifu & Țuțuianu, 2016), and by some actors are subject to plans to use such instruments as weapons in an informational operation (Chifu & Nantoi, 2016).

One of the most important informational warfare weapons is *character assassination*. It is most effective because it targets the personality situated at the critical point of a network in the decision-making system of a country. Character assassination is both old and new, having appeared together with humanity and language, from the very first gossip and lies, and has evolved until being theorized in 1950 (Davis, 1950), during the Cold War. It is used not only in informational wars but also in political campaigns, the source of a multitude of such instruments. At the same time, social media has relaunched itself as having the capacity to act as a more effective and harmful weapon than before, with the ability to reach a well-chosen public.

As we have seen, character assassination is the targeting of a personality in a decision-making system. An attack using this method relates both to the individual at stake and to the public targeted by the attack, and this duality makes the weapon difficult to build and calibrate, extremely nuanced and sophisticated but with a high level of effectiveness when used (Simons & Chifu, 2017). Moreover, to identify, prevent and defend against such a weapon is complicated. The current analysis is aimed at identifying what is and what is not character assassination and to establish the characteristics and typologies of such a weapon used in informational warfare.

It should be mentioned that character assassination is completely linked to some psychological behaviours, uses and practices that can fall under the ideas related to selective exposure and confirmation bias. One of the most widely accepted principles in sociology and social psychology is the principle of selective exposure (Sears, 1967). Exposure to media occurs predominantly selectively, so media effects can only occur if individuals choose specific media content for further consumption. In fact, media users often select media messages to instigate desired media effects on themselves (Knobloch-Westerwick, 2015). Different patterns of exposure to the news can cause people to develop different impressions of what is happening in the world around them. Without a common information base, it is hard to imagine that citizens can agree on public policy issues. People engage in selective exposure once they select media that match their beliefs and predispositions (Palmera, 2008).

Far from a limited effects perspective, selective exposure can serve as an important predictor of media effects (Stroud, 2010). It is often said that people tend to "hear what they want to hear". Similarly, it seems to be a common assumption that people tend to seek out information that supports their opinions and avoid information that contradicts them (Smith et. al., 2008). The effects of selective exposure are not limited to cognitive dissonance research. The principle of selectivity in information seeking (i.e., certain specific conditions will motivate people to seek certain types of information) emerged as a phenomenon of interest in other contexts not traditionally conceptualized in terms of dissonance theory.

Confirmation Bias comes from an obvious difference between impartially assessing the evidence to reach an unbiased conclusion and building a case to justify a conclusion already reached (Nickerson, 1998). First, evidence is sought from all sides, assessed as objectively as possible, and the conclusion drawn that the evidence, on balance, seems to dictate.

In the second case, one selectively gathers or gives undue weight to evidence that supports one's position, neglecting or disregarding evidence that contradicts it. In recent years, confirmation bias (or "my bias"), the tendency for people to seek out information that supports their beliefs and to ignore or misrepresent data that contradicts them, has been frequently discussed in the media, in science and in philosophy. This bias has been mentioned, for example, in debates on the spread of 'fake news', the 'replication crisis' in science, the impact of cognitive diversity in philosophy, the role of values in research and the evolution of human reasoning (Peters, 2020).

Offensive Weapons in Informational Warfare Depend Equally on the Target and Audience

Informational warfare is the creation of alternative realities by perverting the objective truth – made on the basis of facts and concrete arguments – its misinterpretation by using a combination of elements: selected and interpreted facts and pieces of truth, combined with reasoning and altered by the use of syllogisms, sophisms, propaganda, forced interpretations, and all embedded within a multitude of lies. The general purpose of the informational war is to determine, control or even alter strategic decisions, foreign policy, security and defence. It also involves the perversion or hindrance of instruments relating to the military component of a state, to impede its functioning and even block elements related to its security. The instrument and mechanism to achieve this goal are used to identify the public, citizens, prepared and organized pressure groups, who can apply pressure on the authorities to change objectively identified solutions if they lack the support of, and are even opposed by, the population.

The battle of alternative narratives, of "alternative truths", has become the most insidious way of constructing beliefs. However, it is based on the groups and publications targeted for each operation, on a vast fabric of knowledge of the training, inclinations and expectations of the target audience, as well as its propensities and frustrations. Obviously, the tools do not only concern information as such, but also refer to other more subtle components related to fundamental emotions, context and the value of the opportunities offered by ongoing events and the mood created at the level of a target population, in order to protect against a certain type of reporting on the subject.

Alternative reality perverts the perception of a target population in a combination of psychological operations – PSYOPS, along with misinformation and propaganda, using fundamental beliefs, feelings and impact images to bring the target audience into a pre-defined perception. And finally, as the audience already has an opinion, its perception has replaced reality (Stern, 1999) and, no matter what argument or proof of truth they have been presented with, they would face the barrier of an already-formed perception.

One of the most important weapons in the early informational war is character attack (the attack on character or reputation), or alternatively, character assassination (destruction of reputation). As we have seen above, the attack does not necessarily rely on real data (Samoilenko, 2016) and is dependent upon, at the same time, the target person (or a group of people, an institution or country) (Icks & Shiraev, 2014) and by an audience targeted to change their mind about the individual. The goal is to remove the opponent from the public sphere, eliminate it from the competition or at least make a fundamental mark with an engram, or with a psychological vulnerability to make it react to a particular topic in a way that will alter its decision-making capacity, its ability to compete or fight, even if it resists an attack on its prestige and reputation.

What is character attack/ assassination? The individual's characterization is based on the impactive power of the perceptions created by words and images (Icks & Shiraev, 2014) in injuring, devastating, or even completely destroying the status, character, reputation or prestige of someone. At the first two levels with a different degree of impact, we are talking about *character attack*, while complete destruction or elimination of someone from the competition space means *character assassination*. The non-elimination from combat of a stubborn or unpleasant enemy does not mean that a character-type action has failed. As long as the target audience reacts and shapes its opinion, that opinion, based on the results of the attack, leading to a withdrawal of support for the target person, means that the attack is successful; that is, competitive ability is completely ruined, even if the person remains in the battle for office. The only difference is that there is a possibility that the victim has a chance at rehabilitation.

There are cases of personalities who have survived a successful action of *character assassination* and who, despite the impact of the action, have remained in office. They have managed to capitalize on the exposure given to them by *character assassination* and taken the opportunity to exploit the situation to their professional advantage.

The classic case is that of the former mayor of New York, Rudolph Giuliani, who was associated with the Mafia, and involved in corruption cases investigated by the tax authorities, while going through a divorce. However, despite these difficulties, he managed to handle the tragic events of 11 September 2001 with great professionalism, and to make the right decisions regarding the city's societal rebuilding in the face of the terrorist attacks. This was achieved through mass crowd gatherings in Times Square to mourn the dead and rebuild hope – a political gesture with a massive risk should a new attack have occurred. The gesture turned him into the 'Mayor of America', and gave him a solid political base to run for the US Presidency, and today he is part of the Trump team. But the conditions for such a jump are exceptional, and even some would say that Giuliani was not the subject of an informational operation of *character assassination*, but only a subject of his own acts and deeds before September 11. For such a characterization, the intention, planning and existence of the attacker (not only somebody who takes advantage of the operation) (Sigelman & Kugler, 2003: 45–62) is absolutely necessary.

We can, therefore, say what is NOT *character assassination*: the natural mass media reaction, the media, the facts, and the facts of facts, the results of investigations, the actions of state institutions that prove actions and deeds in the courts of a public or political character. It is not *character assassination* or *character attack* if the authentic action is related to the current developments, where real facts, even if these accounts are accompanied by subjective interpretations, forced, or political options and opinions are involved in the actual press reports.

Character assassination is about intention, planning, pursuit of a plan, and its application with various, though not always real, elements of false and *fake news*, with insinuations, rumours and anonymous artificial constructions, all of this is meant to be in conjunction with the specific target, utilizing a spectrum of tools from the above. And above all, the focus is not on the arguments, facts, and combating the ideas displayed by the target person, but to the source of the attack on the person who communicates the ideas, to the elements that define and validate those ideas, which are not even discussed in such a process.

A Concept that Unites Multiple Fields

The character, in the sense that it is used in the term character assassination, refers to the moral aspects of the experience and behaviour of an individual (Danzinger, 1997). If human emotional mood has a psychological relevance and is rather natural, it is biased, biology, personality is the sum of the stable characteristics, reaction and behaviour of an individual, and the character has to do specifically in this definition and understanding of character assassination, ethical and moral aspects, those key elements that give substance to the credibility and legitimacy of a personality.

As far as moral issues are concerned, their impact relies on the public perception of actual standards, or the ability to absorb such news. For example, in France, the news about Francois Hollande's "parallel families" did not affect his election, while the same information in the US would have led to a definitive exclusion from political life. It is through generally accepted standards that appearances and rules, public expectations, and the habits of the community are viewed – with the hint of hypocrisy and snobbery that these standards can contain (we are told "to do what the Pope says, not what the Pope does"). It's about respecting appearances, doing "what is appropriate", avoiding gestures and actions that "shouldn't be done" because "it's a shame", "your neighbours know you", "the good world knows us and we will be in an embarrassing situation". And for a leader, for an aspirant to public confidence, the standards required and society's exigencies are higher than those of ordinary people who are entitled to be wrong, but they do not want to lead the people and have no desire to be adored, believed and followed.

Public figures occupying important positions of authority, are representative and visibile – emperors, kings, presidents, generals, bishops, opinion leaders, teachers, moralists themselves – and must "adhere with utmost ostentatiousness and obstinacy to general morality and standard behavioural norms and to the expectations of the public" (Shiraev, 2010). An actor or a creative person has greater freedoms than an orthodox Bishop, a judge, or a member of a party ethics committee. They must maintain their good reputation for a long time, and in any case as long as they are in office. That is the only way they maintain the support of the relevant public, not only in the role being performed, but for their standing.

Moreover, character assassination does not just refer to the target person, or to a type of audience. First of all, the public or the decision makers depend on the effective maintenance and exercise of the function by the person in question, but perhaps more importantly, for the offensive weapon in an informational war is represented by character assassination, the public who supports the person concerned, the reference public, whose loss causes a loss of self-esteem, support in the professional class and the elite to which it belongs.

The concept itself is present in many fields of study, from psychology, psycho-sociology, political science, political and public communication. Within each of these areas, relevant elements may be examined to know exactly how the offensive weapon of character assassination is constructed in the informational war. In addition, from each element comes the nuances, subtleties and sophistication needed to plan an attack in order to avoid victimization of the target or indolence or indulgence of the public; there may even be support for the target personality of the attack should it become too ostentatious which could lead to unwanted side effects. Here are found some of the sources of mistakes in the use of character assassination as a weapon in the informational warfare, but also the elements that present a survival opportunity to such an operation and the lessons learned to counteract such a destructive attack.

It is suggested that a successful operation involving *character assassination* is when the facts and the attack are not made public, but the prospect of the attack is used to constrain the target to act in a certain manner. Conditioning or extortion, or the exposure of important elements, may result in the acceptance of the targeted person of the expectations and directions suggested/imposed by the attacking group. In addition, the target's resignation achieves the same result, and has the same relevance; that is, the character assassination attack has reached its target even if it is not visible in the public space. Careful research and evaluation of previous events in a person's past professional life may create vulnerabilities and make performance of another public function untenable.

The Concept of Character Assassination

Character assassination is the deliberate destruction of the reputation of an individual, institution, group or country (Shiraev, 2010; Icks & Shiraev, 2014; Samoilenko, 2016). Targets of actions that fit the typology for character assassinations are political leaders, officials, celebrities, scientists, public figures, and athletes. In operations, they are targeted by actions that relate to individual elements such as personal life, behaviour, values, identity and self-confidence; their beliefs and aspirations may be shaken by such attacks. The tools used are: bibliographic details are altered, interpreted or manufactured; intimate elements and personal moments are made public; their professional and personal achievements, family context, ethnicity or sexual behaviour examined; their good will being questioned as well as their good faith and intentions behind the action, using exaggeration, false allegations, insinuations and lies. The purpose is to affect (in the case of *attacks)* and to destroy (in the *character assassination*) the moral position of the person (group, institution, state) in the eyes of the public and generate a negative emotional response to the target of such an attack.

A nuance needs to be highlighted here between the process of *character assassination* against a group, movements, ethnicities, institutions, countries and the process of creating the image of the enemy (Keen, 1988; Bruckner, 1996), elements used in the conflict polarization phase. The instrument could be sensibly identical, but the intention, purpose and planning may be different. In the first case we are in the classical situation of destroying reputation, in the second we want to polarize society, coagulate public support for an emerging conflict, or use the image of the existing enemy to blur internal failures and divert attention from other subjects and stringent themes to target the attacker.

Defamation associated with the attack on the person aims at undermining the credibility of the opponent, based on a rhetorical strategy (Walton, 1985, 2007) and targets its own public and its mobilization for the coming conflict, as in the case of *character assassination*. On the contrary, in the second case, there may be an interest in maintaining the image of a fierce, threatening enemy in order to be able to mobilize the public, which is not in the same category as destruction of reputation (where the objective is one of complete deterioration of the target's image).

Character assassination, in terms of essence and content, was introduced into political communication in the United States in the mid-1920s.[5] The concept as such was introduced in the book of the same title written by sociologist Jerome Davis (Davis, 1950), based on deconstructed fundamental emotions (Moisi, 2009), such as fear, ignorance, envy and jealousy, which are used to propagate rumours and lies about a targeted person.

The major turning point in the 1960s was the emergence of television. The use of television in political debates and electoral competitions rendered candidates powerless in the face of public communication tools, the attacks and defence formulas in electoral strategies and the preparation of public debates (which included negative propaganda and attacks against the person), even attacks on the character (morals) of the candidates. Sociology and psychology have brought new elements to persuasion mechanisms in politics, using emotions and perceptions (Graber, 2009). Negative campaigns have prompted the creation of a large number of instruments and tactics (Flower, 2007) used in character assassination.

Character Assassination: A Weapon of Contemporary Informational Warfare

As in the case of hybrid war or informational warfare, a whole conceptual dispute between military and security theorists and strategic studies or international relations creates differences by looking at issues using both new and old instruments. Some believe that all tools have historically been used for military purposes during conflicts. Others believe that these are all tools and methods of the twenty-first century (Simons & Chifu, 2017).

[5] Although the notion of character assassination appeared in the US, there are equivalents on the European continent; Rufmord in Germany and Karaktermoord in the Netherlands. There are also studies in Russia concerning attacks on the reputation of a person.

The mere fact that any potential attacker has access, as a result of the emergence of the Internet and social media, to anyone who is online or has a social media portal is a qualitative difference that is a huge jump from radio and cinema to television in the 1960s. Indeed, the massive leap forward to the Internet and vast connectivity has allowed social media to emerge; a tool whereby communication borders no longer exist, globalization is achieved, and which allows anyone to target the public space and connected citizens of a state or any other group aim. We have seen how sophisticated the informational wars have become (Chifu & Nantoi, 2016), and how fast the successive generations of informational warfare develop (Chifu, 2017a: 20–27).

In the case of *character assassination* operations, the reality of the twenty-first century informational warfare allows the use of social media tools without measure. Thus, in closed societies, rumour, insinuation and gossip have played a major role in *character assassination* formulas (Ficeac, 1996; Wierzbicki, 1996; Volkov, 2009) (as well as a way to reduce tension within the society), in the contemporary era, with the existence of social media providing major freedoms to anonymous sources through third parties to launch and multiply *character assassination* attacks. During the Communist era, someone who was gossiping rarely knew or heard about gossip unless it had directly affected his family, reputation or position. Today, using established social media formulas, consisting of closed groups and through anonymity formulas determined by avatars and false accounts, all categories of attacks in the *character assassination* sphere can be promoted and distributed without the knowledge of the target, but with greater impact, and especially from some distance away.

Moreover, the psychological operations of informational warfare can determine the maximum effect of character assassination attacks, particularly when the aim is to destroy the target person by threatening a campaign against them or using similar tools. Removing the competitor without the attack being made public is a successful assassination operation which has the greatest impact, even if the general public only sees the result of this operation, not the content of the attack (a credible threat of attack).

A more in-depth approach to the systematic study of character assassination is being undertaken with the growth of institutions, such as the International Society for the Study of Character Assassination.[6] In July 2011, researchers and teachers from nine countries – a group of historians, social science specialists, psychosocial and political psychology experts – gathered at the University of Heidelberg to discuss the art of

[6] The International Society for the Study of Character Assassination (ISSCA), https://characterattack.wordpress.com/.

defamation in both the past and present. The result was the creation of the aforementioned society. The fundamental elements and lessons learned in this area are: the interdisciplinary nature of the tactic; the difficulty of detecting and counteracting *character assassination* operations; the varied instruments used; and the ability to reach different audiences and those whose own aspirations, values and options were freely disclosed by them on social networks.

Moreover, the third generation of informational war (Chifu, 2017a: 20–27) allows for individual targeting on a large scale – for example, all voters in Britain for Brexit, or all voters in The United States – with messages of interest which directly affect them, or causes for which they have empathy, all elements that allow the *character assassination* operation to exponentially multiply its target impact, following individual calibration of the targeted audience.

However, it should not be forgotten that considerations for *character assassination* operations target the groups and decision makers who determine the perspective of choosing and maintaining the target public person and especially the public that counts for the target person, their personal and individual, professional and emotional characteristics. Targeting this component may have a greater impact than targeting the general public.

However, the degree of sophistication must be high, and the attack must be properly calibrated, because we are talking about important public figures for the target personality and a reference public that knows the person, has lasting emotional ties and has perceptions, beliefs and convictions regarding the person which are well rooted and difficult to shake. Seeding some incipient doubts or mistrust is a sufficiently ambitious objective, because in this type of attack any exaggeration can have disastrous side effects for the attacker; it may create the impression of victimization of the target personality that elicits supportive public emotions, thereby damaging the reputation of the attacker.

How Do We Alter Reality in Informational Warfare?

The fourth and fifth generations of informational warfare are directly linked to the evolution of perception and a journey leading us from post-truth via post-factual approaches to post-humanity. If the post-truth[7] is

[7] Post truth – Relating to or denoting circumstances in which objective facts are less influential in shaping public opinion than appeals to emotion and personal belief. *'in this era of post-truth politics, it's easy to cherry-pick data and come to whatever conclusion you desire', 'some commentators have observed that we are living in a post-truth age'*, in Oxford Dictionaries, https://en.oxforddictionaries.com/definition/post-truth.

about emotions that are influencing perceptions and the attempt to alter perceptions on a large scale, especially due to interpretations of reality, the post-factual approach already deals with altering the facts and changing the past, so that the very reality is at stake.

Moreover, if we move further, in the use of artificial intelligence[8] (Poole, Mackworth & Goebel, 1998), we reach a moment where our mind is at stake and the very processes of thinking, perceiving, understanding and knowing reality are altered via the direct impact on our minds, when facing an intelligence that plays with enormous groups of big data[9] (Laney, 2001; Reichman, Jones & Schildhauer, 2011: 703–705), that knows each of us in detail and is playing with all of us, individually, in real time, with effective psychological tools, in order to twist everything that we know. It is no longer about perception or even facts, but about our own minds.

However, we can defend ourselves from the future, to protect our sanity of mind. We need to adapt, but costs are going to be heavy for a majority of our population. Sustainability (Jones, 2017) and resilience (Chifu, 2018: 23–30) of humankind, at the whole level, is going to be tested, as well as other characteristics of adaptation to the reality in a post-human challenge about the reality. We need to adapt more and more to this life in between worlds, the real one, the virtual one and even several alternative worlds that we are soon to be confronted with at the same time, in the very near future. Welcome to the post-human world!

Informational warfare continually evolves (Chifu & Nantoi, 2016), with the fourth generation and the thesis of a fifth generation approaching fast. The theme of informational warfare shows us the developments that are taking place these days about the perspectives and pressures faced by the human mind, in our everyday internal world in the near and predictable future. Already the informational war has passed to the fourth generation and intensive technological studies are being

[8] Artificial intelligence (AI) is a term for simulated intelligence in machines. These machines are programmed to "think" like a human and mimic the way a person acts. The ideal characteristic of artificial intelligence is its ability to rationalize and take actions that have the best chance of achieving a specific goal, although the term can be applied to any machine that exhibits traits associated with a human mind, such as learning and solving problems. At www.investopedia.com/terms/a/artificial-intelligence-ai.asp. A.I. research is defined as the study of "intelligent agents": any device that perceives its environment and takes actions that maximize its chance of successfully achieving its goals (Poole, Mackworth & Goebel, 1998).

[9] Big data is a term used to refer to the study and applications of data sets that are so big and complex that traditional data processing application software are inadequate to deal with them. Big data challenges include capturing data, data storage, data analysis, search, sharing, transfer, visualization, querying, updating, information privacy and data source. See Laney (2001). See also Reichman, Jones and Schildhauer (2011).

conducted towards the fifth generation of mind games, with what a man perceives, with what he understands, with what he knows and how these stages can be altered to impact his opinions, convictions and choices.

If the first generation explicitly refers to misinformation, propaganda, active measures and elements of psychological operations (Chifu, 2015, 2016), the second aims at objectification of the abstract construct in the virtual space (Chifu & Ivan, 2013: 221–239) and the communication between the real space and the virtual space, the third refers to the individual access but based on models – social masses, based on big data, to target, alter or direct options.

The fourth generation directly targets the unpredictable personalities of the third-generation Cambridge Analytica instrument, the 3–15 per cent whose likelihood of being influenced is nil, through specific instrumentation tailored to the individual concerned following analysis, the clearest method being *character assassination* (Chifu, 2017b: 11–17).

Finally, the outlook of a fifth generation, which will occur in about two years, is already being drawn, when technology will allow "real images" to be assigned to every public figure, and that can transmit any pre-set audio or audio message. Imagine the confusion created when Donald Trump declares war in a fake video, his image being manipulated instead of using an actor or a look-alike, or Emmanuel Macron resigns live (Green, 2018). Vision, our richest sense and 87 per cent of the information reaching the human brain, becomes relative and without the prospect of authentic selection, to discern the fake from the real, the prefabricated from reality. And when there is not only one such enormity present in the public space, but seven distinct and nuanced messages on a theme, confusion is perfected and playing with the human mind reaches its peak. In this way, we will have reached post-truth, post-factual, and, further down the road, post-humanity.

Post-Truth: Theorized Origins, Tools of Validation in Modern Informational Wars

The *Post-Truth* is a term that has been used since 1992 by Steve Tesich (Tesich, 1992). This is according to the Oxford Dictionaries[10] which also notes that it was the most searched word in 2016, when nearly all post-truth and informational warfare appeared in various forms. Tesich then maintained that the history revealed by the opening of the archives related to Watergate, the Iran-Contras scandal and the first Gulf War

[10] Oxford Dictionaries, *"A Government of Lies"*.

demonstrated that "we, as free men, have accepted and voluntarily decided to live in a kind of post-truth" (Tesich, 1992). His references discuss how unfamiliar ordinary people are with the great dossiers and themes of contemporary operations – special, classified, military – and how great the distance is between what is publicly presented and what we absorb and understand from a more complex and sensitive reality which is denied in all its details.

The subsequent definition of the term post-factual, or post-reality policy, refers rather to a political culture in which the debate calls for emotions[11] (Oxford Dictionaries), unrelated to the details of concrete policies, in which the propaganda is summed up in talking points (general ideas that are established by the communication strategy of ideas being repeated unanimously by a group), yet completely ignoring the facts and the surrounding reality. Post-truth does not mean challenging or falsifying the facts and expert opinions, but putting everything in full subordination to public emotions that ignore the truths and facts presented objectively, being blinded by emotions.

There is a whole debate about how old or new the informational war is, as well as the populist policies that use emotions. Some say that elements of it date back to when there were debates in the Agora of Ancient Greece, but the explosion of post-truth is inseparably linked to the Internet and social media, to anyone's ability to convey, to a broad, uncontrollable but connected public network, ideas and deep emotions that affect their views.

The origins of post-truth – classical philosophy, technology, populism – can be, in part, found in the research of Bernard-Henri Lévy who was the one who tried to explain the paradigm of post-truth, with all its components. The French philosopher speaks about the three components of this development of human reality that we will have to get used to; there is "a philosophical suspect, a technological culprit and a very responsible politician" (Levy, 2018) in present-day life and this hideous post-truth.

Therefore, everything has its origins in modern philosophy. All those who question 'the Truth' have their origins in Friedrich Nietzsche's philosophical thinking, which questions the Truth as an absolute, considers it relative, and furthermore, places it as an equal to the Lie. Nietzsche wondered why Truth had greater value than Lie (Nietzsche, n.d.). This is a great sign of a question that still exists today, and the effects of this philosophical syllogism exclude ethical, moral, or complex axiological reporting, and lead to the crazy world of today.

[11] Oxford Dictionaries, available at https://en.oxforddictionaries.com/definition/post-truth.

But, in a way, there is a return to the cycle of a classical idea of Greek philosophy. It is the renaissance of the oldest idea of philosophy – the school of thinking of the sophists (Braunstein & Pepin, 1997) – and perpetuates the battle of ideas between the Philosophers and the Sophists. The Sophists questioned the owner, the absolute Truth holder. Their doctrine speaks of the fact that Socrates and Plato are not entitled, nor cannot claim to have, the Truth. So, through social media, we are all the reincarnation of the Sophists in the modern world.

Technological interference, with the advent of the Internet and social networks, has led to absolute democratization in a virtual space, but able to directly attack the mind and perception of real people connected to this space. Democratization has led to an explosion at enormous cost by promoting a generous concept that every truth and idea has equal value; however, ideas compete against each other to be the most listened to and therefore take on the greatest importance.

It is not a question of the verification of human nature, rather an absolute check of the way in which a human really looks on average, and which is the nature of the human spirit, how individualism and the spirit of the herd matter. It is influenced by education, habits, traditions, individual characteristic features, but also from society, such as relationships, community spirit and its natural hierarchy.

The Internet and social media have made each sentence and every opinion of equal importance. An absolute democratization of the Truth has emerged, which has pushed far beyond Nietzsche and the relativization of the Truth. It is no longer about assigning an ethical value superiority of the Truth in relation to Lying, here is the equation of all nuances – which hold a similar value in the competition for perception on the Internet – between the extreme lie and the Absolute Truth. Moreover, it relativizes the source credibility of these contested ideas, the idiot put together with the genius, the connoisseur, the expert, the professional. It's a disaster. Not necessarily absolutely. Levy (2018) claims this is part of human nature and becoming, that a human being is forced, in his becoming, to confront and adapt to this stage of evolution in order to resist the future.

Finally, populist politicians are the third source who are guilty of the problem that is being discussed. Here we are already entering the realms of Kafka or Eugen Ionescu, a theatre of the absurd. Populists resort to emotions, but their main idea is that "Truth is not the Reality, Truth is what I say." This is how Nazism, Modern Absolutism and Authoritarianism emerged. It is the new idea that pushes, through consequences, to the absurd and to the destruction of freedom, and to the absolute control of the centre, whatever it may be, on peoples, societies, nations, the world.

From Post-Truth to Post-Factual and Post-Humanity

Post-truth means relativizing the value of Truth, questioning the owner of the truth (Sophists), equalizing as value of Lying with Truth (Nietzsche), democratizing Truth (social media, all sentences are equal, regardless of the issuer and the connection with Truth), about the connection of human perception, human understanding of the Truth with the reality (Chifu, 2015, 2016). We talk about interpretations, emotions that alter the understanding of truth, or obstruct and guide our individual Truth. But it does not affect the facts.

The post-factual begins from where we speak of the paradigm, '*What you see is not True, What you perceive with the senses is, in fact, not true*', namely, the alteration of the facts through altering perception. It is no longer about the interpretation, but about the fundamental alteration of the facts. The populist approach is *Truth is what I say is the Truth* (Goebbels, 2004: 13; Giuliani, 2018) which turns the fact into non-fact or its opposite.

This type of approach, combined with media censorship, blocking press freedom, free circulation of ideas and freedom of expression, pushes the targeted society towards authoritarianism or dictatorship. If we look today at the official presentations from the Kremlin by its spokespersons – Russia's aggression in Ukraine, the shooting down of flight MH17 over Donbas, poisoning with Novichok of Sergei Skripal, interference in the US elections and those of other states – we see how the truth can be manipulated, through the construction of alternatives to reality, altering the simple facts for the Russian and captive audiences. Or, as Bernard Henry Levy said, "You cannot refrain from thinking that Putin is the commander of his country, he's chief historian, chief journalist, and wants to erase, rewrite, reinvent the history of the past" (Levy, 2018).

Here the reference to Francoise Sagan is clear and the saying "No one ever knows what the past has in store for us" – a direct reference to the alteration and rewriting of future history (Sagan, 1965). This is about populist and post-factual regimes. The facts are rewritten, not reinterpreted, but dramatically changed. We also know the history of Ceausescu's heroism, rewritten after his rise to power in 1965. History is also being rewritten in present-day Poland, with the role of the opposition Solidarity. It destroyed the image and role of Lech Walensa and puts Geremek in the foreground. That is what populist politicians are doing; the post-factual approach (Levy, 2018).

Post-humanity is even more tragic, and it links directly to another human creation, artificial intelligence (AI). This is an automatic, independent machine-learning mechanism capable of infinitely manipulating more data than a human being in a fraction of a second, is infinitely more efficient and capable of reproducing the reality of any human sense, altering the human mind in a programmed manner.

Everything began from the over-use of AI in efforts to detect *fake news*. It taught cars to corroborate language elements, taught them to translate, to read, to write lyrics (not poems!!!), to understand grammars and syntaxes, to create correct sentences in any language, more correct than real people's speech. Moreover, it taught them to recognize faces, to interpret gestures, to create grimaces, and today the program by which a text and an image of anyone, together with his voice, can create "authentic" factual statements that are transmitted on it will take up to two years for improvement (Buterin, 2018; David, n.d.). Then watch out for informational warfare!

We have already stated above what it will be like to wake up to an artificial intelligence engine that creates the life-like image of a leader making a speech purely from algorithms and mega-data, starting only with text, and with an image, voice and behaviour traits more authentic than the original. We have now reached the stage, following more than two years of technological development, where we have a clear case of a mechanism superior to humans in terms of memory, speed and access to information, and that affects the personal internal universe of each individual and psychologically alters us. This is post-humanity.

This is not an isolated case; it is not just for the experts, but a whole human population that will have to confront and adapt to the stress of this post-human reality, which affects us deeply on an individual level. It will also show the content and capacity of humanity in its entirety to adapt, and we will know how the average person, humanity in its entirety, struggles with these real future mental difficulties of living in parallel virtual spaces; a concurrent schizophrenia caused by the constant shift between computer-generated realities and everyday life, factuality, truth and humanity.

Conclusion

Of course, there are methods of reacting to post-reality, now and in the future. Every human construct also has its antidote, even if it is already used as a weapon against humanity itself. We see today the reactions to generations of past informational warfare, and the constant antidote accompanying the responsible construction of new technical discoveries, including artificial intelligence in action.

Niall Ferguson, the well-known historian at the Hoover Institution, Stanford University, is the most ardent supporter of Kondratiev's historical cycles and the repeating history of interpretation. It is the task of an applied historian, he argues, to choose from the possible multiple futures the relevant, condemning and stating that he is shocked at how often politicians and journalists are going to make the wrong mistakes for the future. Ferguson argues that "Technology is important, but less important, in fact, in the future, it changes very quickly and creates frequent ruptures, so the life of a technology is much smaller than it used to be. Computers and social media have a major impact, but they are eliminated more quickly than the 50s" (Ferguson, 2018).

What is new is the decentralization of communications. Decentralization has happened with print media already and social media has brought about changes in their impact, but a fairly limited influence over time. As an analogy it is not sufficiently relevant today, even if it supports democracy and blocks the authoritarianism of that time. The difference is not only almost free Internet access to virtually infinite databases, but also access to large numbers of influential people within any society, whose personal data and intimate experiences are contained within social networks.

Ferguson argues that the impact of decentralization of information and the strengthening of capacity and impact on people's opinions has led to the recent upgrading of power. Even artificial intelligence can mean that over time an instrument may leads to increased state control over the individual.

However, I do not believe in this hypothesis. The difference is in creativity, with the clearest model coming from China, where enormous databases copied from all Internet traffic create knowledge offered on a silver platter to the Communist Party. Political control leads to fundamental consequences, I believe, and we find it in every centralized society and politicized institutions that lose the capacity to act once the rule is that political truth is imposed from the top down, not through knowledge that comes from the base of social organizations and professional institutions.

Likewise in post-humanity, the fundamental consequence of control over Chinese society is repeated; Chinese control over networks leads to implosion by blocking freedom of thought and opinion, free experimentation and creativity. The technological difference, to paraphrase Ferguson, applying a real world example of post-humanity to our analysis, is the fantastic ability of artificial intelligence to create a situation where chess is played better by machines. But are they able to interpret, be creative and perform better that in a chess game, such as in all types of life situations? The answer is No (Ferguson, 2018).

No matter how powerful computers become, the greatest determinants are the human mind and the human condition. This is the decisive factor, regardless of technology. There are some human constants, and it is suggested that the post-humanist era, as we have it defined above, cannot destroy people, human society, the human condition, or the human mind. Even if it does not create a major impact, it fundamentally changes what we know today about our way of life, learning, knowledge and adaptation. It is, indeed, a major challenge for the future, which will collaterally make many contemporary victims who will no longer understand anything and will isolate themselves in comfortable bubbles, incapable of adapting to future shocks.

References

Air Force Doctrine Document, 2-5.3 (27 August 1999), *Psychological Operations*.

Andreescu, C. (18 August 2016), EurActiv.com: SUA își mută armele nucleare din Turcia în România, DC news, www.dcnews.ro/euractiv-com-sua-i-i-muta-arme-nucleare-din-turcia-in-romania_513671.html.

Anderson, T. (2012), Learning Technology through Three Generations of Technology-enhanced Distance Education Pedagogy, *European Journal of Open, Distance and E-Learning*, n2.

Barry, E. (7 April 2009), Protests in Moldova Explode, with Help of Twitter, The New York Times, www.nytimes.com/2009/04/08/world/europe/08moldova.html?mcubz=3.

Bhalla, M. R. & Bhalla, A. V. (2010), Generations of Mobile Wireless Technology: A Survey, https://citeseerx.ist.psu.edu/viewdoc/download?doi=10.1.1.206.5216&rep=rep1&type=pdf.

Braunstein, F. & Pepin, J. F. (1997), *Marile Doctrine. Politice. Economice, Filosofice*, Bucharest: Antet.

Bruckner, P. (1996), *Melancolia democrației: Cum să trăiești fără dușmani*, Bucharest: Antet.

Buterin, V. (13–15 September 2018), Will Blockchain Change the World?, at 15th Yalta European Strategy, Annual Meeting, Kiev, "The Next Generation of Everything", available at https://yes-ukraine.org/en/photo-and-video/video/15-a-shchorichna-zustrich-yes/chi-zminit-blokcheyn-svit.

CCA Report (2016), Chishinau (CCA – Comisia Centrală a Audiovizualului - Audiovisual Central Commission).

Chifu, I. (2020), The Fourth Generation Informational Warfare, *Romanian Intelligence Studies Review* (Revista Română de Studii de Intelligence - RRSI), RRSI nr.22/2019, Bucharest, pp. 67–105.

(2018), *Războiul hibrid și reziliența societală. Planificarea apărării hibride*, Infosfera, no. 1, pp. 23–30.

(2017a), *Trei generații de război informațional. Nivelul dezvoltării domeniului în partea sa publică*, Infosfera, Year IX, no. 3, pp. 20–27.

(2017b), *Character Assassination, Armă ofensivă în războiul informațional*, Infosfera, no. 4, pp. 11–17.

(2016), *Războiul informațional: componentă a războiului hibrid și instrument de acțiune agresivă cu relevanță strategică*, in Infosfera, no. 3 pp. 3–11.

(2015), *Război hibrid, lawfare, război informațional. Războaiele viitorului*, in Universitatea Națională de Apărare, Strategii XXI. Complexitatea și dinamismul mediului internațional de securitate, UNAP "Carol I" Publishing.

Chifu, I. & Ivan, C. (December 2013), *Rolul democrației participative, al meta-narațiunilor și autorului colectiv în secolul XXI. Mesaj din viitor*, Bucharest: Revista Română de Studii de Intelligence, RRSI no. 10, pp. 221–239.

Chifu, I. & Nantoi, O. (2016), *Război informațional: tipizarea agresiunii informaționale a Federației Ruse*, Bucharest: ISPRI.

Chifu, I. & Țuțuianu, S. (2016), *Torn Between East and West: Europe's Border States*, London: Routledge.

Concept of the Foreign Policy of the Russian Federation, Ministry of Foreign Affairs of the Russian Federation (12 February 2013), http://static.kremlin.ru/media/events/files/41d447a0ce9f5a96bdc3.pdf.

CORUPTIA UCIDE: Peste 35.000 de oameni au ocupat strazile Capitalei. Proteste de amploare au avut loc in toata tara (4 November 2015), http://stirileprotv.ro/stiri/actualitate/noi-proteste-sunt-anuntate-si-pentru-miercuri-seara-ce-vor-manifestantii-sa-se-intample-dupa-caderea-guvernului.html.

Danzinger, K. (1997), *Naming the Mind: How Psychology Found Its Language*, London: Sage.

David, M., Artificial Intelligence and Deep Fake, available at https://yes-ukraine.org/en/photo-and-video/video/15-a-shchorichna-zustrich-yes/shtuchniy-intelekt-ta-deep-fake.

Diacu, L. (3 February 2017), "Noaptea, ca hoții!"- peste 300.000 de români, în stradă împotriva mutilării Codurilor penale, Epoch Times, http://epochtimes-romania.com/news/ne-vedem-la-guvern-ziua-a-patra-de-proteste-impotriva-mutilarii-codului-penal—257156.

Digi24, #Ro-mânia, ziua a șasea | Protest istoric: 600.000 de oameni au fost în stradă Referință Ordonanța 13 (5 February 2017), available at www.digi24.ro/stiri/actualitate/justitie/legea-gratierii/cel-mai-mare-protest-fata-de-oug-13-anuntat-pentru-duminica-in-capitala-663703.

EurActiv.com: SUA mută arme nucleare din Turcia în România / MAE și MApN neagă (18 August 2016), EurActiv.ro, available at www.euractiv.ro/extern/sua-muta-arme-nucleare-din-turcia-in-romania-5170.

EurActiv: SUA isi muta armele nucleare din Turcia in baza militara de la Deveselu. MAE neaga informatiile (18 August 2016), Știrile PROTV, http://stirileprotv.ro/stiri/actualitate/euractiv-sua-isi-muta-armele-nucleare-din-turcia-in-baza-militara-de-la-deveselu.html.

Ferguson, N., The Next Generation of Everything. A Historian's and an Investor's View, at 15th Yalta European Strategy, Annual Meeting, September 13–15 2018, Kiev, "The Next Generation of Everything", available at https://yes-ukraine.org/en/photo-and-video/video/15-a-shchorichna-zustrich-yes/maybutnye-pokolinnya-vsogo-poglyadi-istorika-ta-investora.

Ficeac, B. (2014), *Tehnici de manipulare*, ediția VII, Bucharest: C. H. Beck.

(1996), *Tehnici de manipulare*, Bucharest: Nemira.

Flower, H. I. (2007), *The Art of Forgetting: Disgrace and Oblivion in Roman Political Culture*, Chapel Hill: University of North Carolina Press.

Giuliani, R. (2018), www.politico.com/story/2018/08/19/giuliani-truth-todd-trump-788161.

Graber, D. (2009), *Mass Media and American Politics*, Washington DC: CQ Press.

Green, Y., Demonstration. Hacking an Online Conversation, at 15th Yalta European Strategy, Annual Meeting, 13–15 September 2018, Kiev, "The Next Generation of Everything", available at https://yes-ukraine.org/en/photo-and-video/video/15-a-shchorichna-zustrich-yes/demonstratsiya-nezakonne-vtruchannya-u-besidi-v-merezhi.

Hale, T. N. & Slaughter, A. M. (2005), *Hardt & Negri's Multitude: The Worst of Both Worlds*, Occasional Paper, Woodrow Wilson School of Public and International Affairs at Princeton University.

Hammes, T. X. (2007), Fourth Generation Warfare Evolves, Fifth Emerges, *Military Review* (May–June), pp. 14–23, www.hsdl.org/?view&did=482199.

Hardt, M. & Negri, A. (2005), *Multitude: War and Democracy in the Age of Empire*, Edmonton: Canadian Committee on Labour History.

Holland, M. (2006), The Propagation and Power of Communist Security Services Dezinformatsiya, *International Journal of Intelligence and Counter-Intelligence*, 19(1), pp. 1–31, DOI: 10.1080/08850600500332342.

Icks, M. & Shiraev, E. (2014), *Character Assassination throughout the Ages*, Basingstoke: Palgrave Macmillan.

Jerome, D. (1950), *Character Assassination*, New York: The Philosophical Library.

Jones, R. G. (29 March 2017), An Applied Approach to Psychology of Sustainability, available at http://psychology.oxfordre.com/view/10.1093/acrefore/9780190236557.001.0001/acrefore-9780190236557-e-76?print=pdf.

Jula, C. (4 September 2014), *Uniţi salvăm Roşia Montană: un an de la "toamna românească"* (Think Outside the Box), available at www.totb.ro/uniti-salvam-rosia-montana-un-an-de-la-toamna-romaneasca/.

Keen, S. (1988), *Faces of the Enemy: Reflections of the Hostile Imagination*, San Francisco: Harper.

Klayman, J. (1995), Varieties of Confirmation Bias, *Psychology of Learning and Motivation*, 32, pp. 385–418.

Knobloch-Westerwick, S. (2015), *Choice and Preference in Media Use: Advances in Selective Exposure Theory and Research*, 1st ed. Taylor and Francis. www.perlego.com/book/1561320/choice-and-preference-in-media-use-pdf.

Laney, D. (2001), 3D Data Management: Controlling Data Volume, Velocity and Variety, *META Group Research Note*, 6(70).

Levy, B. H., *Who Killed Ms. Truth?*, at 15th Yalta European Strategy, Annual Meeting, 13–15 September 2018, Kiev, "The Next Generation of Everything", available at https://yes-ukraine.org/en/photo-and-video/video/15-a-shchorichna-zustrich-yes/hto-vbiv-mis-istinu.

Lind, W. S. (2004), Understanding Fourth Generation War, *Military Review*, September–October, www.hsdl.org/?view&did=482203.

Lobbying Reform: Background and Legislative Proposals, 109th Congress (PDF) (2006), Order Code RL33065. The Library of Congress, Congressional Research Service. 2006-03-23.

Moisi, D. (2009), *The Geopolitics of Emotion*, New York: Anchor Books Publishing.

Morrell, S. C. & Kosal, M. E. (2021), Military Deception and Strategic Culture: The Soviet Union and Russian Federation, *Journal of Information Warfare*, Summer, 20(3), pp. 134–152, available at www.jinfowar.com/journal/volume-20-issue-3/military-deception-strategic-culture-soviet-union-russian-federation.

Nickerson, R. S. (1998), Confirmation Bias: A Ubiquitous Phenomenon in Many Guises, *Review of General Psychology*, 2(2): pp. 175–220, DOI: 10.1037/1089-2680.2.2.175.

Niezsche, F., *On Truth and Lie in an Extra-Moral Sense*, available at www.kth.se/social/files/5804ca7ff276547f5c83a592/On%20truth%20and%20lie%20in%20an%20extra-moral%20sense.pdf.

Original Joseph Goebbels, in Koonz, C. (2004), *The Nazi Conscience*, London and Cambridge, MA: Belknap Press of Harvard University, p. 13. Quote from March 1933.

Palmera, E. J. (2007), Media Use and Political Predispositions: Revisiting the Concept of Selective Exposure, at www.uky.edu/AS/PoliSci/Peffley/pdf/Stroud%202008%20PB%20Revisiting%20the%20Concept%20of%20Selective%20Exposure.pdf.

Pădure, R. (4 March 2017), *Marele SECRET din spatele Victoriei lui Trump Insectarul DIGITAL de Oameni. O uluitoare OPERAŢIUNE de Penetrare a CREIERELOR americanilor*, Evenimentul Zilei, available at http://evz.ro/marele-secret-din-spatele-victoriei-lui-trump-i.html.

Peters, U. (2020), What Is the Function of Confirmation Bias? *Erkenntnis*, 87, pp. 1351–1376

Poole, D., Mackworth, A. & Goebel, R. (1998), *Computational Intelligence: A Logical Approach*, New York: Oxford University Press.

Popp, G. & Canna, S. (2016), The Characterisation and Conditions of the Grey Zone. A Virtual Think-tank Analysis, NSI Team, Winter, at http://nsiteam.com/social/wp-content/uploads/2017/01/Final_NSI-ViTTa-Analysis_The-Characterization-and-Conditions-of-the-Gray-Zone.pdf.

Reichman, O. J., Jones, M. B. & Schildhauer, M. P. (2011), Challenges and Opportunities of Open Data in Ecology, *Science*, 331 (6018), pp. 703–705.

Rudy, G. (23 August 2018), "Truth Isn't Truth": So Is That True?, The Guardian, available at www.theguardian.com/books/2018/aug/23/truth-isnt-truth-so-should-we-expect-lies-from-donald-trump.

Sagan, F. (1965), *La Chamade*, Julliard.

Samoilenko, S. A. (2016), Character Assassination, in Caroll, C. E. (Ed.), *The SAGE Encyclopedia of Corporate Reputation*, pp. 115–118. New York: Sage Publications.

Sandru, A. (4 November 2015), Ponta si-a dat demisia. Klaus Iohannis, asteptat sa numeasca un premier interimar, available at www.ziare.com/victor-ponta/demisie/romania-in-strada-ponta-demisioneaza-live-1391492.

Sears, D. (1967), Selective Exposure to Information: A Critical Review, *Public Opinion Quarterly*, 31(2), DOI:10.1086/267513.

Shiraev, E. (2010), *Character Assassination: An Interdisciplinary Approach*, George Mason University, USA, available at https://characterattack.files.wordpress.com/2010/10/character-assassination-an-interdisciplinary-approach.pdf.

Sigelman, L. & Kugler, M. (2003), Why Is the Research of the Effects of Negative Campaigning So Inconclusive? Understanding Citizens' Perceptions of Negativity, *The Journal of Politics*, 65(1), pp. 45–62.

Simons, G. & Chifu, I. (2017), *The Changing Face of Warfare in the 21st Century*, London: Routledge.

Smith, S. M., Fabriger, L. R. & Norris, M. E. (2008), Reflecting on Six Decades of Selective Exposure Research: Progress, Challenges, and Opportunities, *Social and Personality Psychology Compass*, 2(1), pp. 464–493. https://doi.org/10.1111/j.1751-9004.2007.00060.x.

Stern, E. (1999), *Crisis Decision Making. A Cognitive Institutional Approach*, Stockholm: University of Stockholm.

Stroud, N. (2010), Polarization and Partisan Selective Exposure, *Journal of Communication*, 60(3), September, pp. 556–576.

Tesich, S. (1992), A Government of Lies, *The Nation*, 254(1), pp. 12–14.

The International Society for the Study of Character Assassination (ISSCA), https://characterattack.wordpress.com/.

"The Military Doctrine of the Russian Federation", Approved by the President of the Russian Federation on 25 December, 2014, available at www.scrf.gov.ru/security/military/document129/.

Uniţi Salvăm Roşia Montana! Prima lună de proteste (2013), available at www.youtube.com/watch?v=obPcorM2wuM.

Volkoff, V. (2009), *Tratat de Dezinformare*, Bucharest: Antet.

Walton, D. N. (2007), *Media Argumentation: Dialectic, Persuasion, and Rhetoric*. Cambridge: Cambridge University Press.

(1985), *Arguer's Position: A Pragmatic Study of Ad Hominem Attack, Criticism, Refutation, and Fallacy*, Westport: Greenwood Press.

Weiner, T. (2020), *The Folly and the Glory: America, Russia and Political War 1945–2020*, New York: Henry Holt and Co.

Wierzbicki, P. (1996), *Structura minciunii*, Bucharest: Nemira.

Zachmann, S. (5 February 2017), Şedinţa de Guvern s-a încheiat: ordonanţa de urgenţă a fost abrogate, Adevărul, available at http://adevarul.ro/news/politica/ora-1400-guvernul-reuneste-retrage-ordonanta-urgenta-modifica-legislatia-penala-abrogare-prorogareo-1_5896e8585ab6550cb825d49e/index.html.

5 The Culture and Language of Contemporary Armed Conflict

Greg Simons

Introduction

Temes notes the dualistic nature of war, and with it the contradictions and dilemmas. "It combines what we fear the most – death, violence, destruction of our homes and communities with that we love most – courage, loyalty, passion" (Temes, 2003: 3). He goes on to state that wars are the "ultimate expression of a nation's, and a man's, moral nature" (Temes, 2003: 4). This is because the morality and ethics of the cause are on display as to why and how we fight wars. These events of brutality and destruction are depicted in language of constructed realities of binary opposition, influenced by the culture(s) of the warring parties in order to convince the supposed righteousness of 'our' cause in contrast to the lack of it in the 'other' opponent. The language and discourse in discussions concerning geopolitical imaginations and practices sets the tone for the resulting relationships, created and/or maintained, in international relations (Ek, 2000). Therefore, these aspects are essential elements in shaping the dualistic public perceptions of 'us' versus 'them'.

Lasky (2017) and his analysis of journalism's use and misuse of language is not only a measure of the declining standards of journalism, but also the cultural and political health of contemporary society. The aim of actors is to dominate the information space as a means to increase the effects of influence and persuasion. "In an uncompetitive information environment citizens are likely to accept facts that go largely unchallenged" (Johansen & Joslyn, 2008: 604). This can be seen when a government makes a case for war, such as the Iraq War in 2003, where the lack of challenged 'facts' or genuine public debate, that could have and should have been facilitated by the mass media, laid the foundations for the escalation of the Global War on Terrorism (GWOT). Maltby and Thornham (2016) note that the digital mundane, the daily routine digital mediations or practices, begins to influence individuals and how they position themselves within the context of institutional and cultural identities. Although social media and the Internet are seen as the front line of the current global information war owing to the immediacy and convenience of the medium, the role of traditional mass media in helping to influence and shape the cognitive domain should not be dismissed.

Information strategy that results from the above needs to synchronise and take into account technological and non-technological factors, which then needs to be coordinated with what an actor wishes to do to an enemy force and wishes to do for themselves (Arquilla & Borer, 2007). This is a rather difficult balance to successfully manage that spans the physical, informational and cognitive domains. It includes not only knowing oneself, but also knowing one's opponent or enemy that is to be encountered on the physical and informational realms of the battlefield. In the twenty-first century within the political and geopolitical context of an evolving global order, from the US-led unipolar order and towards a more non-Western multipolar order, international tensions and conflicts are increasing. This leads to the questions as to how the West has got itself into so many wars of choice and what have been branded as Endless Wars within the GWOT, and why does it continue to choose to do so?

This chapter is broken down into various sections that shed light on different aspects and facets in order to attempt to answer and bring clarity to these questions. The first section concerns the use of language as a means to manufacture consensus and create the necessary environmental conditions that can support the creation and sustaining of warfare. The second section concerns the significance and use of intangible symbolism and meaning to prime and mobilise target audiences through conceptual and theoretical mechanisms to try to justify and legitimise the choice of armed conflict as policy. In the third section, the topic concerns the Western philosophy about war and politically conceiving the management of the genesis and development of armed conflicts. The final section involves a discussion and analysis of the role played by culture and identity in war and armed conflicts.

Use of Language in Conjunction with Armed Conflicts and War

The information age of the twenty-first century has enabled information to develop into a strategic resource, which is integral to playing critical roles in comprehensive national power through national security and social stability (Shen et al., 2007). Conversely, when meaning and perception are manipulated through language and communicational concepts, it can undermine these things. It has been observed that people can unlearn long-held views and learn new ones if information is presented in an attention-catching and resonating manner, although substantial change is rare. It is hard for politicians (and others) to change what people think and then mobilise them to action based on newly acquired information. There is the need to overcome the resulting cognitive dissonance (Hochschild & Einstein, 2015–16: 623–624). However, the 2003 Iraq War proved this is, on occasion, a possible feat to accomplish.

War is on paper a highly legally regulated human activity that is also governed by conventions and expectations as to why and how it is physically conducted, which also includes its informational representation. This informational representation can and is subject to persuasive organised communications as a means to make the appearance of a legitimate political and moral case. O'Brien (1966: 590) notes that propaganda falls within the purview of international law in the following categories: "(1) the law regulating the threat or use of force; (2) the law governing dictatorial intervention; (3) the law of international responsibility for acts injurious to sovereign states originating with another state's jurisdiction; and (4) the corpus of positive legal obligations to promote world law, order, and justice." From a legal perspective, it may appear to be a clear-cut issue, but it is far from the only perspective as other perspectives can create fog around the apparent clarity. "The most fundamental obstacle to regulation of international propaganda results from philosophical-ideological rifts which divide the world. Men differ profoundly on the definition of ultimate reality and, consequently, on the interpretation of human events" (O'Brien, 1966: 595). These interpretations of human events are shaped by the nature of communication and the resulting relationships this activity creates and maintains.

Communication has been increasingly made easier in terms of the potential audience reach and the immediacy of communicating that moves from passive one-to-many monologic to active many-to-many dialogic models. On the eve of the twenty-first century, the Internet and later social media provide a means of engaging audiences and publics physically near or far, in a manner previously unimaginable, in dialogic communications for creating human relationships. Keeping in mind that this is a time-consuming and dynamic process to create reciprocal relationships, symmetrical or asymmetrical in nature (Kent & Taylor, 1998). However, traditional media (newspapers, radio and TV) still play an important function in agenda-setting, priming audiences and in framing conflict news and international relations events. These have shown the gulf that exists between news and its reflection of the 'truth', the gap widening under the conditions of a government being able to dominate the information space with their agenda (Iyengar & Simon,

1993; Wanta, Golan & Lee, 2004; Johansen & Joslyn, 2008). This is owed, at least in part, to the fact that "the news media can show the public how vitally important countries [and issues] are to the United States [and other countries] and how negatively the countries should be viewed" (Wanta, Golan & Lee, 2004: 375). One of the key ways of achieving these ends is through a specific and deliberate use of language and meaning.

Language is a crucial tool, not only in uniting society against division and violence, but also in dividing it to enable division and violence. Bugarski (2001) uses the example of the break-up of Yugoslavia, where competing nationalisms that hated each other and sought an ethnically 'pure' state territory, all agreed on the premise that the maintenance of the Yugoslav federation was no longer feasible. The use of language proved to be both an instrument for the violence and a casualty of the deliberate manipulations. Framing can be used to bridge the divide between instrumental and social-psychological approaches in conflict studies, in order to understand how mobilisation is achieved in ethnic violence through the mobilisation of ethnic solidarity and producing the will to act in the event of conflict. Desrosiers (2012) demonstrates that actors seeking to prime and mobilise audiences need to communicate in a specific contextual manner, which may likely evolve with time and circumstances, in order to resonate with their target audience. The actor that evolves the most appropriately according to the circumstances is more likely to succeed in influencing and persuading their respective audience. A deliberate use of language can also be made to unite a country's population to enable warfare. Presidents George Bush and Barak Obama both used a projected negative image of terrorism and a positive projection of counter-terrorism to try and rally public support for wars of choice through "carefully selecting emotionally charged vocabulary and expressions" (Sarfo & Krampa, 2013: 378). This situation points to the weaponising of language in order to enable political actors to create extraordinary circumstances that will then logically require extraordinary measures to 'cure', which is visualised through the aid of conceptual and theoretical constructs.

Speech and language are powerful mechanisms of the human world as they are used to define and provide a framework of value judgement to people, places and events to differing degrees of objectivity versus subjectivity. This ensures that it is highly contested by those actors that seek to define a 'reality' that supports their agenda; warfare as an extreme and deadly form of contested politics ensures vigorous competition as the spoils go to those in terms of operational latitude and perceived legitimacy that are able to influence the information flows around an armed conflict event. In effect, it creates the weaponising of speech that intends to produce an uneven informational domain that in turn creates a disproportionate advantage to an actor; it does this through limiting the speech possibilities of the opponent or competitor by constraining the allowable/permissible limits of the debate or discussion. For example, the idea of the "information bomb" was used by the Commander of Guantanamo Bay prison when a number of inmates committed suicide as the device of their choice in place of the unavailable explosive vest or by declaring a war on terrorism (Hartmann-Mahmud, 2002; Stahl, 2016; Flusberg, Matlock & Thibodeau, 2018). One may also add the current attempt to cultivate the metaphor the New Cold War. These kinds of metaphors are intended to pass a culturally informed value judgement by a society on the person/people (and their supporters) against which it is applied.

Metaphors play an important role in political communication through their ability to explain a particular policy stance, helping to persuade people towards a particular point of view, and by affecting public opinion on the issue in focus. Therefore, metaphors provide a useful mechanism for political persuasion (Boeynaems et al., 2017). Metaphors are an integral part of the everyday use of language that includes media texts; their potential ideological or rhetorical power is the subject of study. A corpus linguistics study of metaphors in British media concludes, "metaphors are more common in newspapers than in fiction but less common in academic texts. They are more abundant in hard news than in soft news and the science pages" (Krennmayr, 2015: 543). This certainly applies to the media coverage of war as a source of hard news. Flusberg, Matlock and Thibodeau (2018: 1) make the case that "war metaphors are omnipresent because (a) they draw on basic and widely shared schematic knowledge that efficiently structures our ability to reason and communicate about many different types of situations, and (b) they reliably express an urgent, negatively valenced emotional tone that captures attention and motivates action." Although these aforementioned examples do not all explicitly involve literal war, the application potential is clear. The expression that "metaphors can kill" (from George Lakeoff (1991)) was clearly demonstrated in the 2003 Iraq War. This led to what Lule (2004) referred to as an unjust and unjustified war that has killed many thousands of civilians and others through metaphors that dramatised the prelude to war and managed public perception and expectations by imposing *timetables*, dismissing the *games of Saddam Hussein*, while speaking of the Bush administration *losing its patience*. The uncritical media coverage allowed the presidential administration to make its case and sell its plan almost uncontested through mainstream media using the weapons of mass destruction and links to terrorism as metaphorical and rhetorical means to legitimise that, which was not legitimate (Sahlane, 2013). The war was the result of an almost uncontested mainstream media space, which permitted the pro-war viewpoint to dominate the information space.

One of the other key techniques to eloquently express one's opinion and vision for an argument or to support or oppose an idea or proposition with language, is through the art of rhetoric. Aristotle divided the art of persuasion into three subsidiary categories – projecting a good impression, expounding the facts of the case, and creating the desired disposition among the intended audience (Aristotle, 2004: 50). Aristotle argues for the division of rhetoric into three distinct forms: *logos* (the use of logic and reason, including false logic); *pathos* (the use and role of emotions, both positive such as patriotism, and negative such as hate); and *ethos* (the use of legitimacy or recognition of the speaker as the source of persuasion) (Tuman, 2010: 38–39; Matusitz, 2013: 185–186). The connection between rhetoric and warfare is the role of politics to use language to persuade or dissuade the audience to go to war. Politics is the art of persuading and influencing an audience through the use of symbolic speech and acts that are intended to prime and mobilise an audience and give the perception of legitimacy to the discussed political cause at hand. As noted in an earlier chapter, by Sun Tzu, von Clausewitz and others, war is another means of waging politics, albeit through an act of overwhelming (physical or psychological) force that is intended to compel the victim to act in a certain way and probably against their self-interest.

Rhetoric provides a framework for harnessing legitimation for government strategy, including war. This framework includes how a government defines national interest and how it identifies threats as a means to offer a policy option through mobilising the audiences. How successful this is depends upon what is said in terms of the legitimation content, the technique with which it is said, and in the context which it is said (Goddard & Krebs, 2015). War rhetoric has been rather successful in initially priming and mobilising public audiences for wars of choice, where the public perception was that wars have definite beginnings and should have a foreseeable end. However, the rhetorical approach to the GWOT using the above-mentioned means, created a gradual and incremental escalation of commitment. These wars are, however, lacking in any definite and certain end; a situation that uses rhetoric and metaphors to try and deflect or operationalise public fear to keep on course in what are being referred to as features of permanent war or Endless Wars. Therefore, when it comes to 'democratic' states entering and staying in a continuous state of war the questions of how often lead to questions concerning the effective use of manipulative rhetoric to persuade audiences, even if it is against the objective national and public interests (Engels & Saas, 2013; Achter, 2016). Words matter greatly in public discussions, especially ones that have high stakes, such as warfare, owing in no small part to the subjectivity potential of words and the meaning that they denote in political matters. Therefore, language is highly deliberative and not something that is 'stitched together' in haste or without a great deal of forethought in order to send the desired and intended symbolic signalling in the messaging.

> The powerful set the terms of the public debate. Media, including independent, privately owned media, usually confirm the political and social agenda of governments. Even when they challenge politicians' spin on events, the media usually report on what the government says is important. The level of recognition that politicians give to an issue usually matches the level of coverage given to that issue by the media. (Moeller, 2009: 8)

Words create assumptions and associations among the audience, which in turn persuade and influence the audience in arriving at certain, pre-determined understandings of 'reality' through the resulting cultivated opinions and perceptions. Therefore, the three modes of rhetoric (rational, intellectual and legal) are used in speech to support their case and cause (Matusitz, 2013: 186). Words come with their very own embedded value judgements, even if the subject or object is not directly known to the consumer of the communicated ideas. The political actor that wishes to wage warfare requires, at a basic minimum, the façade of a sense of legitimacy for this intended risky political act. In terms of the relationship between language and worldview, language tends to reflect the worldview and in turn the worldview is shaped by language use. "We can see this relationship very clearly in the language around contested issues. Consistent use of words in discourse on a particular issue does not simply reflect ideology; their use also helps circulate and spread ideology by framing issues and events in the world" (Marcellino et al., 2017: 38). Therefore, words are a key element of the information domain that influences the outcomes occurring in the cognitive domain concerning events and people in the physical domain.

This creates an information domain-based rhetorical trinity of war[1] – where the rhetoric of ethics and legitimacy is represented by ethos, the rhetoric of logic and reasoning represented by logos, and the rhetoric of emotion, priming and mobilisation represented by pathos. Taking a

[1] Based loosely on Clausewitz's "Trinity of War", for more information refer to www.clausewitz.com/readings/Bassford/Trinity/TrinityTeachingNote.htm.

closer look at the role of ethos and how this looks from an operational sense, an actor can attack the ethos of an actual or potential opponent as a means of defending one's own narrative from attack (either pre-emptively or reactively). One of the tools with which to carry this out is by invoking the label (as opposed to description) of *conspiracy theory* to try and discredit them without touching the message in order to stymie or disarm a public debate that could harm the narrative and the success of advocating policy. "Conspiracy theories generally attribute extraordinary powers to certain agents – to plan, to control others, to maintain secrets, and so forth" (Sunstein & Vermeule, 2008: 5). The impact of labelling something or someone as a conspiracy theory/theorist has a rather dismissive undertone that lacks an articulate and reasoned rebuttal and can be a form of character assassination. This paper was picked up by investigative journalists, they identified in particular the proposal "that the US government employ teams of covert agents and pseudo-independent advocates to cognitively infiltrate online groups and websites – as well as other activist groups – which advocate views that Sunstein deems false conspiracy theories about the government."[2] In refining methods and tactics to destroy reputations and control online discourse through deceptive means, Edward Snowdon's expose had revealed. The Joint Threat Research Intelligence Group of the British signals intelligence agency GCHQ had two core "tactics: 1) to inject all sorts of false material onto the internet in order to destroy the reputation of its targets; and 2) to use social sciences and other techniques to manipulate online discourse and activism to generate outcomes it considers desirable."[3] These black informational tactics are intended to negate any informational threat by directly attacking the messenger and thereby discourage others, and to create an infodemic with an atmosphere that creates distrust of 'alternative' information to the official line. This is critical in protecting one's own informational channels in supporting proposals advocating for further warfare by attempting dominance of the information space that limits the opponent's operational opportunities.

Careful choice of wording and language is an integral part to creating the façade of a relative moralisation of warfare, how and why it is fought, and the manner in which it is fought. Situations arise when grey informational tactics are used to create an impression or an association of

[2] G. Greenwald, *Obama Confidant's Spine-chilling Proposal*, Salon, www.salon.com/test/2010/01/15/sunstein_2/, 15 January 2010 (accessed 21 May 2020).

[3] G. Greenwald, *How Covert Agents Infiltrate the Internet to Manipulate, Deceive, and Destroy Reputations*, The Intercept, https://theintercept.com/2014/02/24/jtrig-manipulation/, 25 February 2014 (accessed 21 May 2020).

moral or ethical judgement to prepare public perception and opinion for possible future conflicts through using the maxim of accusing your opponent of something of which one is guilty. It is a tactic increasingly used within the era of the New Cold War metaphor, being applied to China and Russia's projected prowess (as a threat) in political warfare[4] or sharp power[5] within the context of 'influence wars'. This is aimed at creating the binary opposing impressions of legitimacy for 'our' cause and illegitimacy for the 'other's' cause. It can even happen simultaneously in the same information space concerning different battles occurring in different parts of the physical space. A good example of this phenomenon is coverage and depictions of the good (us) battle for Mosul in Iraq versus the bad (other) battle for Aleppo in Syria, both against Sunni terrorist groups occupying major urban areas (Simons, 2016). This politically subjective representative distinction between the good us versus the bad them plays out in mass media. For example, the newspaper headline *Russian presence in Libya more dangerous than ISIS, says US Africa Command.*[6] There is no mention of the role played by the US-led NATO regime change operation in Libya in 2011 that led to a failed state, let alone the dubious 'logic' of the comparison of a recognised state of the international community with a non-state terrorist organisation. This type of opinion and advocacy in the mass media supports the theses of Payne (2005), that media as an instrument of war, and Zollmann (2019), that the propaganda activity of liberal democracies, have vastly increased. The lack of Western mass media outlets and journalism in managing due diligence in the coverage of warfare has been noted in academic research, such as the political elite being sheltered from blame for civilian casualties in drone strikes through their assessments. As noted by the authors (Davies, Schulzke & Almond, 2018: 477), this has "the potential to undermine the ability of the public to constrain elites from launching military operations using drones." The power of language in potentially shaping the worldview of media audiences are understood and practiced and exploited for short-term political and operational advantages and gains.

[4] F. Hoffman, *On Not-So-New Warfare: Political Warfare vs. Hybrid Threats*, ISN, https://css.ethz.ch/en/services/digital-library/articles/article.html/182335, 4 August 2014 (accessed 8 August 2014).

[5] C. Walker, S. Kalathil & J. Ludwig, *Forget Hearts and Minds*, Foreign Policy, https://foreignpolicy.com/2018/09/14/forget-hearts-and-minds-sharp-power, 14 September 2018 (accessed 25 September 2018).

[6] A. Mahshie, *Russian Presence in Libya More Dangerous Than ISIS, Says US Africa Command*, The Washington Examiner, www.washingtonexaminer.com/policy/defense-national-security/russian-presence-in-libya-more-dangerous-than-isis-says-us-africa-command, 29 April 2020 (accessed 30 April 2020).

Language has been used successfully to make the case for war, either through the offensive role of advocacy or the defensive role of accommodative strategies. By the offensive role of advocacy, it is meant that language is used to make the case for war where the role of mass communication presents an interpretive and symbolic representation of people and events to shape target audience perception and opinion in favour of war (regardless of the distinction or deviation of subjective representation from 'objective' reality). Defensive accommodative strategies refer to where the legitimacy or cause for war or continued war is under threat and needs to be defended by such tactics as a justification or explanation, a distraction (red herring) or an information vacuum. These become evident within the context of armed conflicts and how actors and actions are characterised, such as: the Free Syrian Army (Turkish controlled, initially a UK proxy); the Syrian Defence Force (Kurdish proxies for the US); or the naming of military operations as Operation Iraqi Freedom or Operation Olive Branch (in Syria). The names denote and suggest one reality and the results demonstrate an entirely different purpose and reality.

When undertaking an examination of some examples of defensive accommodative strategies, there are many to choose from; this is partly owing to the global publics' war wariness with the Endless Wars phenomenon that has put governments that want to continue wars of choice as a policy on the defensive and need to 'protect' (gate keeping) the public information space from further delegitimising materials. However, sometimes such potentially damaging revelations are briefly exposed to the public. One such example is the current US prosecutors' indictment against Wikileaks founder Julian Assange under the Espionage Act, which mentions that the lives of Coalition troops were put at risk by the public release of information and yet fails to mention the exposure of US war crimes and lies through the Collateral Murder video.[7] Sometimes stories appear in mainstream media outlets that expose the 'curation' of a pathos-based narrative to support the regime- change ambitions in Syria as humanitarian and an 'inevitable' development into a Western-style democracy, while ignoring the obvious unpleasant geopolitical reality of targeting the only Arab ally of Syria. The ugly details are covered in favour of value and norm-based

[7] P. Daley, *Julian Assange Indictment Fails to Mention Wikileaks Video That Exposed US 'War Crimes' in Iraq*, The Guardian, www.theguardian.com/media/2020/jun/15/julian-assange-indictment-fails-to-mention-wikileaks-video-that-exposed-us-war-crimes-in-iraq, 14 June 2020 (accessed 16 June 2020).

judgements and predictions or ignored,[8] although sometimes media content does allude to the hypocrisy of 'humanitarian interventions' by the West.[9] On occasion, a journalist from the mainstream media defects on the basis of ethical principles and exposes the inner workings of the manipulations; the dilemma of earning a steady salary and knowingly suppressing important facts on the Syria War proved too much for Tareq Haddad. The deception conducted by the mainstream media was deliberate, systematic and in support of the US-led policy of warfare and regime change.[10] On other occasions, media watchers attack the 'humanitarian' narrative on Syria, such as the disastrous results of earlier regime change operations[11] or the exposure of non-public documents of policy makers that contradict their public statements and thereby exposing the lies, complicity and heavy costs of this type of warfare.[12] The 'unpleasant' geopolitical realities that are masked by the humanitarian façade are exposed, such as the geopolitical motivations for regime change in Syria that are publicised by platforms, such as Wikileaks.[13] Language is the mechanism that is able to inject a sense of symbolism and meaning to people, places and events located in the physical domain.

The Significance and Use of Intangible Symbolism and Meaning

Intangible symbolism and meaning in international relations and affairs has likely tangible significance and consequences, which is observable in how (and why) state actors position themselves in the marketplace of global (regional and local) roles. This positioning determines foreign and security policy stances in order to live up to the promised brand and

[8] R. Fisk, *Everything You Were Told about Syria Was Wrong – Until Now*, The Independent, www.independent.co.uk/voices/ghouta-siege-syria-death-toll-civilians-armed-attack-rebel-aleppo-latest-a8221086.html, 7 November 2019 (accessed 8 November 2019).

[9] R. Fisk, *Western Howls of Outrage Over the Ghouta Siege Ring Hollow – We Aren't Likely to Do Anything to Save Civilians*, Independent, www.indepdendent.co.uk/voices/ghouta-siege-syria-death-toll-civilians-armed-attack-rebel-aleppo-latest-a8221086.html, 21 February 2 018 (accessed 23 February 2018).

[10] T. Haddad, *Lies, Newsweek and Control of the Media Narrative: First-Hand Account*, Tareq Haddad, https://tareqhaddad.com/2019/12/14/lies-newsweek-and-control-of-the-media-narrative-first-hand-account/, 14–I15 December 2019 (accessed 18 December 2019).

[11] I. Eland, *Courting Catastrophe in Syria*, Consortium News, http://consortiumnews.com/2013/02/28/courting-catastrophe-in-syria/, 1 March 2013 (accessed 1 March 2013).

[12] D. Lazare, *The Memo That Helped Kill a Half Million People in Syria*, Consortium News, http://consortiumnews.com/2019/01/13/the-memo-that-helped-kill-a-half-million-people-in-syria/, 13 January 2019 (accessed 17 January 2019).

[13] US Department of State, Case No. F-2014-20439, Doc No. C05794498, 30 November 2015. Available at https://wikileaks.org/clinton-emails/emailid/18328 and www.foia.state.gov/Search/results.aspx?searchText=C05794498.

reputation, for example, as a hegemon or a challenger or a nicher (specialised role, such as non-aligned movement). Since the end of the Second World War, and especially after the Cold War, global politics has been highly influenced and shaped by two constraints: the US position as the single most powerful state; the US choice to be heavily engaged around the globe. This has created three core objectives: reducing threats and challenges by shaping and managing the security environment in key geographical areas; spreading a liberal political and economic order for generating its prosperity and therefore power (hard and soft); and sustaining global networks that work in favour of maintaining US interests and power. This mixture of historical event and contemporary logic has been used as an excuse to continue US global hegemony as a "natural duty" that should not be neglected or abandoned (Brooks & Wohlforth, 2016). Therefore, there is the expedient rational logic (though not necessarily true) used that it is better the devil you know, the symbolic meaning intending to serve as the rhetorical legitimacy for continued US global hegemony.

The public are being conditioned to accepting certain states of being as normal when they are not, such as the normalisation of wars of choice as being not only a necessity, but a 'moral' obligation. A number of dilemmas and contradictions are evident to a dispassionate observer. For example, on the one hand the US projects a self-image that cultivates a belief of being a force for good in the world and an ingrained sense of self-righteousness, which is contradicted by educational, mass media and cultural representations of militarism. One of the major ingredients of conditioning is fear, and the presence of a constant threat or danger that must be countered by extraordinary means and efforts. Therefore, any signs of US aggression are represented as being purely 'defensive' and 'justified' as being urgent and necessary. When these individual elements are brought together, it "suggests, anti-intellectualism seems to be an important ingredient in conditioning the American mind-set for war, with the system relying on a population motivated by fear and forgoing critical thinking".[14]

In making an argument to go to war, by priming and mobilising support and minimising active opposition to the proposal, a speaker can use reasoned logic or emotional logic. In the short-term at least, emotional logic is more effective in priming and mobilising a trusting or non-suspicious audience. Two conceptual constructs are available to

[14] D. Niose, *You've Been Conditioned for War*, Psychology Today, www.psychologytoday.com/us/blog/our-humanity-naturally/202001/youve-been-conditioned-war, 11 January 2020 (accessed 15 January 2020).

frame the appeal of the cause, *Just War* and the *Responsibility To Protect* (R2P). These options are also representing the two argumentation choices, where the reasoned logic of Just War encounters the emotional logic of R2P. In the short term, it is easier to elicit the desired output and impact of enabling the façade of legitimacy to facilitate regime change for politically expedient goals, through the perception of 'legitimate' military means, by creating an ethically and morally ambiguous fog around a military act that is likely to have significant negative consequences that would be measured in a Just War approach.

Just War is the result of the why and how we fight wars (to be defensive and not to build an empire), often narrated in abstract terms (GWOT, The War to End All Wars, in the name of or against a religion, political ideology or an empire, as a clash of civilisations …etc.), together with the of the moralisation of the conflict as being necessary and just (in the minds of individuals) in its aims and results (Temes, 2003; Hoffmann, 2006; Fisher & Wicker, 2010). There are three basic and yet enduring principles of Just War thinking, which has its origins as a tradition some two thousand years ago: "a Just War sanctifies human life and treats all life as equally precious; Just War is about the future, not the past; and a Just War preserves and strengthens the principles of the rights of the individual, based on the notion that the legitimacy of government derives from the consent of the governed" (Temes, 2003: 198). When applied to *jus ad bellum*, Just War prioritises the following set of principles: 1) it is a just cause; 2) the correct authority makes the decision for war; 3) war is waged with the right intention; 4) the proportionality of ends applies; 5) that war is a last resort only; 6) that war is waged with a reasonable prospect for success in attaining the stated goals (Heinze & Steele, 2009: 5–6). Although there may be a logic (logos) used to justify warfare as being just, it does not necessarily mean the logic is sound and true, let alone a shared vision and perception by all the parties in the armed conflict as to what is just and how to measure what is just and not just. Just War is also the cue to start a struggle for the competitive communicational positioning of the clashing political actors (state and non-state) as to who is moral and just and who is not.

Wars can be narrated as just by state and non-state actors alike, in keeping with the moralisation of an activity that is intended to subjugate an opponent by the use of deadly force, but this is a matter of perception and what individuals understand and process as being just. Wars and armed conflicts can be fought or 'justified' in the name of a religion, where an actor attempts to use the sacred legitimacy of the religion in question to perform an act of political violence on another people. This is evident in the narratives beginning and continuing the GWOT, with

George Bush's gut reaction to the attacks in declaring a "Crusade Against Terrorism", and Barack Obama's use of language influenced by the Christian faith urging Americans in the fight against radical elements of Islam. In this case, religion is simultaneously seen as an inhibitor to peace and a solution to conflict. Even though, theoretically, the West is cherishing secularism as an ideal, in cultural, political and military practice it is influenced by historical experience and religious roots of the country in question (Durward & Marsden, 2009). However, there is the counter-argument that religious-based terrorism by non-state actors is not the gravest threat in the twenty-first century, but rather state-based and sponsored terrorism and proxy wars are the biggest security threat to the world, where one side is fighting a war against terrorism and the other is defending their religion (Hashmi, 2014). It is a question of the sum of the outputs and the impact of Unjust Wars. Another conceptual mechanism for considering whether to commit to warfare is the value- and norm-based R2P as opposed to the reasoned logic of Just War.

Moreover, R2P is a post-Cold War concept that is an evolution in the approach to the idea and practice behind the humanitarian intervention justification. A set of eight fundamental principles, guidelines and procedures were drawn up for R2P: 1) prevention and preventive diplomacy; 2) rigour in pushing for peaceful means; 3) the use of force, including the exercise of responsibility to protect, must always be authorised by the Security Council; 4) the authorisation of the use of force must be limited and strictly governed by international law; 5) the use of force must produce little violence and instability; 6) the use of force must be judicious, proportionate and limited to the objectives established by the Security Council; 7) these guidelines must be observed throughout the entire length of the authorisation; 8) the Security Council must ensure the accountability of those to whom authority is granted to resort to force (Kassim, 2014: 3–4). However, there is a great deal of qualitative variation and departure from the conceptual ideal combined with the messiness of politicised operational practice.

In 2005 the UN adopted the concept of R2P and in 2011 used the cover of the concept as a means to ostensibly protect civilians against the use of deadly force in Libya that was sponsored by the countries that were leading the push for regime change (Morris, 2013; Thakur, 2013; Kassim, 2014: 2). A problem of R2P is the imprecise nature of the concept when implemented beyond theoretical exercises in humanitarian intervention when the military is tasked by policymakers to practically implement the mission to meet their political will. A number of significant aspects need to be considered as the specific nature of the object of protection: who they are being protected from, the means of that protection and who is tasked to physically execute the mission (Thakur, 2013). The disastrous impact of Western-led regime change in Libya, that was enabled by the invocation of R2P, caused a fundamental rethinking of the subjective instrumentalisation of the humanitarian concept for geopolitical purposes (Kassim, 2014; Nikitina, 2014). This created the situation and consequence that an attempt to apply R2P to the Syrian conflict was soundly rejected by the UN Security Council (UNSC) (Morris, 2013; Kassim, 2014). There was also evidence of anti-branding the 'Arab Winter' in a rhetorically oppositional approach to the expectation management exercise of the 'Arab Spring' brand (Feldman, 2020). Another such example is the NATO-led operation's results in Libya by referring to NATO as "Al Qaeda's Air Force".[15] As noted by Thakur (2013: 73), even if R2P against Libya was 'sincere' it tangibly demonstrates that "good intentions do not automatically shape good outcomes."

Western attempts to use such conceptual instruments to superficially legitimise their war of choice does not make it just or responsible; it is however, a deeply political exercise. The dilemma is that public support for military intervention is critical to secure and difficult to obtain. This invariably leads to a governmental 'informational' campaign intended to affect the cognitive domain to secure their support by providing a compelling and yet short, simple and succinct narrative. Colley (2015: 67) notes that the British government failed to achieve this in the regime change operation in Libya: "The government claimed it was simultaneously going to war to: reduce the threat of terrorism; protect Libyan civilians; safeguard the economy; minimise immigration; protect against organised crime; remove a tyrant; uphold international law; demonstrate global leadership; support its allies; spread freedom and democracy; and because it was morally right to do so." In effect these represent moral and ethical relativism, and the attempt to operationalise and exploit the emotional use of fear (of terrorism and immigration for example). The impact of the military operation has in effect ensured that none of these goals have been fulfilled, and in fact the situation is much worse in Libya and the region as a result of the war of choice.

When academic or philosophical theoretical creation and development is the initial stage of the process, from academic knowledge to political policy to military practice, it is this initial stage that can potentially

[15] A. J. Kuperman, *America's Little-Known Mission to Support Al Qaeda's Role in Libya*, The National Interest, https://nationalinterest.org/feature/americas-little-known-mission-support-al-qaedas-role-libya-73271, 13 August 2019 (accessed 29 September 2020).

influence the way that society thinks, perceives and reacts to people and events in its environment. The combination and interaction between language, theory and symbolism can shape the resulting quality and nature of relationships that govern the purpose of dysfunction of individuals and collectives as a result. This becomes apparent within the context of political and military thinking on conflict management and about the decision as to whether or not to wage war.

Western Philosophy about War and Conflict Management

There is a risk and a tendency to take an overly Western-centric understanding to the imagining and practice of war, especially given their centuries of being a dominant global military and political force that has exerted a great deal of influence in its messianic-like spread of ideas and thinking. The understanding and approach to war is guided by the nature of the strategic culture of the nation or civilisation in question that is informed by historical experience and debates. Therefore, when Western countries conduct military operations in foreign countries, ignorance of the local history, culture and politics can be costly with possibly different rules within the same broad game of warfare (Zaman, 2009). Western philosophy on managing war and armed conflict is rooted in historical experience and traditions. This is reflected in the intellectual and cultural approach, which differs from other cultures and civilisations. Milevski (2019: 14) observes that "the Western tradition seems to emphasize the difficulties of war and of strategy in war [...] The Chinese tradition differs. Instead of dwelling on the difficulties, it seeks to avoid them by preparing the environment in advance." Therefore, the Chinese approach concerns the engineering of the environment (physical, informational and psychological) where the conflict is to take place so that it yields them advantages while disadvantaging the opponent.

In an era of increasingly dangerous and risky war-like rhetoric between major global powers, which can be understood as being linked to the context and specific circumstances of the relative decline of US military and economic power, it is political influence that is tied to these resources. As noted elsewhere, the US attempts to project its self-constructed international brand image as a force for good in the world, and as the most single most militarily powerful nation in the world. There is an emphasis on the difficulties and strategy for war but communicated with a sense of 'inevitability' of the outcome. Marsella (2011: 714) notes that "at the core of US pursuit for empire is a historic commitment to the beliefs in 'manifest destiny' and American 'exceptionalism.' It is proposed that the process and consequences of much of this are rooted in the existence of a US 'culture of war' that dominates US domestic and foreign policies, and their destructive implementation around the world." This is in line with Zaman's (2009) idea of a nation's strategic culture of war.

This stance does inevitably clash with the rapidly rising power of China, which is seen in the understanding and communication of the possibility of an upcoming armed conflict, that has intensified with Obama's 'Asia Pivot' and Trump's trade war, which stresses the need for preparation of the physical, informational and cognitive domains. As a case in point, is an article penned by Lieutenant General He Lei in the Global Times.[16] One of the points made by the author was that the Korean War (1950–53) demonstrated that the US was not an invincible military force, where the outcome of the war was not a loss for the tangibly weaker military forces facing the better-equipped US military and its allies. It is noted that COVID-19 has produced changes in the international strategic landscape, which has created severe challenges for China. As such, a revisiting of the lessons of the Korean War (as the US did with the Vietnam War) have been prompted and it is found that "to undertake the great struggle in the new era, we must have the will to fight and the confidence to win." At the same time, the massive differences and circumstances between China in the 1950s and the China of the twenty-first century were noted. He highlighted several lessons from the Korean War that are applicable today: (1) the necessity to safeguard national security interests and actively carry out international struggles; (2) to engage in one decisive fight in order to avoid being forced to fight on multiple occasions; (3) to lay the foundations for a long-term peace and that national development required political consensus, will and leadership. Of course, given the current level of rising tensions, such articles and thoughts capture the attention of the opponent who then tries to interpret it from their own political and military strategic culture.[17] There is the resulting attempt of false logic to try to moralise and justify the supposed civilisational 'superiority' (read legitimacy) of the democratic system that is allegedly under scrutiny, as opposed to the lack of scrutiny in the Chinese system. However, the permanent and often covert warfare of the GWOT and Endless Wars of the US-led West tend to negate this argument.

[16] L. He, *China Should Effectively Enhance Ability to Fight, Win Wars*, Global Times, www.globaltimes.cn/content/1202566.shtml, 1 October 2020 (accessed 3 October 2020).

[17] T. Rogan, *Chinese General Says Korean War Shows How to Defeat America*, Washington Examiner, www.washingtonexaminer.com/opinion/chinese-general-says-korean-war-shows-how-to-defeat-america, 2 October 2020 (accessed 3 October 2020).

From a Russian perspective, the events of the past are important to understanding contemporary trends and processes and these lessons should not be forgotten or ignored. The current uncertainties and instabilities have been experienced in the past, albeit in different forms, but they contain salient lessons and warnings. "Times of change are always associated with uncertainties, risks and opportunities. At such moments the price to pay for a mistake increases dramatically" (Barabanov et al., 2015: 1). However, the cognitive clarity in understanding and comprehending the price of mistakes can be obscured by the symbolic interpretations and projections of the informational domain, which is intended to affect the opponent's decision-making capability and capacity but can also adversely affect one's own.

Informational representations of reality are influential in shaping the cognitive domain, and consequently the response to events and processes occurring in the physical domain. Realism and Neo-realism stress the role of rational mankind and the 'cold' logic of political decision making, however in the reality of international relations this does not always hold. There are two broad routes to persuading and influencing an audience to accept war, which operate at two different levels in society. These were visualised in Western philosophy millennia ago. The elite governing classes of a given society are held together by the bonds of pragmatic mutual interests (reasoned logic) that bring them benefits, whereas the mass level of governed society tends to be held together by the use of emotional symbolism (emotional logic) (Grube, 1992). This is also seen in the division of the approach to armed conflict and warfare, where reason versus emotion is used, including in combination with each other.

For example, Löwenheim and Heimann (2008) conceptualise the role played by revenge in international politics. They argue that a state will decide whether or not to take revenge based on a number and/or combination of interrelated and influencing factors. These are: "(1) the degree to which a state emotionally experiences harm against it as morally outrageous, (2) the extent of humiliation the harmed state feels, and (3) the degree to which international retaliation is institutionalised by rules and laws that govern the use of cross-border force." This in turn leads to the goals of the 'revenge', whether it is to inflict suffering per se on the other or whether this concerns correcting the past versus deterrence (Löwenheim & Heimann, 2008: 691). Other implications also exist, in terms of the means of revenge. Various choices include the use of excessive and disproportionate force, the role of symbolism (targeting symbolically significant targets), low sensitivity to material costs and explicitness in revenge (sense of pride); the above is also influenced by the longevity for revenge (Löwenheim & Heimann, 2008: 692–693).

Historically, this played a significant role, such as the end of the First World War paved the way to the Second World War, and contemporarily, where the US in response to the terrorist attacks of 11 September 2001 grew into the GWOT; this can also be seen in the rising tensions between Armenia and Azerbaijan in 2020.

Another emotional line taken in the genesis of armed conflict and warfare is the "prestige motive", where an individual or collective desire and seek public recognition of self-distinctiveness as an end goal (Markey, 1999: 126). The cognitive notion and understanding of prestige can influence the opinions, perceptions and actions of people, both intrinsically and instrumentally. This includes creating the cognitive context that can create the necessary pre-conditions that facilitate the creation of and then the ability to sustain warfare.

> In other words, traditional realists believe that prestige, whether denoted as glory, honour, or reputation, is a motive force behind the behaviour of individuals, statesmen, and nations. It has also been implicit throughout this discussion that prestige motive can inspire competitive behaviour, possibly even violent conflict, between men. (Markey, 1999: 155)

However, prestige is seldom the only reason for an actor to make the conscious decision to go to war; it is usually in conjunction with a number of other reasons. According to traditional realists there are three such broad and essential motivations – profit, security and prestige. Others argue that armed conflict can also be due to the competition for scarce resources (Markey, 1999: 156). Although, the notion of security is open to an extremely broad selection of interpretations and applications, such motivations can be used to try and justify/legitimise warfare while masking the actual and hidden motivations. Markey (1999: 157–160) argues that prestige is linked to conflict in four important ways: 1) That prestige is the ultimate relative (or propositional) good; 2) Prestige is the public recognition of eminence, and thus an assessment of relative position; 3) Prestige is not only a relative good, but must be sought through competition, though the thirst for prestige is not entirely satisfied; 4) Prestige is irrational for two distinct reasons. In the material sense, because it does not yield optimal material outcomes. It is also irrational from a proportionality perspective as it does not always align with the means–ends calculus. The afore-mentioned theoretical and conceptual points are brought to life in the context of the relative decline of US global hegemony and the competing visions of foreign policy, and especially the rhetorical proposition that the US is an exceptional and irreplaceable actor, the 'natural' leader of the international community.

In both the case of revenge and prestige, these are intangible factors that drive policy makers to irrationally go to war or to continue one. This falls within the area of emotional psychology, where influential decision makers construct a desired cognitive projection of an ideal vision of the future. In turn, this emotionally based future vision affects their behaviour, that is, how they perceive and react to events unfolding around them. It also affects the risks that they may be prepared to take, where more rational minds would likely not dare to venture (Payne, 2015). Therefore, emotions in the context of warfare can be used to produce rapid decisions, but they are poor in delivering a much-needed clear and consistent strategy, to not only fight battles, but to win wars.

One view of war as a crisis of the civilisational age, reflecting on the First World War during the period of the Second World War, understood the issue as a general Western phenomenon as opposed to any particular single state. War is viewed as a symptom and as such it is a "war of spirits". Where the "war is holy, not because any intrinsic value attaches to warlike events, but because the war is the execution of a judgement passed on a guilty mankind, because mankind is expiating through it a common guilt" (Voegelin, 1944: 199–200). Intangible and cognitive value assessments and judgements play as important a role (if not more so) as tangible and physical evaluations and careful calculations. A number of trends and aspects developing in Western culture and society has been the cause of the prediction of the decline of Western civilisation as a whole. Some of these points have already been noted decades back in time and others are more recently observed. These, when taken together and as a whole, point to the weakening of intangible elements in the cognitive domain.

> They all share 'family resemblances': a kind of cultural despair that reaches across the centuries of human history to make its case; a suspicion of modern rationality for its capacity to reduce and level the complexity of life in logic, abstraction, and number; a critique of urban culture for its atomisation of individuals, reification of technology, predominance of money, and the appearance of the 'masses'; and finally some sort of proposal of what to do in this state of falleness. The Enlightenment is unmasked as harbouring shadows within its illumination. (Potter, 2017: 85)

The reference to the gradual 'dumbing down' of society and its gradual evolution towards collectivism and a mass society (physically and psychologically) has the potential to create a number of vulnerabilities for the mass level of society and opportunities at the elite level. Cahill (2017) discusses the role of the rise of the neo-liberal ideology, which has been systematically spread in a messianic-like manner that has been assisted through the instrumentalisation of historical revisionism and the wars of 'liberation.' This is witnessed in the contemporary age through the invocation of the 'legitimising' narratives of warfare against various identified and highlighted enemies of the neo-liberal political system; such as the revisionism present in the idea of the clash of civilisations in GWOT that invoke the projected image of a civilised West in an existential fight with an uncivilised, ruthless and determined 'terrorist' other. Alternatively, the 'necessity' to form a united and unquestioning collective front within the context of the New Cold War, where another existential battle is projected as being played out between the ('defensive') so-named democratic neo-liberal versus those state actors that resist conversion and challenge the geopolitical status quo, the ('attacking') authoritarian other (in particular attributed to China, Iran and Russia).

The current weakening global political hegemony of liberalism, which witnessed a large increase in influence after the end of the Cold War in 1991, currently finds itself on the defensive in the twenty-first century. There has been a systematic attempt to brand liberalism as the viable global political option based on notions of bringing peace and wealth (Doyle, 1986; Owen, 1994), as well as the notions of multiculturalism and globalisation as part of being an inclusive society (Fleiner & Basta Fleiner, 2009), as an inevitable consequence of the political ad geopolitical trends of the twentieth century. This can be viewed as being an exercise in reputation management in order for the system to derive the perception of legitimacy for its ideology and for its actions. Therefore, although the theory of democratic peace poses the idea that 'democratic' states do not go to war against one another, they can go to war against 'non-democratic' states rhetorically for idealistic values and normative reasons as opposed to the reason of selfish gain and interests.[18] However, this is a vague concept that is open to interpretations of convenience through the lack of rigorous definition and delineation of the key parts of the logic, evidenced by the following questions: What exactly is 'war' and how can one qualify and quantify it in this context? How does one distinguish a 'democracy' from a 'non-democracy' in the age of hybrid political systems? Is 'peace' the mere absence of 'war', or does it involve something more?

As noted earlier, the US-led liberal international order is under a great deal of competitive pressure; a number of questions stem from this evolution of the balance of power in the advancement of the global order

[18] M. Louis, *Liberalism, between War and Peace: An Interview with Michael Doyle, Part 1*, Books and Ideas, www.booksandideas.net/Liberalism-between-War-and-Peace.html, 3 September 2018 (accessed 15 September 2018).

from a unipolar to a multipolar configuration. One of these questions asks, is the unipolar order truly ended or not? If this order has ended, is it advisable and/or feasible to retake this hegemonic position? What are the possible political and geopolitical consequences of an evolving global order?

Joseph Nye contests the popular perception of the US's global role, arguing that after 1945 the US order was neither global nor always very liberal, excluding the Eastern Bloc and China, and yet included many authoritarian states. In this regard their influence and strength were exaggerated. According to the UN Charter, states can only go to war in self-defence or with the blessing of the Security Council. China and Russia were not in a position to directly contest or balance US power in the 1990s and early 2000s, which was evident when the US chose to override sovereignty in the, at times, unilateral pursuit of 'liberal' values (for example, Kosovo 1999, Iraq 2003, Libya 2011). This had the effect of further alienating states, such as China and Russia, and set stricter limits to liberal interventionism. Nye argues that there should probably not be an attempt to revive and relive what has been lost, but it is necessary to replace it.[19] However, the situation is a delicate one with one order on the wane and another in the ascendency, which will create an extremely volatile environment, fraught with risks and hazards that require a long-term and carefully considered approach. This is something that seems to have been unlearned by leading Western countries that have become accustomed to centuries of global (military, economic and political) dominance and increasingly short-term and politically expedient approaches to managing conflict and warfare.

In attempting to entrench their global hegemonic position, the US-led West is trying to isolate and contain threats or challenges to this privileged position. As the rivalry between the US and China begins to take a more central and dramatic stage in terms of global great power competition, the attempts by the US-led West to isolate and coerce Russian decision-making are gradually decreasing relative to it. The likely result of this misguided strategic game is to create a more sustainable and durable alliance between China and Russia despite historical antagonisms. In the process, this has created a highly capable and formidable combination of state actors placed in a position of competition and conflict with a declining US-led West, coming at a point in time where the West has created and become entangled in a global quagmire of

[19] Nye Jr., J. S., *After the Liberal International Order*, Project Syndicate, www.projject-syndicate.org/commentary/biden-must-replace-liberal-international-order-by-joseph-s-nye-2020-07, 6 July 2020 (accessed 10 July 2020).

irregular conflicts that seem to lack an end.[20] One of the criticisms of the current risky strategy of creating global conflict with other powers is that there is an absolute reliance on short-term solutions of hard power, and the role of the US military as the only means to wage "Great Power" competition (where the invocation of the *New Cold War* slogan is being used to drive conflict against national interests and security), which is at the expense of developing other longer-term solutions (including soft power) through development of social, intellectual and economic resource capacity and capability. This would mean a reduction in the military budget for infrastructure, education and research,[21] an unlikely scenario in an increasingly aggressive and hostile international relations environment driven more by emotion rather than rational logic; a 'game' that will incur a great number of opportunity costs at a time of increasing decline and intractable irregular conflicts.

Aspects of Culture and Identity in War and Armed Conflict

The idea and spread of the concept of the culture of international values and norms is encapsulated in the Western approach to war, which contains a number of embedded false logics and presumptions. One of them is the US-led West as a global force for good and stability, yet with the approach of the reckless use of armed force as a policy of first choice. Furthermore, the questionable logic is that countries shall become more like the US-led West through military intervention and regime change. "On the contrary, there is no humanitarian crisis so grave that an outside military intervention cannot make it worse. The use of military force must always – always – be the option of last resort, not the tool of choice for dealing with threatened or occurring atrocities" (Thakur, 2013: 73). In this case, the West is positioning itself as a saviour in the script of warfare, but not taking into account the background circumstances that have shaped the conflict that may have been gradually building; this is in addition to the medium to long-term impact on both the immediate area and wider region by the use of military force, together with any counterproductive results to national interests and the security of the self-appointed 'saviour'.

[20] M. Dal Santo, *The West Risks Creating a Central Powers 2.0*, ABC News, www.abc.net.au/news/2014-08-04/dal-santo-the-west-risks-creating-a-new-central-powers/5646244, 4 August 2014 (accessed 7 September 2020).

[21] D. Larison, *'Great Power Competition' Is a Cheap Slogan Justifying Cold War*, The American Conservative, www.theamericanconservative.com/articles/great-power-competition-is-a-cheap-slogan-justifying-cold-war/, 5 June 2020 (accessed 8 July 2020).

Yakushik (2019: 89) maintains that resolving inter-civilisational and intra-civilisational tensions and conflicts requires a careful vision and responsible actions from policy makers and practitioners. This includes being "able to consciously, systematically and efficiently develop and implement strategies related to maintaining the required balance between their country's various civilizational layers." This is certainly not understood by foreign actors involved in civil and proxy wars, or adhered to by local actors engaged in those conflicts.

One of the areas that contains strong notions, and contestations of culture and identity, especially in time of armed conflict, is history. It can create identities and causes that are based upon constructed and projected emotionally based senses and perceptions of justice and injustice, problems created by the Other and solutions created by Us. This can create a version of 'national' history that is both essentialist and culturally exclusive in nature in terms of national memory and identity.

> Therefore, history and historians were mobilised to perform a major task: to construct (or to reconstruct) a picture of the past that would explain the present and that would legitimise the new nation state and its titular nation. History (which in many respects was readily and unwittingly confused with collective memory) became a part of civic education of the nationalising state. (Kasianov, 2015: 149)

History is symbolic in nature and conveyed by language that holds special meaning to groups and collectives by offering a sense of culture and identity that can bind them to an interpretation of the past as a means to guide the path to the future. When exploited for the expediency of political causes and operationalised, history can be weaponised through the creation of binary opposing and incompatible versions of history that create the ideological and then tangible basis for conflict between groups and their interpretations. Kasianov (2015: 150–151; Himka, 2015) has identified three major narratives[22] of history/collective memory in Ukraine since the beginning of the 2000s. One is a nationalist version that stresses and emphasizes the exclusive role and position of ethnic Ukrainians in Ukrainian history. It contains myths and frames of a heroic past and victimhood as a means of presenting a continuous struggle for the Ukrainian people for statehood and freedom. A second version involves nostalgia for the Soviet era and is based upon the idea of an eternal struggle for the 'people' for social and national liberation, building upon the effort and deeds of their ancestors in the building of

[22] The first version was based predominantly in Western Ukraine and partly in the centre, the second version in the East and Crimea, the third version in the centre and South Ukraine (Kasianov, 2015: 151).

a great power. The third version is a mixed and ambiguous variant that combines elements of both of the previous versions, where pride does not necessarily contradict the tolerance for things Soviet.

These various versions and interpretations of history managed, until recently, to co-exist relatively peacefully in the same public sphere. The gradual politicisation of history during elections, the Orange Revolution and Euromaidan, changed the dynamics and created a zero-sum game where the hegemony and dominance of one version threatened the survival of the other versions, and set the scene for conflict (Himka, 2015; Kasianov, 2015) in the physical domain via conflict in the information and cognitive domains. This was clearly demonstrated in the 'Battle of the Monuments' as a symbolic representation of history and collective memory, where Soviet monuments were torn down and monuments dedicated to Ukrainian historical figures, such as Stepan Bandera (Organisation of Ukrainian Nationalists) were proposed or erected. Therefore, a sense of cultural identity and belonging has a dualistic outcome when brought into contact with armed conflict and warfare. On the one hand, it unites some individuals together as a collective. On the other, by creating collective groups, the psychological dynamics are altered by potentially creating a herd mentality that can stand in binary opposition to 'other' collective groups of an incompatible political or ideological construction. Identity and culture are intangible, cognitive feelings and intuitions, rather than necessarily a tangible physical 'fact'.

Conclusion

In the beginning of this chapter, in the introduction, the following questions were posed to the reader: how has the West got itself into so many wars of choice and what has been branded as Endless Wars within the GWOT, and why does it continue to choose to do so? The answer to these questions is somewhat complex and evolving with time and circumstance. The current state of being is both related to and influenced by the evolving and transforming processes and trends in both the tangible and intangible realms. The twenty-first-century context is the relative decline of US-led Western military and economic power, which is witnessing a transformation in the global order from a unipolar to multipolar configuration. As such, the unstable era elicits actions and reactions from global actors to defend their privileged global position or, alternatively, for rising powers to challenge for greater influence and power in an increasingly chaotic international system that is increasingly diverging in terms of the understood or agreed upon sets of rules and customs that govern and regulate international relations and affairs.

Language and speech form an essential means of positioning between and within the international system at the national, regional and global levels. It is used both as a means of persuading and influencing domestic and international audiences of the legitimacy or righteousness of certain suggested political courses. However, it is also a means to articulate values, norms, culture and identity and to place this within the context of the wider human environment as to where one's own nation or civilisation is situated in relation to others. In this instance, the US-led West situates itself as the global military and economic hegemon, and the US as an 'indispensable' global actor with numerous self-given privileges to shape international politics, economics and geopolitics to its own advantage.

This self-understanding of its role, which affects its thinking and approach to other international actors and events, functioned more effectively when the bipolar global order ceased, and the unipolar order rose with the Soviet collapse and the end of the Cold War. As US power and influence has continued to decline relative to rising non-Western powers, such as China and India, the old thinking and perceptions of the international relational rules persisted in the cognitive domain in spite of the increasingly obvious effects in the physical domain. Hence the messianic approach to spreading the likeness of the Western image through coercive and non-coercive means (hard power and soft power) was owed to the innate self-belief in the moral and ethical superiority of liberal democracy. This functioned so long as other countries and actors did not feel cognitively able to directly or indirectly challenge this global supremacy, which is a time that has now passed.

The feeling of tangible and intangible superiority over other actors and systems has also led to miscalculations by the US-led West. In part this has been brought about by the flawed Western policymakers' perceptions concerning the capability, capacity and consequences of applying warfare as an expedient policy of choice to 'solve' foreign and security problems. The lack of any critical function played by Western mass media and journalism has merely served to compound the assumption of moral and military superiority over any single opponent that has served to enable the current serious dilemma and hazard for Western military and economic power, and increased threats to national security and public safety. This is seen in the current cycle of permanent and endless wars, which lack sufficient scrutiny and oversight due to the undemocratic, unjust and unjustified roles and actions by politicians from 'democratic' countries for rhetorically legitimate means in the name of the publics that they represent.

This is becoming an increasingly risky and dangerous game as the US-led West draws itself in to tensions and conflicts with state actors

(China, Iran and Russia, for example), in addition to the current array of non-state actors. This is increasingly looking like warfare is being waged in the name of political vanity against various challengers, and that lacks a clear sense of beginning or end, ostensibly to defend an eroding privileged global position. Instead, the effects of the tactics, which are supported through concepts that are intended to cognitively exploit the power of language, symbolism and meaning by emotionally priming and mobilising publics to consent to actions that go against their interests. The likely impact of these misguided and poorly conceived outputs is the hastened decline of Western power and the transformation of the global order towards a non-Western multipolar configuration.

References

Achter, P. (2016), Rhetoric and the Permanent War, *Quarterly Journal of Speech*, 102(1), pp. 79–94.

Aristotle (2004), *The Art of Rhetoric*, London: Penguin Classics.

Arquilla, J. & Borer, D. A. (Eds.) (2007), *Information Strategy and Warfare: A Guide to Theory and Practice*, Oxon: Routledge.

Barabanov, O., Bordachev, T., Lukyanov, F., Sushentsov, A., Suslov, D. & Timofeev, I. (2015), *War and Peace in the 21st Century: International Stability and Balance of the New Type*, Valdai Discussion Club Report, Moscow: Valdai Club.

Boeynaems, A., Burgers, C., Konijn, E. A. & Steen, G. J. (2017), The Effects of Metaphorical Framing on Political Persuasion: A Systematic Literature Review, *Metaphor and Symbol*, 32(2), pp. 118–134.

Brooks, S. G. & Wohlforth, W. C. (2016), *America Abroad: The United States' Global Role in the 21st Century*, New York: Oxford University Press.

Bugarski, R. (2001), Language, Nationalism and War in Yugoslavia, *International Journal of the Sociology of Language*, 151, pp. 69–87.

Cahill, D. (2017), The Spectre of Collectivism: Neo-Liberalism, the Wars, and Historical Revisionism, in Sharpe, M., Jeffs, R. & Reynolds, J. (Eds.) (2017), *100 Years of European Philosophy since the Great War*, Cham: Springer, pp. 183–197.

Colley, T. (2015), What's in It for Us, *The RUSI Journal*, 160(4), pp. 60–69.

Davies, G. A. M., Schulzke, M. & Almond, T. (2018), Sheltering the President from Blame: Drone Strikes, Media Assessments and Heterogeneous Responsibility 2002–2014, *The British Journal of Politics and International Relations*, 20(2), pp. 477–496.

Desrosiers, M.-E. (2012), Reframing Frame Analysis: Key Contributions to Conflict Studies, *Ethnopolitics*, 11(1), pp. 1–23.

Doyle, M. W. (1986), Liberalism and World Politics, *The American Political Science Review*, 80(4), pp. 1151–1169.

Durward, R. & Marsden, L. (Eds.) (2009), *Religion, Conflict and Military Intervention*, Farnham: Ashgate.

Ek, R. (2000), A Revolution in Military Geopolitics? *Political Geography*, 19(7), pp. 841–874.

Engels, J. & Saas, W. O. (2013), On Acquiescence and Ends-Less War: An Inquiry into the New War Rhetoric, *Quarterly Journal of Speech*, 99(2), pp. 225–232.

Feldman, N. (2020), *The Arab Winter: A Tragedy*, Princeton: Princeton University Press.

Fisher, D. & Wicker, B. (Eds.) (2010), *Just War on Terror? A Christian and Muslim Response*, Farnham: Ashgate.

Fleiner, T. & Basta Fleiner, L. R. (2009), *Constitutional Democracy in a Multicultural and Globalised World*, Berlin: Springer.

Flusberg, S. J., Matlock, T. & Thibodeau, P. H. (2018), War Metaphors in Public Discourse, *Metaphor and Symbol*, 33(1), pp. 1–18.

Goddard, S. E. & Krebs, R. R. (2015), Rhetoric, Legitimation, and Grand Strategy, *Security Studies*, 24(1), pp. 5–36.

Grube, G. M. A. (Translator) (1992), *Plato Republic*, Indianapolis: Hackett Publishing Company.

Hartman-Mahmud, L. (2002), War as Metaphor, *Peace Review*, 14(4), pp. 427–432.

Hashmi, T. (2014), *Global Jihad and America: The Hundred-Year War Beyond Iraq and Afghanistan*, New Delhi: Sage.

Heinze, E. A. & Steele, B. J. (Eds.) (2009), *Ethics, Authority, and War: Non-State Actors and the Just War Tradition*, New York: Palgrave.

Himka, J.-P. (2015), The History Behind the Regional Conflict in Ukraine, *Kritika: Explorations in Russian and Eurasian History*, 16(1), Winter, pp. 129–136.

Hochschild, J. L. & Einstein, K. L. (2015–2016), Do Facts Matter? Information and Misinformation in American Politics, *Political Science Quarterly*, 130(4), pp. 585–624.

Hoffmann, J. R. (Ed.) (2006), *The Just War and Jihad: Violence in Judaism, Christianity & Islam*, New York: Prometheus Books.

Iyengar, S. & Simon, A. (1993), News Coverage of the Gulf Crisis and Public Opinion: A Study of Agenda-Setting, Priming, and Framing, *Communication Research*, 20(3), pp. 365–383.

Johansen, M. S. & Joslyn, M. R. (2008), Political Persuasion During Times of Crisis: The Effects of Education and News Media on Citizens Factual Information About Iraq, *Journalism and Mass Communication Quarterly*, 85 (3), Autumn, pp. 591–608.

Kasianov, G. (2015), How a War for the Past Becomes a War in the Present, *Kritika: Explorations in Russian and Eurasian History*, 16(1), Winter, pp. 149–155.

Kassim, Y. R. (2014), *The Geopolitics of Intervention: Asia and the Responsibility to Protect*, Singapore: Springer.

Kent, M. L. & Taylor, M. (1998), Building Dialogic Relationships Through the World Wide Web, *Public Relations Review*, 24(3), pp. 321–334.

Krennmayr, T. (2015), What Corpus Linguistics Can Tell Us About Metaphor Use in Newspaper Texts, *Journalism Studies*, 16(4), pp. 530–546.

Lakoff, G. (1991), *Metaphor and War: The Metaphor System Used to Justify War in the Gulf*, http://philosophy.uoregon.edu/metaphor/lakoff-l.htm.

Lasky, M. J. (2017), *Media Warfare: The Americanisation of Language*, The Language of Journalism, Volume 3, New York: Routledge.

Lule, J. (2004), War and Its Metaphors: News Language and the Prelude to War in Iraq, 2003, *Journalism Studies*, 5(2), pp. 179–190.

Löwenheim, O. & Heimann, G. (2008), Revenge in International Politics, *Security Studies*, 17(4), pp. 685–724.

Maltby, S. & Thornham, H. (2016), The Digital Mundane: Social Media and the Military, *Media, Culture & Society*, 38(8), pp. 1153–1168.

Marcellino, W., Smith, M. L., Paul, C. & Skrabala, L. (2017), *Monitoring Social Media: Lessons for Future Department of Defence Social Media Analysis in Support of Information Operations*, Santa Monica: RAND.

Markey, D. (1999), Prestige and the Roots of War: Returning to Realism's Roots, *Security Studies*, 8(4), pp. 126–172.

Marsella, A. J. (2011), The United States of America: "A Culture of War", *International Journal of Intercultural Relations*, 35(6), pp. 714–728.

Matusitz, J. (2013), *Terrorism & Communication: A Critical Introduction*, Thousand Oaks: Sage.

Milevski, L. (2019), The Idea of Genius in Clausewitz and Sun Tzu, *Comparative Strategy*, 38(2), pp. 139–149.

Moeller, S. D. (2009), *Packaging Terrorism: Co-opting the News for Politics and Profit*, Chichester: Wiley-Blackwell.

Morris, J. (2013), Libya and Syria: R2P and the Spectre of the Swinging Pendulum, *International Affairs*, 89(5), pp. 1265–1283.

Nikitina, Y. (2014), The "Colour Revolutions" and "Arab Spring" in Russian Official Discourse, *Connections*, 14(1), Winter, pp. 87–104.

O'Brien, W. V. (1966), International Propaganda and Minimum World Public Order, *Law and Contemporary Problems*, 31(3), pp. 589–600.

Owen, J. M. (1994), How Liberalism Produces Democratic Peace, *International Security*, 19(2), Fall, pp. 87–125.

Payne, K. (2015), Fighting On: Emotion and Conflict Termination, *Cambridge Review of International Affairs*, 28(3), pp. 480–497.

(2005), The Media As an Instrument of War, *Parameters*, Spring, pp. 81–93.

Potter, J. (2017), The Spengler Connection: Total Critiques of Reason After the Great War, in Sharpe, M., Jeffs, R. & Reynolds, J. (Eds.), *100 Years of European Philosophy Since the Great War*, Cham: Springer, pp. 83–104.

Sahlane, A. (2003), Metaphor as Rhetoric: Newspaper Op/Ed Debate of the Prelude to the 2003 Iraq War, *Critical Discourse Studies*, 10(2), pp. 154–171.

Sarfo, E. & Krampa, E. A. (2013), Language at War: A Critical Discourse Analysis of Speeches of Bush and Obama on Terrorism, *International Journal of Social Science & Education*, 3(2), pp. 378–390.

Shen, C. X., Zhang, H. G., Cao, Z. F. & Huang, J. W. (2007), Survey of Information Security, *Science in China Series F: Information Science*, 50(3), pp. 273–298.

Simons, G. (2016), "Good" Battles versus "Bad" Battles: A Comparative Analysis of Western Media Coverage of the Battles of Mosul and Aleppo, *Tractus Aevorum*, 3(2), Winter, pp. 114–138.

Stahl, R. (2016), Weaponizing Speech, *Quarterly Journal of Speech*, 102(4), pp. 376–395.

Sunstein, C. R. & Vermeule, A. (2008), *Conspiracy Theories*, Paper No. 387, Law & Economics Research Paper, University of Chicago Law School.

Temes, P. S. (2003), *The Just War: An American Reflection on the Morality of War in Our Time*, Chicago: Ivan R. Dee.

Thakur, R. (2013), R2P After Libya and Syria: Engaging Emerging Powers, *The Washington Quarterly*, 36(2), pp. 61–76.

Tuman, J. S. (2010), *Communicating Terror: The Rhetorical Dimensions of Terrorism*, 2nd Edition, Thousand Oaks: Sage.

Voegrlin, E. (1944), Nietzsche, The Crisis and the War, *The Journal of Politics*, 6 (2), pp. 177–212.

Wanta, W., Golan, G. & Lee, C. (2004), Agenda Setting and International News: Media Influence on Public Perceptions of Foreign Nations, *Journalism and Mass Communication Quarterly*, 81(2), Summer, pp. 364–377.

Yakushik, V. (2019), Importance of Inter-Civilizational and Intra-Civilizational Dialogue in National State Consolidation and Development, *Ukrainian Policy Maker*, 4, pp. 85–90.

Zaman, R. U. (2009), Strategic Culture: A "Cultural" Understanding of War, *Comparative Strategy*, 28(1), pp. 68–88.

Zollmann, F. (2019), Bringing Propaganda Back into News Media Studies, *Critical Sociology*, 45(3), pp. 329–345.

6 Different Shades of Information and Communication in Armed Conflict: White, Grey and Black

Greg Simons

Introduction

In executing the art of war and warfare there are a multitude of environments and considerations that need to be taken into account that span the physical, informational and cognitive domains. Tangible and intangible aspects need to be weighed and balanced against each other in order to derive a viable and functioning strategy to maximise strengths and minimise weaknesses, while simultaneously and negatively affecting the decision-making calculus of the enemy with the aim of overcoming the opponent on the physical and informational battlefields. The art of strategy in warfare has been gradually invented, refined and evolved over centuries of experience in organised warfare (Greene, 2006). Innovation and strategy are developed to gain an advantage over opponents, which are impacted in each society by numerous unique sets of factors such as its history, economy, culture, politics, technological development, philosophy and other elements.

A subtle and yet important distinction lies in the use of the terms war and warfare, which have various and multifaceted dimensions, including but not limited to: legal, political, scholarly, communicational, military, cultural and so forth. These aspects influence and shape not only the tangible manner (the way wars are fought), but also the intangible aspects (the way warfare is conceived). This creates different tangible and intangible approaches to warfare across different local and regional settings. For the purposes of this chapter, in the most simplistic understanding, wars are a formal legal and political matter that oblige the respective members of the international community to adhere to certain sets of enshrined principles and rules of engagement in the prosecution of their kinetic military operations. Warfare is broader, and without the formal declaration of war and engagement in military operations, which can involve unrecognised and non-state actors as well as those who are state based.

The nature of wars and warfare has been evolving in the wake of the Cold War, together with the evolutions in global geopolitics, producing a more complex and unstable environment, which has accelerated in the twenty-first century in the wake of the 9-11 attacks and the declaration of the Global War on Terrorism (GWOT). It is the difficult task of information and communication strategies by Western governments and military forces to support and sustain the kinetic military operations (Payne, 2005), although this same problem and dilemma is likely to have been faced by political and military leaderships since the practice of organised warfare (Taylor, 2003). US unipolarity has bred an overconfidence in the understanding of tangible military superiority, but without an understanding of the consequences of the overuse of warfare that is currently occurring.

Even though there are numerous wars and conflicts being fought around the globe currently, the level of transparency and accountability of Western governments is being questioned (Zollmann, 2017). This is seen in the burgeoning number of labels given to types of warfare and armed conflict, such as "Dirty Wars" with the use of small-scale covert (therefore plausibly deniable if 'discovered') global military operations by the US (Scahill, 2013). The US has been involved in a long line of such operations that stretch back over decades, including assassination attempts (with varying degrees of success and failure) on various world leaders that were proving to be a thorn in the side of their foreign and security policy ambitions, such as Cuba's Fidel Castro. However, these were usually conducted clandestinely. The assassination of Iranian General Suleimani in Baghdad was done overtly in a brazen manner that lacked any of the previous legal 'niceties' or regard to human consequences.[1] Increasingly contradictory and confusing signals are given, such as President Trump's earlier denial of any attempt to assassinate Syria's President Assad, yet later stating that he wanted him killed.[2] This disregard for international law and the consequences of actions sets a progressively dangerous precedent in terms of increasing risk, uncertainty and hazards in the waging of warfare in the twenty-first century.

Another label that has emerged in the wake of numerous on-going armed conflicts following the declaration of the GWOT is the notion of "Forever Wars". To date, according to the US Department of Defense,

[1] E. Pilkington, *Suleimani Killing the Latest in a Long, Grim Line of US Assassination Efforts*, The Guardian, www.theguardian.com/us-news/2020/jan/04/us-political-assassinations-history-iran-suleimani, 4 June 2020 (accessed 17 September 2020).

[2] L. Seligman, *Trump Changes Story, Says He Wanted to Assassinate Syria's Assad*, Politico, www.politico.com/news/2020/09/15/trump-assassinate-syria-assad-415093, 15 September 2020 (accessed 17 September 2020).

more than 7,000 US military personnel have lost their lives and another 53,000 wounded in action. This does not take into account the conservative estimate of thirty-seven million people (figures range up to fifty-nine million) displaced by US military operations worldwide.[3] There is also the issue of the results and costs of the GWOT, which saw from 2001 to 2014 an increase in the incidence of terrorism by 6,500 per cent.[4] This is in stark contrast to the implied promises imbedded in the brand name, the Global War Against Terrorism!

However, in spite of the focus on irregular warfare, regular state–state warfare should not be ignored, as the rising border tensions between the nuclear-armed and rising Asian powers of India and China amply demonstrate, with a volatile mix of tangible (territorial control and interest) and intangible (national pride and leadership credibility) elements.[5] In keeping with the 'defensive' narrative and rhetoric, China has been characterised as being a source of economic, political and military threats for NATO to adapt to and counter.[6] This is a façade as the agreed upon rules of international relations are being dangerously rewritten without any thought to the long-term consequences and risks involved. As an illustration, Secretary of State Mike Pompeo openly boasted in a speech at Stanford University's Hoover Institute that assassination of Iranian General Soleimani was part of a broader strategy of deterring challenges to US supremacy, including from China and Russia, and not about the original preventative strike reasoning.[7] Furthermore, the potential for geopolitical expansion of the conflict to a much larger one involving alliances of states, such as India, possibly being tempted to assist the

[3] J. Ismay, *At Least 37 Million People Have Been Displaced by America's War on Terror*, The New York Times, www.nytimes.com/2020/09/08/magazine/displaced-war-on-terror.html, 8 September 2020 (accessed 8 September 2020); D. DePetris, *The Never-Ending War on Terrorism Goes on, 19 Years Later*, Washington Examiner, www.washingtonexaminer.com/opinion/the-never-ending-war-on-terrorism-goes-on-19-years-later?utm, 9 September 2020 (accessed 10 September 2020).

[4] Costs of War project at the Watson Institute of International and Public Affairs, Brown University, https://watson.brown.edu/costsofwar/figures; *The US War on Terror Has Cost $5 Trillion and Increased Terrorism by 6500%*, CS Globe the World Online, http://csglobe.com/the-us-war-on-terror-has-cost-5-trillion-and-increased-terrorismby-6500/, 19 September 2016 (accessed 30 June 2019).

[5] T. Rogan, *The Rising Risk of Conflict between China and India*, Washington Examiner, www.washingtonexaminer.com/opinion/the-rising-risk-of-conflict-between-china-and-india?utm, 9 September 2020 (accessed 10 September 2020).

[6] J. Gehrke, *NATO Turns a Wary Eye toward China*, Washington Examiner, www.washingtonexaminer.com/policy/defense-national-security/nato-turns-a-wary-eye-toward-china, 2 April 2019 (accessed 4 April 2019).

[7] H. Pamuk & J. Landay, *Pompeo Says Soleimani Killing Part of New Strategy to Deter US Foes*, Reuters, www.reuters.com/article/us-iraq-security-pompeo-soleimani-idUSKBN1ZC2I3, 14 January 2020 (accessed 20 January 2020).

US as a proxy to contain China's influence and growth in order to maintain the American global hegemony.[8] There is also a growing risk associated with the direct rhetoric between the US and China. In August 2020 with the failed attempts at establishing détente, China warned the US that it would be prepared to resist any military measures that may be used, although they would not fire the first shot.[9] This came one month after the US had used "a B-52 Stratofortress (strategic bomber) to send a sharp reminder to both China and US allies: American forces can hurt the People's Liberation Army in a hurry."[10] This was a misguided move and a politically symbolic threat given the development of surface-to-air missile technology. Such strategies are fraught with risks and hazards, and unforeseeable consequences.

This chapter intends to explore the use and interplay of information and communication with contemporary armed conflict, which is occurring in an increasingly complex and more politicised global environment. This is also happening in an environmental system that is less transparent and accountable, even if these armed conflicts are being waged rhetorically in the name of national security and the public good. There are various opportunity costs, apparent and hidden, in the price of limitless warfare. These costs are obscured by the manner in which information and communication are used to represent and project the 'reality' of contemporary warfare. The chapter begins with the topic of the role and purpose of managing and manipulating the perception and opinion of the physical domain's 'reality'. The third section engages in the topic of the concepts and roles of information within the context of armed conflict. Next is the theory and practice of communication in armed conflicts of the present day. An account and overview of the different forms and types of warfare and the transformation of armed conflict in the aftermath of the Cold War is in the following section. The final section concerns an attempt to visualise the theory and concepts of the previous sections, by bringing them to life with a case study, the 2010 military offensive Operation Moshtarak in Afghanistan.

[8] J. Gehrke, *Border Violence Could Spur India to Help US Counter China*, Washington Examiner, www.washingtonexaminer.com/policy/defense-national-security/border-violence-could-spur-india-to-help-us-counter-china, 18 June 2020 (accessed 19 June 2020).

[9] T. Rogan, *Why China Is Warning Trump of War*, The Washington Examiner, www.washingtonexaminer.com/opinion/why-china-is-warning-trump-of-war, 9 August 2020 (accessed 11 August 2020).

[10] J. Gehrke, *'They Can't Compete': US Flexes Bombers that Can Strike Mainland China*, The Washington Examiner, www.washingtonexaminer.com/policy/defense-national-security/they-cant-compete-us-flexes-bombers-that-can-strike-mainland-china, 8 July 2020 (accessed 8 July 2020).

Managing and Manipulating the Perception and Opinion of Reality

The role of definition of concepts, theories and terms is inextricably linked to the attempt to define and manage 'reality' itself, or at least its perception. In the current times of increased global political and geopolitical tensions, the motivation of actors to define the 'reality' of the information domain in order to influence the target audiences' cognitive domain increases within the framework of a struggle for hearts and minds of publics that begins with communication and reputation management. Therefore, there is an initial struggle at the stage of definition, which is used as an initial means of manipulating the associated values and meaning of the informational flows that surround people, places and events. This can be viewed from the perspective of the spiral of silence or the practice of censorship, where the selective altering and manipulation or blocking of words and terms is a means to achieve the desired ends of being in a better position to more effectively define how reality is perceived (Simons, 2015). This issue becomes apparent in the important discussion and distinction between the use of definition and labels, where definitions are intended to be a more objective attempt to bring clarity and understanding to a concept or theory, whereas labelling is a more subjective attempt to engage in deliberate reputation or brand management for political or geopolitical purposes (Fridman, Kabernik & Pearce, 2019). Therefore, a more balanced and unemotional approach is needed to decoding and deconstructing the intangible elements of influence and persuasion.

Nancy Snow (2003: 149) notes, "like the Cold War, the Age of Terror is being fought primarily through the information war. Capture and control of the public mind is nothing to fear if we understand that propaganda is a neutral concept – any systematic process of mass persuasion – often misunderstood as censorship or lying." Concepts are not only relevant, but important in managing public perception and opinion of the GWOT; for example, the automatic associations that come with labelling someone as a terrorist and the consequences that stem from such. "The very word 'terrorist' connotes someone who is subhuman. To combat a terrorist, one is tempted to do less-than-human things" (Poole, 2004: 216). Of course, the basic premise is that not every single resistance fighter is a 'terrorist' and may well be motivated to fight what they consider to be terrorism from occupying Western military forces. The resulting situation has the potential to create an endless vicious cycle of violence. Information Operations and Influence Activity are pursued with the intention of outcomes and impacts that assist in securing the operational and strategic goals and objectives of the military and the government that tasked them.

The ability for governments and armed forces to achieve key national security objectives requires that they are able to effectively and credibly communicate with and influence diverse foreign audiences (Paul, 2017: 1). Around the year 2017, the US Department of Defence spent more than US$250 million annually on Information Operations and information-related capabilities to influence target audiences at the strategic and operational levels (Paul, 2017: ix). This provides an approximate aspect within the broader field of the measure of activity; however, the measure of effect is a relatively unknown quantity in the sequence of informing, influencing and persuading. In defining the approach to operational art, there are four suggested steps: 1) understand the strategic direction; 2) understand the operational environment; 3) define the problem; 4) use the results in steps 1–3 to develop a solution – that is, the operational approach (Paul, 2017: 7). However simple this seems on paper and in conceptual practice, the subjectivities present in the physical and especially the informational domain make this an extremely difficult practice.

For example, how can Information Operations be used to support a war that is to be waged on the basis of vanity and/or emotions in a global information environment of decreasing trust in governments and authorities? There are likely to be dilemmas and contradictions inherent in the political realm that must be solved by the military in an era of the so-called Democratic Control of Armed Forces, but without effective oversight and control of those political forces wielding control of the military for intended short-term political gains (and these have the potential to damage both national interests and national security).

Concepts and Role of Information in Armed Conflict

Any actor occupying a competitive and contested information environment has two primary concerns: 1) to identify capabilities and practices that could be adopted and 2) identify adversary capabilities and practices as a means of preparing to counter them in future operations taking place in and through the information environment (Paul et al., 2018: iii). As technology has changed the weapons with which wars are waged, there is a general principle of strategy that applies, and this includes the current information age. This involves the gaining of both freedom to manoeuvre and greater access to achieving goals and interests, while simultaneously trying to deny them to the opponent/enemy (Arquilla & Borer, 2007). With the development of new information communication technologies and the increasing level of international competition and conflict, the determination and control of global flows of information that help to formulate audience opinion and the perception of people and events in the physical domain. For all intents and purposes, the actual and objective 'reality' of what occurs is of much lesser concern than the actual and subjective perception of that 'reality', even if these diverge. Manoilo and Strigunov (2020) on the topic of the role of the genesis, evolution and practice of information technologies in "non-classic wars" facilitate an indirect means of shaping and influencing a target audience's cognitive domain via the information domain. It is clear though, the results can produce significant tangible (physical) results that have operational consequences and policy results that can undermine the target and attain the political/geopolitical goals of the offensive informational operations.

The role and concept of information in warfare is not a homogeneous global idea or approach but is influenced by different aspects in any given country as historical experience, politics, geography, elements of culture and identity and other elements. This creates approaches and concepts that in some respects may seem similar, but upon closer inspection exhibit some uniquely British, German, US, Russian, Chinese or Iranian characteristics. In terms of the development of a viable approach to asymmetric information warfare (defined as the use of surprise by a weaker actor against a stronger, yet vulnerable opponent), China blends traditional stratagems (as Sun Tzu's overcoming the superior with the inferior), more modern concepts (as Mao's People's War) and the development and use of new information communications technology for an operational advantage in the contemporary conflict context (Wang & Stamper, 2002). Innovative approaches to waging evolving forms of warfare require an approach to appraise the existing capabilities and capacities with a view to developing those in the future for conflicts that are yet to be fought.

Three broad categories have been identified that create gaps in the information operations and influence activity capabilities of an actor. The first category relates to the presence of capacity gaps. A second category is the influence of conceptual gaps, where an effective concept has not been adopted. In the third gap category, constraints that have been imposed by higher level authorities or ethical concerns can be the source (Paul et al., 2018: iii, 42). Means and avenues of negating these capability and capacity gaps are constantly sought, especially in the current highly competitive geopolitical age. What has come to be characterised as currently being a global information war is waged on the basis of competing channels of information in the "discourses of hegemonic flow and contra-flow. Hegemonic flow and contra-flow may be considered the play of soft power and contestations of soft power" (Xie & Boyd-Barrett, 2015: 78). Thus, there is an attempt to accumulate informational dominance or at least sufficient intangible advantages over an opponent that can simultaneously increase one's own potential for political and policy manoeuvre whilst limiting the opponent's.

Being able to effectively manage the informational aspects of military campaigns is increasingly central to strategic planning by political and military actors (Waldman, 2019: 163). Information is selectively used in the context of armed conflict, impending and actual, to justify aggressive and/or extraordinary foreign and security policy. This is particularly noticeable within the realm of the politics of intervention, where the use of selective human rights shaming is a common practice. An opportunity for the West to apply more 'liberal' and politically convenient interpretations of whether to intervene or not came with the end of the checks and balances in the international system when the bipolar global order came to an end with the fall of the Eastern Bloc and the rise of a US-centred unipolar global order in the early 1990s. This period witnessed an attempt to shift away from the traditional argumentations for or against military intervention that were more focused on military and political logic, tied to the pragmatic idea of state interests. The logic then moved towards the more imprecise and unclear use of ethical, moral and psychological dimensions to communicate specific information to justify the policy preference (Forbes & Hoffman, 1993: 1–2). Other authors (Durward & Marsden, 2009) argue other factors within the decision to intervene or not in the GWOT were largely ignored, to begin with, such as the role of religion and religious leaders, owing to value and normative differences in Western culture versus other cultures, such as Islamic culture. The longer-term results have been tangible in terms of the outcome of armed conflicts of choice in the twenty-first century.

Powers and O'Loughlin (2015: 177) note that the usual explanations of how information can be used to reduce violence through accurate and real information to decrease misunderstanding, to ensure compliance and to build trust, does not function during periods of sustained conflict. There is no free-flowing information between the sides, although the information domain is saturated, and there is no listening to one another as information is being weaponised by the various sides to support their agenda. They conclude that the priority of information during an armed conflict is to perpetuate violence and not to bring peace. Sources of information and perceived legitimacy can be used offensively to selectively 'justify' regime change policy in a target country, such as Libya in 2011 during the so-called Arab Spring. However, information can also be used defensively to prevent the call for regime change or facilitate friendly force military operations (that are at least equivalent to those in the country being targeted for regime change and the primary reason for the call) in a vassal or allied state, such as the destruction of Fallujah in Iraq in 2004. This is done in an environment of apparent collusion between mainstream mass media and the sitting government that is pursuing the regime change or intervention policy, which requires a façade of public approval and legitimacy, or at least not strong public resistance against the policy (Zollmann, 2017). This is seen in the approach to the selective use and release of information to support or oppose a specific policy.

It was certainly evident in the Libyan regime change of 2011, when information was selectively collected and used in the publicly available motivations by the British government. A report from the Commons Foreign Affairs Committee severely criticised the government of Prime Minister David Cameron for "erroneous assumptions" and an "incomplete understanding" of the ground realities (including Islamist elements in the 'opposition') in a display of opportunism. One of the illustrations of these assessments was "Mr Cameron argued the intervention was necessary to prevent a massacre of civilians but the committee said the available evidence showed that, despite his appalling human rights abuses over 40 years, Gaddafi had no record of large-scale attacks on Libyan civilians."[11] One of the criticisms of the role of media is the lack of critical reporting at the time of the armed conflict, when something could have been changed, with criticism only coming some years after the event when little or nothing can be changed and the damage is done. This was the case with the reflection of Hillary Clinton's role in creating the informational environment that supported the bombing of Libya, when even her critics were slated for understating the catastrophic consequences of this decision that was created and supported in the information domain.[12] The above picture demonstrates a lack of adversarial relationship existing between the government and mass media during periods of armed conflict, which has helped to create and sustain a pro-war agenda.

There is the argument that there is a lack of a clash existing between the executive and legislative branches of government with media, which

[11] *MPs Maul 'Opportunist' Cameron Over Libyan Bombing Campaign*, Express, www.express.co.uk/news/politics/710187/Report-MPs-criticises-David-Cameron-Libya-bombing-campaign, 14 September 2016 (accessed 11 September 2020).

[12] B. Norton, Even Critics Understate How Catastrophically Bad the Hillary Clinton-led NATO Bombing of Libya Was, Salon, www.salon.com/2016/03/02/even_critics_understate_how_catastrophically_bad_the_hillary_clinton_led_nato_bombing_of_libya_was/, 2 March 2016 (accessed 15 September 2016).

clashes with the widely perpetuated myth of the mass media as being a check and balance in society by holding the government to account in the interests of the public in a guardian function. A number of different converging trends have been identified as being possible contributory factors to enable such effective and relatively unchallenged attempts at information management, which enable and sustain armed conflict as a matter of policy choice.

> The development and testing of government information control strategies over the last three decades, and the emergence of a for-profit giant conglomerate media system that lends itself to propaganda due to its structural limitations. The convergence of these two trends has seen a further integration of the media into the military-industrial complex, building upon existing Cold War relationships. As the "war on terror" continues, we can expect the growth of more sophisticated methods of information control and the further curtailment of diversity and debate, unless significant challenges are posed by an informed public. (Kumar, 2006: 49)

Mainstream media has become entangled with mainstream politics and tends to echo the elite consensus, when one exists, in an effort to go to war. Instead of information being accurate, accessible, reliable and timely to enable an informed audience to make the correct decision based on an objective assessment of the topic being covered, information tends to be selective, accessible, manipulative and timed to encourage a faulty and subjective assessment of the situation that encourages the audience to arrive at a convenient and pre-determined conclusion that benefits the originator of the information campaign. In short, there is no coincidence in terms of the news narrative or its precise timing. There are also clear dilemmas and contradictions at play within the military structures too. Planners and practitioners of military deception will cultivate and feed unreliable information through mass media sources in order to mislead the enemy for a short-term operational goal. However, a public affairs officer needs to create and maintain relationships of trust and respect in the long-term management of information where the military is judged not only on who won or lost, but the [ethical] manner in which it is fought (Ross, 1998). It is important to try to win hearts and minds in the cognitive domain by attempting to control the flows in the information domain about events and people in the physical domain.

Theory and Practice of Communication in Armed Conflict

Attempts have been made periodically to streamline the communicated information flows in order to give the appearance of greater message consistency and therefore increase the façade of credibility and believability. It is a matter of trying to obtain a more equitable balance of the measure of effect from the measure of activity. One such attempt was made in 2016 by President Obama with the Global Engagement Centre that was intended to "lead the coordination, integration, and synchronisation of Government-wide communications activities directed at foreign audiences abroad in order to counter the messaging and diminish the influence of international terrorist organisations [...]" (Obama, 2016). This is an illustration of the organisational, technical and cognitive dimensions of the art of attempted influence in competitive communication.

Communication seeks to influence and persuade audiences through the effective and selective use of information using an appropriate communicational concept for achieving the wider organisational objectives. There are different communication mechanisms and avenues available for different tasks and objectives. These are influenced by the political and military culture of the user and the nature of the target to be influenced and/or persuaded. In order to accomplish the desired end state, combinations or selections of various appropriate communicational means can be used by the offensive communicator. For example, when psychological warfare's functions are studied, an interplay between the tangible and intangible is used to create the desired effect.

> Psychological warfare, in the broad sense, consists of the application of parts of science called psychology to the conduct of war; in the narrow sense, psychological warfare comprises the use of propaganda against an enemy, together with such operational military measures as may supplement the propaganda. (Linebarger, 2009: 25)

There is a concentration, not on abstract academic definitions of such communicational means, but rather on how to better apply (more effectively) these mechanisms (operational considerations) in order to achieve the desired goals. Therefore, the definition of psychological warfare revolves around three considerations. Firstly, the specific given situation in which it is to be applied. Secondly, designating the responsibilities and authority in the given tasks. Thirdly, stating the realistic results that can be achieved with the assigned resources (Linebarger, 2009: 37). This is visualised currently in international relations through the new politics of protection (embodied in concepts such as the Responsibility to Protect – R2P), which is applied selectively in order to shape public and governmental responses to armed conflicted that are framed as being humanitarian crises (Kassim, 2014). There is a tendency to communicate in binary opposing sets of constructed realities that are intended to be an interpretive representation of physical domain realities.

Communications theory has identified two primary factors that enable and account for narrative effectiveness. The first is the coherence or probability of the story as being reliable, consistent and convincing. This normally involves a not overly complex central theme rationale, which focuses on a single and clear message with clear purpose. It can include the 'costs' of not following the diagnosis of the narrative. This needs to be at least somewhat aligned with the physical reality in order to create and maintain audience influence and persuasion. The second point is the fidelity or resonance of the narrative with the target audience (Waldman, 2019: 163–164). This is, in part, achieved through selecting and applying an appropriate communication concept to meet the operational tasks and objectives designated by the political and military leadership.

These conceptual lenses are dualistic in use; on the one hand they can be used to assist with the communication of messages and the effectiveness of this activity. On the other hand, the same theoretical lenses can be used to deconstruct messages and reveal the inner workings of persuasion and influence. One such conceptual lens, which has been compromised through its associations with specific actors such as Nazi Germany and the Soviet Union (who used it as a tool for messaging), is propaganda. As a lens for deconstruction, propaganda has been used in the deconstruction of multiple armed conflicts. One of the examples is the 2003 Iraq War, which is still setting the benchmark for deception in the twenty-first century for creating the false pretext for war. The communication campaign against Iraq, described variously between public relations and propaganda, intended to attain operational objectives by subverting the global publics' objective understanding of the physical domain events that would permit other unthinkable military intervention (Samsap, Sung & Jaemin, 2008; Ryan & Switzer, 2009). Early in the aftermath of the opening US-led invasion, Kellner (2004) noted that the false logic of Iraqi weapons of mass destruction and alleged links to terrorism were used by the Bush administration as the basis for a pre-emptive strike. The fickle nature of information and communication within armed conflicts is also noted. "Media spectacles can backfire and are subject to dialectical reversal as positive images give way to negative ones. Spectacles of war are difficult to control and manage and can be subject to different framings and interpretations" (Kellner, 2004: 336). Official reports that include assessments of government propaganda, such as the British Chilcot Report into the nature of the deceptive entry into the Iraq War, also provide moments for the potential for public and government reflection and learning from past mistakes by exposing past abuses (Robinson, 2017). However, the 2011 Libyan campaign seemed to be somewhat of a re-run of those same mistakes from eight years prior.

Other theoretical and conceptual lenses in the sphere of communication are more suited to the task and intention of gaining influence and persuading a target audience (offensively and defensively) rather than providing explanatory power to acts of communication. An example of this type is the concept and practice of psychological operations (PSYOPS). This is a relevant tool in the information age, where social-psychological threats endanger the balance of intangible elements in a target audience that is set in a rapidly evolving international relations environment. Napoleon Bonaparte noted that there are two sources of power in the human realm, the sword and the mind. He noted that in the long run the mind always beats the sword. PSYOPS is an important non-lethal force multiplier in military operations. In a relatively straightforward definition, "psychological operations may broadly be defined as the planned use of communications to influence human attitudes and behaviour, to create in target groups, behaviour, emotions, and attitudes that support the attainment of national objectives" (Narula, 2004: 184). Therefore, PSYOPS engages and deals directly with human behaviour, which includes opinions, attitudes, emotions and behaviour to shape the desired perceptions and elicit the desired response (Narula, 2004: 189; Wall, 2010: 290). However, targets of PSYOPS are not limited to combatants, but also civilian populations within areas of operational interest. Criticism levelled at PSYOPS stems from the 'bordered' nature of the practice prior to cable TV, the internet and social media, which has blurred distinctions and boundaries between segmented audiences and the ability to target only intended audiences (Wall, 2010: 289–291). This is testimony to the effects of new technology in shaping the practice, outcomes and impact of communication efforts.

The rapidly evolving sphere of new information communications technologies is creating new opportunities as well as threats for state and non-state actors alike. Paterson and Hanley (2020) note that the digital age has forever altered the way states conduct political [and other forms of] warfare, which necessitates a rebalancing of security priorities. Machine learning and artificial intelligence merely serve to increase the offensive capability potential of this means of indirect attack. Cyberspace is being currently used by a variety of different state and non-state actors for subversion and destabilisation purposes in foreign and security policy.

There are attempts to increase the communication capability and capacity of governments and authorities; one of the ways to do this has been through the outsourcing of certain tasks and abilities, but also to bring in sought-after external expertise, such as key people from Madison

Avenue or Silicon Valley.[13] Given the current global trend of decreasing trust in authorities, it is an activity that may involve intermediaries to be seen as communications coming from "independent", or at least not state, sources, as a means to increase the possible perception of credibility among the target audience. As a case in point, the US Pentagon paid the British PR firm Bell Pottinger US$500 million to run a top-secret influence programme in Iraq. There were three types of media operations used in Iraq: white (attributed); grey (unattributed); and black (falsely attributed) material. It included black operations, such as the tracking of who was watching certain materials. The PR firm was brought into Iraq soon after the US-led invasion and occupation in 2004 (and lasted until 2011), and one of the first assignments was to promote the idea of democratic elections being held in the country. Media materials were produced, such as TV commercials portraying Al Qaeda in a negative light, news items that were made to look as if they were filmed by and for Arabic TV, and the production of fake Al Qaeda propaganda films (which were placed in different national and international locations, such as Iraq, Iran, Syria and the US). Between 2006 and 2008, some forty different communications companies were being paid for professional communications services.[14] This particular aspect and many previous incidents and instances are having the effect of eroding US credibility as a communicator and as an actor.

A raft of various disclosures concerning non-transparent, unaccountable and ethically dubious military and political operations in media outlets in the twenty-first century severely compromises the logic of the argument that Western governments do not have the operational freedom of their opponents as they must abide by the law and behave in an ethical manner. Here is a small sample of some of those revelations. These subjects include the United Kingdom's covert operation to overthrow Syria's Assad, and as a consequence to prolong and radicalise the conflict.[15] A number of stories also exist on the aspect of the British government funding a covert propaganda war about Syria that may be in

[13] C. Watts, *Silicon Valley Meets the Snake Eaters: State Department's Counter-messaging Goes Back to the Future*, Foreign Policy Research Institute, www.fpri.org/geopoliticus/2016/02/silicon-valley-meets-snake-eaters-state-departments-countermessaging-goes-back-future, 10 February 2016 (accessed 16 February 2016).

[14] C. Black & A. Fielding-Smith, *Fake News and False Flags. How the Pentagon Paid a British PR Firm $500 Million for Top Secret Iraq Propaganda*, The Bureau of Investigative Journalism, http://labs.thebureauinvestigates.com/fake-news-and-false-flags/, 2 October 2016.

[15] M. Curtis, *How Britain Engaged in a Covert Operation to Overthrow Assad*, Middle East Eye, www.middleeasteye.net/opinion/how-britain-engaged-covert-operation-overthrow-assad, 27 April 2018 (accessed 21 May 2020).

breach of the law.[16] These aforementioned issues and aspects of covert British government operations in Syria challenged the media narratives on the nature of the war.[17] Stories have also appeared in some mainstream media outlets calling for readers to be sceptical of government claims and the assumed validity of information on armed conflicts such as Iraq and Syria.[18] Other revelations include what high-ranking government officials, such as former Secretary of State Hillary Clinton, knew of troubling information about conflicts that they were actively engaged in and did not admit publicly.[19] This is about citizens being manipulated into a war[20] or lied to in order to continue an existing war, such as Afghanistan and the Forever Wars.[21] The mass media do not escape investigation and judgement, where stories explore the nature of who supplies the news on armed conflicts,[22] how the press are misinforming

16 I. Cobain, A. Ross, R. Evans & M. Mahmood, *How Britain Funds the 'Propaganda War' Against ISIS in Syria*, The Guardian, www.theguardian.com/world/2016/may/03/how-britain-funds-the-propaganda-war-against-isis-in-syria, 3 May 2016 (accessed 21 May 2020); I. Cobain & A. Ross, *Exclusive: British Propaganda Efforts in Syria May Have Broken UK Law*, Middle East Eye, www.middleeasteye.net/news/exclusive-british-propaganda-syria-war-opposition-illegal-internal-review, 11 May 2020 (accessed 21 May 2020); I. Cobain & A. Ross, *Revealed: The British Government's Covert Propaganda Campaign in Syria*, Middle East Eye, www.middleeasteye.net/news/revealed-british-government-covert-propaganda-campaign-syria, 19 February 2020 (accessed 21 May 2020).

17 M. Curtis, *Revelations About UK Covert Operations in Syria Challenge Media Narratives on the War*, Daily Maverick, www.dailymaverick.co.za/article/2020-05-18-revelations-about-uk-covert-operations-in-syria-challenge-media-narratives-on-the-war/, 18 May 2020 (accessed 21 May 2020).

18 P. Cockburn, *We Should Be Sceptical of Far-Away Governments Who Claim to Know What Is Happening on the Ground in Syria*, The Independent, www.independent.co.uk/voices/syria-assad-uk-government-christian-persecution-isis-foreign-office-a8314616.html, 20 April 2018 (accessed 23 April 2018); P. Cockburn, *This Is Why Everything You've Read About the Wars in Syria and Iraq Could Be Wrong*, The Independent, www.independent.co.uk/voices/syria-aleppo-iraq-mosul-isis-middle-east-conflict-assad-war-everything-you-ve-read-could-be-wrong-a7451656.html, 2 December 2016 (accessed 5 December 2016).

19 P. Cockburn, *We Finally Know What Hillary Clinton Knew All Along – US Allies Saudi Arabia and Qatar Are Funding ISIS*, The Independent, www.independent.co.uk/voices/hillary-clinton-wikileaks-email-isis-saudi-arabia-qatar-us-allies-funding-barack-obama-knew-all-along-a7362071.html, 14 October 2016 (accessed 9 June 2017).

20 R. W. Merry, *Lies They Told Us: A Long History of Being Manipulated Into War*, The American Conservative, www.theamericanconservative.com/articles/lies-they-told-us-a-long-history-of-being-manipulated-into-war/, 21 June 2019 (accessed 25 June 2019).

21 R. Kheel & E. Mitchell, *Bombshell Afghanistan Report Bolsters Calls for End to 'Forever Wars'*, The Hill, https://thehill.com/defense/474005-bombshell-afghanistan-report-bolsters-calls-for-end-to-forever-wars, 11 December 2019 (accessed 16 December 2019); C. Whitlock, *At War With the Truth*, The Washington Post, www.washingtonpost.com/graphics/2019/investigations/afghanistan-papers/afghanistan-war-confidential-documents/, 9 December 2019 (accessed 16 December 2019).

22 P. Cockburn, *Who Supplies the News?*, London Review of Books, www.lrb.co.uk/v39/n03/patrick-cockburn//who-supplies-the-news, 2 February 2017 (accessed 16 February 2018).

the public about their government's role in the world[23] or the media supporting rather than questioning or holding to account Western government violence against foreign countries.[24] This leads some observers to openly state that the media are intentionally misleading the public in support of aggressive Western foreign policy that involves military intervention.[25] These revelations reveal the tip of a larger problem, the misuse of information and communication to manipulate audience perception and opinion of people, places and events by the political elite as a means of implementing risky and/or illegal acts of warfare within the pursuit of foreign policy.

This can be seen within the context of specific categories of situations: 1) the effects of deception upon credibility and 2) the effects of the lost credibility and prestige on policy preferences. In the first category is the accumulating number of wars and warfare that has been initiated by deception or at the very least manipulation, such as Iraq in 2003 (Colin Powell and the "irrefutable proof" of Iraqi Weapons of Mass Destruction (WMDs)), and Libya in 2011 (the mass and systematic rape allegations by Hillary Clinton). In 2018, the then Secretary of Defence James Mattis admitted publicly that there was "no evidence" that the Syrian government used sarin gas on its own people, in spite of a US military strike against Syria that was based on the premise it had done so. This confirmed the doubts cast earlier on the US narrative by experienced arms control experts such as Scott Ritter, Hans Blix, Gareth Porter and Theodore Postol.[26] The second category not only concerns the verbal and visual communication of wars and warfare, but the communicated acts of actual warfare and especially the path of aggressive unilateralism against the interests and wishes of other international actors, together with the results that are reflected in the attitudes and acts of members of the international community. This was seen in the northern hemisphere in the summer of 2020, when the Trump administration attempted to

[23] M. Curtis, *How the UK Press Is Misinforming the Public About Britain's Role in the World*, Daily Maverick, www.dailymaverick.co.za/article/2020-03-09-how-the-uk-press-is-misinforming-the-public-about-britains-role-in-the-world-part-one/, 9 March 2020 (accessed 21 May 2020).

[24] G. Shupak, *Media Support US Violence Against Syria, But Long for More*, FAIR, www.fair.org/home/media-support-us-violence-against-syria-but-long-for-more/, 20 April 2018 (accessed 23 April 2018).

[25] S. Kinzer, *The Media Are Misleading the Public on Syria*, The Boston Globe, www.bostonglobe.com/opinion/2016/02/18/the-media-are-misleading-public-syria/8YB75otYirPzUCnlwaVtcK/story.html, 18 February 2016 (accessed 13 April 2018).

[26] I. Wilkie, *Now Mattis Admits There Was No Evidence Assad Used Poison Gas on His People*, News Week, www.newsweek.com/now-mattis-admits-there-was-no-evidence-assad-using-poison-gas-his-people-801542, 8 February 2018 (accessed 28 February 2018).

garner international support for further anti-Iranian measures at the United Nations. Instead of gaining the desired support, it was criticised for the unilateral stance in the Iranian nuclear deal that was rendered void on the basis of not serving US interests. Thus, instead of the US reasserting itself on the world stage, it was shown just how exposed and isolated it had become.[27] This has further exacerbated the contradictions and political conflicts in US domestic politics concerning the current and future path of US foreign and security policy.

It is sometimes communicated through official governmental channels, that the underlying logic and reasoning of aggressive US military deployments abroad is to facilitate American interests and security, as well as to bring peace and stability to the country and region in which they are deployed. However, in contradiction to these communicated logics are the hard policy maker and practitioner facts of heavy-handed and often illegal military occupations (according to international law), the support of repressive governmental regimes or dubious 'freedom fighters', flooding conflict zones with arms, driving instability and resentment that threaten US interests. This pattern of forever wars is increasingly leading to calls to break the cycle by ending these various armed conflicts and no longer making the same mistakes as in the past, which has created the endless succession of armed conflicts and hostilities. However, it becomes difficult to reverse the trends because of a combination of bureaucratic inertia and simplistic shaming based on the false logic that US troops are needed in the region because of the insecurity that requires their continued (without any timeline) presence (Pillar et al., 2020). Given that 2020 was the year of the US presidential elections, a sharp divide occurred between the political establishment that supported continuing armed conflicts around the globe and President Trump who rhetorically stood against the "Endless Wars" as part of his "America First" policy platform.[28] This is demonstrative of the important role played by politics in the outcome of wars rather than the military influence and outcome of individual battles in a given armed conflict.

An argumentation that is often used in the discourse of defending the quality of Western governments' communications, with the application

[27] C. Lynch & R. Gramer, *World Rebukes US Over Iran*, Foreign Policy, https://foreignpolicy.com/2020/06/30/world-rebukes-us-iran-arms-embargo-united-nations-diplomacy-pompeo-middle-east/, 30 June 2020 (accessed 3 July 2020).

[28] J. Hunter, *Trump's 'America First' Redefining of Republican Foreign Policy Might Be His Greatest Legacy*, The Washington Examiner, www.washingtonexaminer.com/opinion/trumps-america-first-redefining-of-republican-foreign-policy-might-be-his-greatest-legacy, 24 July 2020 (accessed 25 July 2020).

of false logic that they cannot tell lies or untruths, need to act within the law, unlike the enemy. This is an attempt to create an aura of ethos around the communicating actor while simultaneously denigrating the named enemy other and the target of those 'truthful' communications. The role of fake news and false flags are not a new means of manipulative form of communication in the history of warfare. In the contemporary age, new information communication technology enables the spread of manipulative persuasive communication further, faster and to more people. However, a number of trends are emerging that are likely to further impede the effectiveness of US and Western information and communication strategies in warfare. There are various evolving political, economic and military processes and trends in the US and globally, which are having the effect of transforming the global balance of power. The global hegemony of the US is being gradually eroded by various internal and external trends and actors, which leaves it in a difficult position both in terms of the physical realities of politics and practice, and also the psychological realities of politics and conceptualisation.

> The US, then, finds itself in the worst of all possible worlds. Its external power and internal cohesion are manifestly in decline. Its psychic hold on the rest of humanity is not. The result is a level of scrutiny that no other nation, not even the other superpower [meaning China], has to experience – or, perhaps, could withstand. The tighter the inspection, the more numerous the revealed flaws, and the more compromised is America's ability to command global deference. The whole process is viciously circular.[29]

As a result of the above, there has been the emergence of a narrative concerning the end of US global hegemony and the ruining of a superpower. There are growing rival conceptions as to how the global order should be conceived; new transnational networks and regional organisations now contest US influence. There has been a return of great power competition to international relations, which has seen an end to the US patronage monopoly. It is noted that currently, the system that upholds US global hegemony based upon projections of economic and military power is unravelling.

> The United States lacks both the will and the resources to consistently outbid China and other emerging powers for the allegiance of governments. It will be impossible to secure the commitment of some countries to US visions of international order. Many of those governments have come to view the US-led order as a threat to their autonomy, if not their survival. And some governments

[29] J. Ganesh, *US Is Stuck in the Worst of All Worlds*, Financial Times, www.ft.com/content/af8e1f30-290c-4181-858e-0bf2f11a0c32, 1 July 2020 (accessed 3 July 2020).

that still welcome a US-led liberal order now contend with populist and other illiberal movements that oppose it.[30]

Even though the US decline as a superpower was noticed some years ago, the recent challenges, such as the mismanagement of the Coronavirus pandemic, seems to be "killing it off". Walt contends that the collapse of the Soviet Union came as a strategic surprise for the US, which resulted in creating a "dangerous combination of hubris and complacency". He notes three categories of strategic mistakes committed by the US during the unipolar era. The first error has been the messianic-like spread of liberal democracy globally, which has created backlashes, proved to be an economic drain and created countless costly wars. The second mistake is allowing public institutions to deteriorate, which has reduced the US capability and capacity to act. The third mistake is the weaponisation of domestic partisan politics in the US, where slander, derision and deadlock have replaced constructive, respectful debate and the art of compromise.[31] In a comparative look at the underlying reasons for the collapse of the Soviet Union and reflecting on the current US decline, a similarity is noted. The process of collapse is a gradual one, where the state is able to see and understand the faults in others, but not its own threats, risks and problems. A final blow came to the Soviet Union when the country's political elite separated their interests and themselves from the national political project.[32] Thus, the decline that leads to eventual collapse is internally influenced and grows as a creeping crisis, therefore less easily detected until the final moment, even though there are early warning signs of an impending failure. Bearing in mind the current state of US politics as noted in the third mistake by Walt, the similarity is an intriguing one.

Shades of Armed Conflict

The physical battlespace (and accompanying information space) of war is divided into three different interacting levels: the strategic, operational and tactical. At the strategic level, there is a focus on defining and supporting national policy and it relates directly to the outcome of the

[30] A. Cooley & D. H. Nexon, *How Hegemony Ends: The Unravelling of American Power*, Foreign Affairs, www.foreignaffairs.com/articles/united-states/2020-06-09/how-hegemony-ends, July/August 2020 (accessed 3 July 2020).

[31] S. M. Walt, *How to Ruin a Superpower*, Foreign Policy, https://foreignpolicy.com/2020/07/23/how-to-ruin-a-superpower/, 23 July 2020 (accessed 24 July 2020).

[32] C. King, *How a Great Power Falls Apart: Decline Is Invisible from the Inside*, Foreign Affairs, https://foreignaffairs.com/articles/russia-fsu/2020-06-30/how-great-power-falls-apart, 30 June 2020 (accessed 3 July 2020).

war or conflict as a whole through strategy, planning and preparations. Operational is the next level down; it concerns the deployment of military force in a theatre of war/operations to gain advantage over the enemy and accomplish a common goal in a given space and time through orchestrated political, economic and military operations. Tactical is the lowest level and concerns translating power into individual battlefield outcomes as a means of achieving goals at the operational level (USAF, 1997). As an example, Operation Overlord (D-Day in June 1944) was at the strategic level part of the plan to defeat Nazi Germany and open a second front in the West for the Soviet allies at the operational level, by beginning the liberation of France at the tactical level. These delineations (which can become blurred in case of armed conflict) are intended to facilitate the understanding of battles and wars for greater clarity in planning strategy for operations.

Wars are in part rational, especially from the elite perspective and the calculations towards achieving concrete tangible goals with war as the political policy means. Warfare is waged for different reasons and goals, which can change and evolve within a given armed conflict, for example ideational goals for society versus a quest for maximising territory or wealth (Rochlin, 2020). However, wars are also emotionally driven, especially at the mass level, with the use of emotional symbols and myths to stir public passion and animosity towards the declared other (enemy) figure (Kaufman, 2001). Short of the physical exercise of war or warfare, the perception of relative military strength related to the show of strength (actual and projected) is a means of what Luttwak (2001: 218–233) refers to as being "armed suasion". This is the measure of ability by an actor to keep allies reliant and compliant through persuasion, and to deter enemy actions that may harm their national interests through dissuasion. Although sometimes this is not enough and warfare occurs.

The nature with which wars and armed conflicts are fought and why they are fought is influenced by the development of technology, the economy, and the political and geopolitical imperatives of the era. These factors also have an influence on how wars and warfare are imagined and communicated to stakeholders and audiences (friend, foe and neutral) (Betz, 2015). On various occasions in history, it is the spectre of military defeat that causes armed forces to innovate their tactical strategy, such as the situation for the Imperial German Army in 1917 (Poole, 2004: xvii) that led to the March 1918 offensive on the Western Front during the First World War. This has been witnessed throughout the history of organised warfare and is still readily apparent today.

In 1991 the sudden and final collapse of the Eastern Bloc and the Soviet Union that saw the final act of the Cold War, caused some to rethink the orthodoxy of knowledge on the role and practice of war and warfare in the new international era. Moaz and Gat (2001: 1–3) give an overview of the three main arguments that followed the end of the Cold War and the attempts to predict how this would impact on wars and warfare. One argument was that wars are obsolete as they have lost their original function (as slavery had done historically). As a result, they are likely to disappear, in a form of Darwinian evolution, as they are unable to adapt to the new environment, such as expressed in the ideas of Francis Fukuyama. A second line of argumentation was that the basic structure of war does not change. This was proven historically with periods of relative peace. The third argument hypothesized that war shall not disappear; it will fundamentally transform in scope, distribution and character.

Kaldor (2006) has argued for the third variant, and has divided wars and armed conflicts as being "old" and "new", which has been influenced by globalisation that has resulted in the state–state-based monopoly of violence being challenged and conflicts becoming more transnational in character with various elements and actors (military, law enforcement, criminals, insurgents and terrorists, etc.). There is a division in the qualitative nature of warfare, between conventional and unconventional wars. The risk to national survival is more acute in conventional conflicts, which "imply a state of open belligerency between nations and a direct confrontation of their armed forces", and unconventional conflicts are more blurred in terms of who is fighting whom (for example insurgencies and guerrilla campaigns) (Department of the Army, 2004: 1–1). In practice, the end of the Cold War has brought an era of less high-intensity and regular conflicts (inter-state wars), and more low-intensity and irregular conflicts (insurgencies and terrorism). This gives rise to the need to define and visualise the different types of armed conflicts, their nature and consequences. After all, not all conflicts are equal in nature and effects. Kassimeris and Buckley (2010: 1) note that there have been a number of new enemies to confront and new security threats emerging that have resulted in a qualitatively and quantitatively diverse array of wars and warfare across the planet in the post-Cold War era.

Wars and warfare are often thought of as being something that are overt and obvious in nature, not least owing to the sheer destructive potential of modern military munitions and hardware. However, this is not always the case owing to the consequences and transparency concerns of openly declaring war and following the international rules of war. Transparency can have the potential to breed accountability for one's policy and its consequences. It should also be noted that there has been a considerable growth in the use of covert military operations since the declaration of the Global War on Terrorism (GWOT or otherwise known as Overseas Contingency Operation post-9/11). The global prevalence of covert kinetic military operations, and particularly those used by international actors such as the United States, are increasingly covered in popular science publications (for example, see Scahill, 2013). Irregular warfare has presented Western political and military leaderships with a number of complex problems and dilemmas to resolve in the twenty-first century, which has in turn created the need to think about and reflect on the strategy as to how to resolve this kind of warfare from a practitioner and policymaker perspective. The current practitioner and policymaker focus is on the management and prosecution of low intensity conflicts, irregular and unconventional warfare, which are a product of the long history of various antagonisms and more recently the GWOT and Arab Spring.

It was noted from a Western perspective that, during the period of the Cold War, insurgencies stemmed from two major sources – the rise of nationalism in colonial territories, and Communist subversion (Galula, 2006: 95). In the contemporary era of the twenty-first century, these sources are no longer the issue. The first issue is the result of creating and abandoning non-state actors that served as proxy forces to fight the Soviets in the Cold War, such as the Mujahedeen in Afghanistan that evolved into Al Qaeda. Short to medium term tactics of convenience have evolved to become strategic problems. Another issue is the poorly conceived and executed Western foreign policy, and that of regime change in particular, and the resulting occupations of "humanitarian" interventions.

The tangible asymmetry of irregular conflicts does not always mean that the most physically powerful actor (often state-based) shall automatically prevail against the actor with lesser kinetic military force projection. Different historical experiences in counterinsurgency are studied in the current times of the intractable insurgencies that Western armies have found themselves fighting across different parts of the globe, such as the possible lessons that could be derived from the British experiences in the Malayan Emergency (1948–57) and the US experience in Vietnam (1950–72) to be applied to the GWOT (Nagl, 2005). As history has taught us, there have been gradual, costly and humiliating defeats including: the British retreat from Afghanistan in 1839; the Soviet Army in Afghanistan by the Western-supported Mujahedeen (Sunni Islam) in 1989; or the forced end to the Israeli occupation of Southern Lebanon by Hezbollah (Shia Islam) in 2000 (Poole, 2004: xix). The study and restudy of Vietnam has been important in order to derive lessons for the present, where Vietnam is seen as a traumatic past marker of humiliating political and military failure that is to be avoided from repetition in present or future armed conflicts (Marks & Bateman, 2019; Phillips, 2019). These forces and experiences are now being encountered by the US-led GWOT and the aftermath of the Arab Spring.

As a result of a combination of various historical experiences (political, religious, social, technological and other factors), different military strategies have evolved. Western military forces tend to favour a top-down leadership approach of set-piece battles, with the use of stand-off weaponry, whereas various Eastern military traditions involve the use of avoiding set-piece battles, preferring the bottom-up use of tactical deception and close-quarter combat (Poole, 2004: xxvi–xxvii). A significant puzzle is thus created; where the strategy of Western military thinking is to "cut off the head" of the enemy to win the battles and wars, it is not effective against the tactics described, which may imply the need for non-kinetic strategies in order to assist in a more positive military outcome (Poole, 2004: 211). Another element of the asymmetry of irregular and unconventional warfare is that it is relatively cheap for the insurgent forces and expensive for the counter-insurgents; insurgents also have greater leeway in terms of military tactics that can be applied on the battlefield in comparison to their counter-insurgency opponent (Galula, 2006: 6–8; The US Army/US Marine Corps, 2007: 3–4). This has an impact upon the conceived timescales, short or protracted warfare, used in strategy calculus. There is also the basic dilemma where the use of violence begets violence in an incompatible understanding of the value of time, or ethical, moral, political and military factors of the physical and psychological battlefields.

The prerequisites needed for an insurgency to occur are: the existence of a vulnerable civilian population; insurgency leadership in order to give direction; and a lack of government control (Department of the Army, 2004: 1–2). From a practitioner's perspective, politics and political power are as central to unconventional and irregular conflicts as insurgency and counterinsurgency (Department of the Army, 2007; The US Army/US Marine Corps, 2007: 2). Military action tends to play a subordinate role to political priorities and imperatives of insurgent forces (The US Army/US Marine Corps, 2007: 9). Insurgents are able to draw upon a number of different sources and means for the purposes of mobilisation, for example: persuasion; coercion; reaction to abuses; foreign support; and apolitical motivations (The US Army/US Marine Corps, 2007: 15). Insurgency and counterinsurgency (COIN) are seen as employing the full spectrum of military operations (offensive, defensive and stability operations) (Department of the Army, 2007; The US Army/US Marine

Corps, 2007: 34–36), which tend to be situationally determined by on-the-ground realities and situational awareness in meeting the political objectives.

> Insurgency and counterinsurgency (COIN) are complex subsets of warfare. Globalisation, technological advancement, urbanisation, and extremists who conduct suicide attacks for their cause have certainly influenced contemporary conflict; however, warfare in the 21st century retains many of the characteristics it has exhibited since ancient times. Warfare remains a violent clash of interests between organised groups characterised by the use of force. Achieving victory still depends on a group's ability to mobilise support for its political interests (often religiously or ethnically based) and to generate enough violence to achieve political consequences. (The US Army/US Marine Corps, 2007: 1)

In spite of the apparent 'simplicity' expressed in the various military manuals, it is not an easy task to coordinate and align political and military aspects in an evolving operational setting. The US and the West have been caught in a malaise that has resulted from unclear thinking about war, war's political essence, and strategies that can align these aspects. As such, this has spawned a somewhat superficial attempt to discover or rediscover the art of irregular warfare, including a raft of new terminological concepts. However, some have called for a need to challenge the underlying political and theoretical orthodoxy of current knowledge and practice (Ucko & Marks, 2018). One of the ways to move forward in COIN operations, in an era of public aversion to casualties and the resulting political sensitivity to the issue that impacts upon the nature of military operations, is to try and build partner (state and non-state actors) capability and capacity in order to reduce the number of US troops exposed to risks and hazards on the battlefield (Guido, 2019). The result of these aspects combined is a US government that is poorly equipped (tangibly and intangibly) to successfully engage in irregular warfare owing to the problematic search for means of lower cost and lower risk in COIN operations that is in need of a significant reform to manage, although, owing to the nature of the current political and military culture, this change is unlikely to occur (Ucko, 2019). Thus, minor tactical victories and measure of activity (body counts, captured enemy equipment and combatants, holding of elections in occupied territories, building of schools and hospitals and so forth), will be used as a means to try and convey to a war-wary public that some level of progress (although this is a false logic) is being made within the cycle of endless wars in order to stave off the needed fundamental change to a political and military system under a great deal of stress and strain.

One of the growing criticisms is that the wrong irregular warfare strategy is currently being used, and another approach is required to secure victory.[33] Edwards (2014) notes the contradictions and tensions that exist between political and military strategies in warfare, and how entry into a conflict influences and defines the exit from that conflict. All of this contradictory mix influences the longer-term impacts on warfare in general and the outcome of any particular conflict. One of the tactical means of attempting to lower the cost and risk of armed conflict is to wage armed conflicts indirectly. This has led to the greater attraction of tactics of surrogate warfare, which is defined as "the conceptual umbrella for all forms of externalisation of the burden of warfare to supplementary as well as substitutionary forces and platforms" (Krieg & Rickli, 2018: 115). Although the concept and practice extends back to ancient times, surrogate warfare is noted as becoming the norm and not the exception during the contemporary Western management of warfare (Loveman, 2002; Marshall, 2016; Krieg & Rickli, 2018: 116). Surrogate warfare, which includes proxy warfare, is defined as "conflicts in which a third party intervenes indirectly in order to influence the strategic outcome in favour of its preferred faction" (Mumford, 2013: 40). Mumford (2013) notes that there are four reasons why this kind of warfare is likely to be increasingly used: 1) decreased public and political will for more drawn-out military campaigns within the context of a global economic crisis; 2) increase in the prominence and role of private military companies in contemporary warfighting; 3) increasing use of cyberspace as a means for waging indirect warfare and 4) the rise of China as a superpower. Proxy wars are attractive tactics for state actors owing to being relatively cheap and plausibly deniable.

Moreover, US COIN/CT has increasingly relied on the strategy of semi-proxy wars to achieve political and military objectives. However, this strategy is creating security risks and hazards by virtue of the actual nature of the 'rebel' forces being supported in regime change operations, such as Libya and Syria. This has created large flows of radicalised foreign jihadist fighters travelling between the battlefield and their European or US homes, which threatens the viability of counterterrorism efforts as well as when a multitude of proxy wars can occupy the same battlespace and destabilise the target country and the surrounding area. There is often a lack of long-term strategy and understanding of the consequences of proxy forces, such as shown when the US abandoned the Mujahedin after the Soviet defeat in Afghanistan and parts of this gradually morphed in to Al Qaeda and contributed to the current

[33] T. Harshaw, *The Terrorists Can Be Beaten, but Not the Way We Are Trying*, Bloomberg, www.bloomberg.com/opinion/articles/2019-05-19/the-terrorists-can-be-beaten-but-we-re-doing-it-wrong, 19 May 2019 (accessed 22 May 2019).

GWOT (Hughes, 2014; Cragin, 2015; Waldman, 2017; Phillips & Valbjörn, 2018). This situation creates a dilemma and a significant problem in the management of proxy wars. In warfare, the elements of politics and identity can influence and motivate actors to fight as proxies for a sponsor. However, experience has shown and amply demonstrated that ideology is not a substitute for the operational control of proxy forces (Biberman & Genish, 2015). The Syrian conflict has proven to be particularly illustrative of the dangers of covert and indirect engagement in military operations owing to the lack of transparency of accountability to the larger political system and the general public; for example, the US $1 billion CIA secret programme that trained proxy forces and funnelled weapons to the 'rebels', including groups linked to Al Qaeda.[34] There was also the equally disastrous and dangerous US Department of Defence Train and Equip Programme in Syria that suffered from large budgetary 'discrepancies' in the audit (Inspector General, 2020). As noted by Clarke (2015: 4), "when wars come to an end, even a decisive military victory does not necessarily translate into a decisive political outcome." This is very discernible in the current GWOT world.

Irregular and low-intensity conflicts, some of which are waged covertly, create an environment of lower transparency and therefore political accountability. Increasingly tough questions are being asked, such as how many wars is the US fighting? The question is made more difficult to answer with the presence of such military entities as the Special Operations Command (SOCOM). In 2015 it was officially announced that US special operations forces had been deployed to 135 countries around the globe (under the Bush Junior Administration the figure was sixty countries). From 2001, SOCOM increased the number of personnel from 33,000 to nearly 70,000 in 2015. At any time in 2015, there were some 11,000 of these personnel deployed beyond the US borders in a range of various military-related activities.[35] Geopolitical positioning between powers and the resulting warfare with and between state and non-state actors, directly and indirectly, has had the effect of blurring the distinctions between war and peace (Bălăşoiu, 2016; Perot, 2019). These acts and measures that fall short of an officially declared state of war, but nonetheless have the potential to subvert an opponent create a grey zone conflict. Grey zone conflict is nothing novel or new, being used actively

[34] M. Mazzetti, A. Goldman & M. S. Schmidt, *Behind the Sudden Death of a $1 Billion Secret CIA War in Syria*, The New York Times, www.nytimes.com/2017/08/02/world/middleeast/cia-syria-rebel-arm-train-trump.html, 2 August 2017 (accessed 8 August 2017).

[35] N. Turse, *How Many Wars Is the US Really Fighting?*, The Nation, www.thenation.com/article/archive/how-many-wars-is-the-us-really-fighting/, 24 September 2015 (accessed 25 September 2015).

in the geopolitically charged atmosphere of competition during the bipolar order of the Cold War (and earlier) in attempts between the US and Soviet Union to defend their own vulnerabilities and leverage their opponent's weaknesses, and the practice continues unabated in the contemporary context of an evolving global order (Sauliuc et al., 2014; Votel et al., 2016; Oakley, 2019; Hughes, 2020). Importantly, it is noted that a difference exists in the measure of success between classic forms of conflict and grey zone conflict.

> A grey zone "win" is not a win in the classic warfare sense. Winning is perhaps better described as maintaining the US government's positional advantage, namely the ability to influence partners, populations, and threats toward achievement of our regional or strategic objectives. Specifically, this will mean retaining decision space, maximising desirable strategic options, or simply denying an adversary a decisive positional advantage. (Votel et al., 2016: 108)

However, what can count as being political or military "success" may be viewed differently by the general public. The results of the lack of successful political outcomes of wars waged by Western powers in the twenty-first century is becoming an increasingly hot topic. This can result in some historical reflections as a means to try and make sense of the present situation and find potential remedies. The Pentagon celebrated the fiftieth anniversary of the American victory in Vietnam, which in reality was a decisive political defeat (the individual battles were won militarily, but the war was ultimately lost politically). It was a conflict that demonstrated: an American military that could not manage effective military and political operations in a protracted irregular war (insurgency and occasionally regular conflict); the mistreatment of proxy forces and allies; the long-term impact of the willingness to become entangled in wars of choice for the country concerned and the wider region; and the impacts of war crimes on the national image and credibility. However, the issues and same mistakes are repeating themselves in the contemporary armed conflicts that collectively constitute the GWOT. A situation that repeats calls for a major rethink on the philosophy and approach to waging armed conflicts.[36] Historically and contemporarily, the US knows how to perform the initial opening military strikes, but lacks the political and military culture to be able to endure the rigours of a

[36] G. Adams, *The Varnish of Vietnam: The United States Still Hasn't Stopped Trying to Win Unwinnable Wars*, Foreign Policy, https://foreignpolicy.com/2014/10/15/the-varnish-of-vietnam/, 15 October 2014 (accessed 17 October 2014); D. Tierney, *Why Has America Stopped Winning Wars?*, The Atlantic, www.theatlantic.com/international/archive/2015/06/america-win-loss-iraq-afghanistan/394559, 2 June 2015 (accessed 8 June 2015).

drawn-out irregular war to the end and achieve victory.[37] The lack of political foresight and accountability in the use of military operations of choice as foreign policy is illustrated by the disaster in Libya after the 2011 NATO-led regime change that has brought severe instability to the country and region to the point where another debate on the necessity of military intervention began in response to the chaos caused by the first military intervention only five years afterwards (Reeve, 2016). Harsher criticism has come from long-term members of the security community, such as Anthony Cordesman, who states that successive political administrations have myopically focused on the tactical victories as a measure of effect in the broader operational or strategic picture. Cordesman refers to the process of "losing by winning", where tactical victories can end with major strategic losses that hurt US interests and security.[38] The case of Iraq from 2003 until 2011 also illustrates the succeeding to fail thread, where the lack of adequate preparation, a dismal understanding of the human terrain, and short-sighted political and military strategies combined to produce a disastrous political and military result (Godfroy & Collins, 2019). The seemingly endless cycle of wars has created a political environment that is becoming increasingly sensitive and adverse to the notion of Forever Wars.

One of the noted problems of Forever Wars is the argument that the US is not at war formally (along the definition posed in the beginning of this chapter), so war denialism adds to the problem of recognition.[39] This is compounded by the normalisation of war and the denial of empire as being complementary and creating an era of continuous war.[40] In the GWOT-era of warfare, successive presidential administrations across political party lines, leadership style and priorities have either expanded the number of on-going armed conflicts or at least continued the existing ones, where warfare has been conceived and executed from the presidential office (Ettinger, 2017). The unilateral practice of an 'energetic executive', where

[37] D. Maye, *We Know How to Strike, but Can We Achieve Victory? A Primer on the American Way of War in the 20th and 21st Centuries*, ISN, www.isn.ethz.ch/layout/set/print/content/view/full/24620?lng=en&id=190839, 25 May 2015 (accessed 30 May 2015); M. Thompson, *Why the US Military Isn't Winning*, Time, http://time.com/4058520/american-military-losing-wars/, 1 October 2015 (accessed 2 October 2015).

[38] A. Cordesman, *Losing by 'Winning': America's Wars in Afghanistan, Iraq and Syria*, Centre for Strategic and International Studies, www.csis.org/analysis/losing-winning-americas-wars-afghanistan-iraq-and-syria, 13 August 2018 (accessed 23 August 2018).

[39] D. Larison, *War Denialism and Endless War*, The American Conservative, www.theamericanconservative.com/larison/war-denialism-and-endless-war/, 16 December 2019 (accessed 26 December 2019).

[40] D. Larison, *We Can End Forever War*, The American Conservative, www.theamericanconservative.com/larison/we-can-end-the-forever-war/, 13 December 2019 (accessed 26 December 2019).

presidential decisions for going to war have proven to be neither wise nor necessary, has been shown by the examples of Libya and Syria (Edelson & Starr-Deelen, 2015). Forever Wars have become increasingly politically sensitive issues with an increasingly war weary public in the leading Western countries that are militarily committed to continuing the fight. Wars all have a beginning and are expected to have an end, for example George Bush's premature announcement in 2003 that the mission was accomplished during a photo opportunity on an aircraft carrier with the closing of the conventional and regular armed conflict phase in Iraq. In 2018, Trump tried a similar communication tactic when he declared "mission accomplished" in Syria as a pretext to try and withdraw US troops illegally stationed there. However, the pertinent question was posed, what exactly is the mission in Syria?[41] This sensitivity is seen in the lead-up to the November 2020 US Presidential elections when President Trump's view, whose political platform in 2016 included ending the Endless Wars, was reiterated again, and where the political mainstream on different sides of the political spectrum supported the politico-military status quo in foreign policy (or to increase military aggression).[42] As has been noted, Western engagement in warfare in the twenty-first century tends to be poorly considered, especially at the political level of decision making.

At lot is at stake; the US is in the process of being characterised as "the sick man of the 21st century",[43] a named coined for the Ottoman Empire on the eve of the First World War. Retired General Stanley McChrystal characterised the armed conflicts in Afghanistan and Iraq as neither understanding the problem nor the objectives when initiating the conflicts, and with little time to consider effectively the approach to military operations in Afghanistan. "Iraq was different, we had time to think about it. Iraq was a war of choice versus a war of reaction. And yet, interestingly enough, we didn't understand the problem there either."[44] This demonstrates a need to go beyond the superficial rhetoric of "good

[41] P. Baker, *'Mission Accomplished!' But What Is the Mission in Syria?*, The New York Times, www.nytimes.com/2018/04/14/us/politics/trump-syria-policy.html, 14 April 2018 (accessed 17 April 2018).

[42] W. J. Antle III, *Trump Troop Drawdown an Attempt to Keep the Campaign Promise to Finish 'Endless Wars'*, Washington Examiner, www.washingtonexaminer.com/news/trump-troop-drawdown-an-attempt-to-keep-campaign-promise-to-finish-endless-wars?utm, 10 September 2020 (accessed 10 September 2020).

[43] D. Klion, *The American Empire Is the Sick Man of the 21st Century*, Foreign Policy, https://foreignpolicy.com/2019/04/02/the-american-empire-is-the-sick-man-of-the21st-century/, 2 April 2019 (accessed 4 April 2019).

[44] M. Miklaucic, *An Interview with Stanley McChrystal*, Centre for Complex Operations, PRISM 6(3), http://cco.ndu.edu/PRISM-6-3/Article/1020271/an-interview-with-stanley-mcchrystal/, 7 December 2016 (accessed 27 January 2017).

intentions" and to make a much more thorough and scientifically based analysis of the pros and cons of military operations on a case-by-case basis in order to try and reduce the likelihood of creating more harm than good (Gleditsch, 2019). This is clearly seen in the evolution of the hasty war of choice on Libya by NATO in 2011 that turned the country into a failed state and increasingly as a zone of proxy war by various regional and international powers.[45] Historically and contemporarily, there is the aspect of the US leaving its soft networks and allies in a vulnerable position and at the mercy of the enemy. "Abandoning these networks endangers USA legitimacy in conflict zones, and risks the success of future operations" (Miska & Romano, 2019: 218).

In spite of being the recipient of the Nobel Peace Prize, Obama has one of the most controversial and corrosive legacies of warfare by choice; for example, supporting the Saudi's war of attrition in Yemen as a means of "reassuring" an ally of the US in spite of the catastrophic humanitarian consequences of the conflict. Another major critique of his administration is the normalisation of permanent war and the role of presidential war-making without the necessary legislative input and role in the process.[46] Similar criticism has also been aimed at the United Kingdom for 'blindly' following the US, but with even less tangible reason for doing so. The UK has been continuing to fight the post-9/11 wars without end, often covertly, with neither victory nor end in sight. These continuous wars being fought by the US and UK are not posing any existential threat, with an increasing lack of public and military personnel support for continuing these armed conflicts of choice. Where attempts to convert other countries and cultures to an image of the West has failed, the political establishment chooses to stay and fight lost wars in order to try and save face rather than to withdraw.[47] The current situation is at least in part to be blamed upon a messianic-like fervour in spreading idealist notions of political utopia that accelerated after the end of the Cold War and spawned the likes of Clash of Civilisations and the End of History.

[45] F. Wehrey, *Is Libya a Proxy War?* Monkey Cage, The Washington Post, www.washingtonpost.com/news/monkey-cage/wp/2014/10/24/is-libya-a-proxy-war/, 24 October 2014 (accessed 26 October 2014); *Arab Media Supports Egypt's Threat to Intervene Militarily in Libya: Turkey's Attempts to Become a Regional Superpower Must be Stopped*, MEMRI, Special Dispatch No. 8828, www.memri.org/reports/arab-media-supports-egypts-threat-intervene-militarily-libya-turkeys-attempts-become, 2 July 2020 (5 July 2020).

[46] D. Larison, *Reviewing Obama's Flawed Foreign Policy Record*, The American Conservative, www.theamericanconservative.com/larison/reviewing-obamas-flawed-foreign-policy-record/, 26 January 2017 (accessed 26 December 2019).

[47] S. Jenkins, *The US and Britain Face no Existential Threat. So Why Do Their Wars Go On?* The Guardian, www.theguardian.com/commentisfree/2019/nov/15/britain-and-us-wars-conflicts-middle-east, 15 November 2019 (accessed 26 November 2019).

This messianic mission to spread 'democracy' began as a theoretical academic exercise, but soon spread to policy maker circles owing to the expediency for 'legitimising' questionable policy. The brief overview of the logic of this utopian academic proposition is that democracy encourages peaceful interaction among states ("democratic peace"). Furthermore, the idea is that democratic states tend not to fight one another (Gleditsch, 1992). With the end of the Cold War, and by default the end of the bipolar global tensions, came the notion of war's obsolescence and the need to renounce it as an instrumental means of foreign policy (Westing, 1990). However, the twenty-first century has seen the political operational expediency of the use of regime change and military conquest as a means of spreading 'democracy' and 'peace' (Cederman & Gleditsch, 2004). This is seen in the narrative logic of the armed conflict that is on-going in Afghanistan.

Case Study: 2010 NATO Offensive in Southern Afghanistan

According to the lessons from the First World War and the involvement of the Committee for Public Information in getting the United States actively engaged in the conflict, information pertaining to an impending war should already be up and running before the first shots are fired in anger. This was certainly the case of a much-hailed anti-Taliban offensive by NATO forces in Southern Afghanistan (focused initially on the town of Marjah in southern Helmand province). In terms of operational security, a calculation can exist that announcing a military operation in advance can expose the troops to additional dangers that would not possibly exist under the conditions of operational security. However, examples have shown that by announcing a coming offensive, people can leave the immediate area of military operations and reduce civilian casualties in addition to the poor publicity that comes with noncombatant deaths in the media headlines. This can have the effect of increasing the latitude of military operations and weapons employed in the operation. This is a means of offensive military forces being in a position to shape the battlefield (physical and informational), such as affecting enemy morale through information operations and pressuring opponent propaganda with greater mass media attention that can potentially challenge their narrative.[48] There was a hint that the coming offensive

[48] C. Lee, *Warning Orders: Strategic Reasons for Publicising Military Offensives*, War on the Rocks, http://warontherocks.com/2016/10/warning-orders-strategic-reasons-for-publicizing-military-offensives/, 28 October 2016 (accessed 14 November 2016).

(Operation Moshtarak – from the Dari language and meaning *together*) had far more than purely military objectives in mind.

> Military operations usually are intended to catch the enemy off guard, but for weeks U.S. and allied officials have been telling reporters about their forthcoming assault on Marjah, a Taliban-held town of 80,000 and drug-trafficking hub in southern poppy-growing Helmand province.[49]

The news of the impending offensive was announced well in advance of the actual attack. From the point of view of the governments whose forces are engaged in Afghanistan, there seem to have been a number of different themes that addressed a number of different concerns of different countries and groups with vested interests in the conflict. I shall highlight four of the messages that were highlighted in and through the mass media: 1) combating the crime concerns of a narcotic trade system (including destination countries); 2) emphasizing the scale and possible results of the offensive to domestic publics to demonstrate progress in the long war; 3) the care taken to try and reduce the number of civilian casualties and the resulting negative publicity for NATO ISAF; and 4) PSYOPS against enemy combatants in the area of military operations. There are a variety of different stakeholders that are simultaneously being communicated to as it is problematic attempting to segment target audience groups owing to the level of interconnectedness in the information domain.

Each one of these points shall be taken in turn, owing to the fact that they are intended for different recipients. One of the intended recipients of the spin of Operation Moshtarak was the Afghan people. One of the sources of friction is the issue of civilian casualties, the exact number being difficult to accurately discern owing to vague and misleading systems of classification (Taliban "Suspects" for instance) and the ability to get to remote areas in order to count the number killed.[50] Numerous references and statements were issued by the military officials and spokespeople commenting on the planned offensive and the care taken to remove civilians from the danger zone. There was also a communication frame involving Afghan civilians in the process of restoring 'normality' to the country.

[49] *Why Are U.S.' Allies Telling Taliban About Coming Offensive?*, McClatchy, www.mcclatchydc.com/news/nationworld/world/article24572932.html, 5 February 2010 (accessed 16 May 2010).

[50] See Afghanistan Conflict Monitor, Human Security Report Project (administered by the School of International Studies at Simon Fraser University), www.eldis.org/organisation/A7764, accessed 16 May 2010.

> Minivans piled high with mattresses and clothing lined up at checkpoints Sunday as hundreds of civilians fled a Taliban-controlled area ahead of a planned NATO offensive in southern Afghanistan.[51]
>
> Facing stiff resistance from Taliban fighters, the Marines radioed for permission to call in an airstrike on the insurgents at midday Monday. It appeared to be the sort of clear opportunity that would have prompted a rapidly executed bombing run during the Iraq war, or even in the first seven years of this conflict.
>
> But not anymore: Officers at the Marine headquarters deemed the insurgents to be too close to a set of houses. In the new way the United States and its NATO allies are waging the Afghan war, dropping a bomb on or near a house is forbidden unless troops are in imminent danger of being overrun, or they can prove that no civilians are inside.[52]
>
> What's really important ... is that if there is a conversation before the operation between the Afghans and the maliks, or the village leaders, on the ground, and it is explained to them what will happen when the government asserts control and authority over those areas, we often find the Afghans don't fight - but they will welcome you. (Major General Nick Carter British Army)[53]

The messages can also be interpreted as communicating to different Afghan audiences that NATO ISAF did not intend to 'abandon' them and therefore to keep up the belief in the cause to continue the fight against the insurgency. However, the attempted projected image of civilians being safe from NATO's military operations evaporated when it came into contact with the realities of the way and means that modern war is prosecuted. News soon started to arrive of a mounting civilian toll. Concerns over the recent elections in Afghanistan also seemed to erode confidence that the government had the concerns of ordinary people at heart.

Since the beginning of the involvement of international troops in Afghanistan in 2001 until early 2010, some 1,624 international troops had been killed (including 984 American troops).[54] In that time there has been relatively little to show for all of the financial resources, aid and

[51] N. Khan & K. Gamel, *Afghanistan Civilians Flee Ahead of U.S. Military Offensive*, The San Diego Union-Tribune, www.sandiegouniontribune.com/sdut-hundreds-flee-south-afghan-town-ahead-of-offensive-2010feb07-story.html, 7 February 2010 (accessed 16 May 2010).

[52] R. Chandrasekaran, *U.S. Curtails Use of Airstrikes in Assault on Marja*, The Washington Post, www.washingtonpost.com/wp-dyn/content/article/2010/02/15/AR2010021500774.html?nav=rss_email/components, 16 February 2010 (accessed 16 May 2010).

[53] J. Borger & R. Norton-Taylor, *British and US Troops to Launch New Afghan Offensive*, The Guardian, www.guardian.co.uk/world/2010/jan/25/new-afghanistan-helmand-offensive, 25 January 2010 (accessed 16 May 2010).

[54] *Why Are U.S., Allies Telling Taliban about Coming Offensive?*, McClatchy, www.mcclatchydc.com/news/nation-world/world/article24572932.html, 5 February 2010 (accessed 16 May 2010).

troops committed to Afghanistan. The rate of casualties is on the increase too, which means that public patience and tolerance of those nations contributing troops to the conflict, will be tested. If it is eroded, there will be significant pressure mounted against those governments to disengage from the conflict there. The military superiority of NATO forces over their Taliban opponents was played up, plus the humanitarian aspects considered by NATO commanders.

> In the coming days, you will see a demonstration of our capability in a series of operations, led by the Afghans and supported by NATO, in southern Helmand. (NATO Secretary General Anders Fogh Rasmussen)

> If they want to fight, then obviously that will have to be an outcome. But if they don't want to fight, that's fine, too [...] We'd much rather have them see the inevitability that things are changing and just accept that. And we think we can give them that opportunity. And that's why it is a little unconventional to do it this way. (General Stanley McChrystal)[55]

> U.S.-led NATO troops launched an offensive on Saturday against the Taliban's last big stronghold in Afghanistan's Helmand province, a test of President Barack Obama's troop surge strategy.[56]

These statements were neither counter-balanced nor challenged. They give the impression of positive military progress being made in Afghanistan by NATO forces. There are no statements that begin to call into question the veracity of the statements in spite of years of setbacks or at least no progress before Operation Moshtarak. A clear attempt is being made to frame this as a gradual "indigenisation" of the hard security burden in Afghanistan in order to imply the logic that Western troops will be less exposed to risks and perils in combat operations in order to soothe domestic public opinion in the core Western countries that committed combat troops.

Much has been made of the goal of squashing drug operations in the Helmand province, which has been a source of tension between NATO and Russia. A number of accusations have been made by Russia, stating that NATO has not been taking the drug production and distribution problem seriously, and that Russia has been paying the price.[57] A number

55 C. Whitlock, *NATO Ministers, Commanders Advertise Planned Offensive in Southern Afghanistan*, The Washington Post, www.washingtonpost.com/wp-dyn/content/article/2010/02/05/AR2010020502554.html, 6 February 2010 (accessed 16 May 2010).

56 M. Georgy, *NATO Launches Major Afghanistan Offensive*, Reuters, www.reuters.com/article/idUSTRE61B1ZJ20100212, 12 February 2010 (accessed 16 May 2010).

57 A. Fedyashin, *Russia and NATO Divided Over Afghan Opium*, RIA Novosti, http://en.rian.ru/analysis/20100325/158312107.html, 25 March 2010 (accessed 16 May 2010).

of statements were made that implied the importance of disrupting the drug trade in the offensive.

> The offensive in Marjah – a farming community and major opium-production centre with a population of 80,000 – will be the first since President Barack Obama announced he was sending 30,000 additional troops.[58]

These statements were from NATO officials, which were taken and published without question (of the track record to date for example). They emphasize the scale of the problem and the alleged resolve in correcting it, rhetorically at least. The reasons for publicising the military offensive match the logic given by Lee in the *War on the Rocks* article in the opening of this section. In part, this is a public affairs communication, a persuasion and influencing exercise with different audiences in Afghanistan and internationally. It is also evidence of the element that involves the intention of affecting the decision-making calculus of the insurgent forces in the area; to win without firing a shot by an implied forthcoming of a tactical level of "shock and awe" combat tactics. It certainly underlines and stresses the increasing influence of political priorities further into the Western approach to waging warfare in the twenty-first century. This is done even down to the tactical level of micromanagement to try and demonstrate the illusion of potential minor tactical victories as a measure for success at the operational and strategic levels of the current Endless Wars. The likely result is the intention to avoid any of the needed and necessary innovation or changes to be made to why and how the West fights, but fails to win, the wars and warfare of choice.

Conclusion

Information and communication are connected links in the operationalisation of the informational domain aspects of parallel kinetic warfare as they can influence and shape the cognitive domain of the various actual and potential stakeholders (friendly forces, enemy forces, neutral actors, civilians – local, regional and global, regional and international institutions, etc.). Furthermore, there are links in the chains of information and communication, where different actors fulfil different functions and goals. At the beginning of this chain are the scholars and academics who seek to create the underlying theories and concepts (to advance the

[58] N. Khan & K. Gamel, *Afghanistan Civilians Flee Ahead of U.S. Military Offensive*, The San Diego Union-Tribune, www.sandiegouniontribune.com/sdut-hundreds-flee-south-afghan-town-ahead-of-offensive-2010feb07-story.html, 7 February 2010 (accessed 16 May 2010).

science and their own peer prominence); the policymakers selectively use these conceptual or theoretical foundations to advance a specific foreign or defence policy agenda (to advance an often ideological agenda) and the practitioners seek to operationalise the same conceptual and theoretical base in order to execute and meet the goals of the political agenda within the framework of the allocated resources and other constraints (time, money, manpower, rules of engagement, etc.) imposed upon them by the policymakers. These roles and goals are sometimes not in harmony, which reduces the effectiveness of the information operations and influence activity.

This situation of interaction and incompatibility between the different constitutive parts of the system, from conceptual birth to pragmatic implementation, has in part contributed to the current era of Endless Wars and Dirty Wars that involve non-conventional and irregular conflicts around the globe in various geographic geopolitical hot points concentrated in the Middle East and North Africa, Central Asia and South Asia. This is a trend that comes at a point in time when the geopolitical balance is shifting from a Western-centric and US-led unipolar global order towards a more non-Western and multipolar global order. It is a globally transformative shift that creates another set of related political and security problems for the US-led West as the relatively declining hegemonic power that seeks to retain its position with a narrowing set of options that focus on economically cheap and low-casualty risk strategies, that in effect erode its tangible and intangible assets and work against its objective security and interests.

At a point in time when the price (in blood and treasure) of maintaining empire (US global hegemony) is likely to increase, the public appetite and hence political ability is decreasing for Endless Wars of the twenty-first century. This has created an unviable and unsustainable political strategy (hence military strategy by default) of waging warfare that is least costly in terms of blood and treasure in the short term to manage long-term and enduring warfare. The fact that there are now openly observable signs of relative decline in the US and West's economic and military power creates an international relations and affairs environment that encourages challenges to this hegemony. Mostly to date, this has been via indirect and covert means, but is moving more openly as demonstrated in conflicts as Syria and Iraq, where there is a gradual move from proxy and surrogate warfare to something that is beginning to involve the risk and actual occurrence of clashes and hostilities between the US and other state actors, such as Russia and Iran. These clashes increase the risk for conventional and regular warfare that tends to be more destructive and costly at a time when the West can least afford it.

The level of functioning of the logic driving the irregular warfare and risking an outbreak of regular warfare creates a worrying situation, given that the more emotional tone and intangible nature (warfare for the sake of vanity, to demonstrate political 'leadership' or for notions of idealism) is masked by the quality and quantity of information and how it is communicated. Mass media has in effect, as was stated by Payne (2005), become an instrument of war by supporting and enabling warfare to occur by not critically examining the false logics given by governments. It is something that has been repeated on numerous occasions, whether Iraq in 2003 or Libya in 2011 and any number of other instances. The basic state of play is that information and communication are more often used to emotionally mobilise publics for war rather than work towards building peace.

An obvious problematic fact that needs to be considered is the trend of the growing influence of politics on the outcome of wars and armed conflicts, where the outcome of individual battles is decided by military factors. This is simultaneously a problem of politics and information, which then creates a burden for military planners and military operations, one that requires a critical rethinking of the current orthodoxy of why and how the West goes to war. The current creeping crisis is masked (or ignored) by attempts to project various tactical and minor individual battlefield 'wins' as a vindication of the West's (and US's) tangible hard power strength and continued façade of global hegemony. It is difficult to judge and scrutinise accurately the scale of US-led warfare around the globe, as it is often indirect and covert in nature. A lack of adequate mass media questioning of governments' words and acts compounds the problem as it lacks transparency and therefore accountability. Governments are increasingly using information and communication in a selective and manipulative manner, in pursuit of the policy of wars as a choice that runs against the objective interests and security of the West and the regions where these armed conflicts are fought.

References

Arquilla, J. & Borer, D. A. (Eds.) (2007), *Information Strategy and Warfare: A Guide to Theory and Practice*, New York: Routledge.

Bălășoiu, N. (2016), *Revenirea la Agenda de putere. Realități sumbre developate de cursa înarmării* (Return to the Power Agenda. Grim Realities of the Arms Race), Bucharest: Editura Institutului de Științe Politice și Relații Internaționale Ion I. C. Brătianu al Academiei Române.

Betz, D. (2015), *Carnage & Connectivity: Landmarks in the Decline of Conventional Military Power*, New York: Oxford University Press.

Biberman, Y. & Genish, O. (27 September 2015), The Problem With Proxies: Ideology is no Substitute for Operational Control, *Small Wars Journal*, accessed at: http://smallwarsjournal.com/print/30939.

Cederman, L.-E. & Gleditsch, K. S. (2004), Conquest and Regime Change: An Evolutionary Model of the Spread of Democracy and Peace, *International Study Quarterly*, 48(3), pp. 603–629.

Clarke, M. (2015), The Ending of Wars and the Ending of Eras, *The RUSI Journal*, 160(4), pp. 4–9.

Cragin, R. K. (2015), Semi-Proxy Wars and US Counterterrorism Strategy, *Studies in Conflict & Terrorism*, 38(5), pp. 311–327.

Department of the Army (2004), *U.S. Army Counter-guerrilla Operations Handbook*, Guilford: The Lyons Press.

Department of Defence (2007), *U.S. Army Counterinsurgency Handbook*, New York: Skyhorse Publishing.

Durward, R. & Marsden, L. (Eds.) (2009), *Religion, Conflict and Military Intervention*. Farnham: Ashgate.

Edelson, C. & Starr-Deelen, D. G. (2015), Libya, Syria, ISIS, and the Case Against the Energetic Executive, *Presidential Studies Quarterly*, 45(3), pp. 581–601.

Edwards, L. M. (2014), How the 'Entry' Defines the 'Exit': Contradictions Between the Political and Military Strategies Adopted in 2001 and How They Have Deleteriously Affected the Longer-Term Possibilities for Stabilisation in Afghanistan, *Conflict, Security & Development*, 14(5), pp. 593–619.

Ettinger, A. (2017), The Patterns and Implications of American Military Interventions in the Post-September 11 Era, *Canadian Foreign Policy Journal*, 23(3), pp. 205–217.

Forbes, I. & Hoffman, M. (Eds.) (1993), *Political Theory, International Relations and the Ethics of Intervention*, New York: St. Martin's Press.

Fridman, O., Kabernik, V. & Pearce, J. C. (Eds.) (2019), *Hybrid Conflicts and Information Warfare: New Labels, Old Politics*, Boulder: Lynne Rienner.

Galula, D. (2006), *Counterinsurgency Warfare: Theory and Practice*, Westport: Praeger Security International.

Gleditsch, K. S. (2019), An Ever More Violent World? *Political Studies Review*, 17 (2), pp. 99–114.

Gleditsch, N. P. (1992), Democracy and Peace, *Journal of Peace Research*, 29(4), pp. 369–376.

Godfroy, J. & Collins, L. (2019), Iraq, 2003–2011: Succeeding to Fail, *Small Wars & Insurgencies*, 30(1), pp. 140–175.

Greene, R. (2006), *The 33 Strategies of War*, New York: Penguin Books.

Guido, LTC, J. (2019), The American Way of War in Africa: The Case of Niger, *Small Wars & Insurgencies*, 30(1), pp. 176–199.

Hughes, G. (2020), War in the Grey Zone: Historical Reflections and Contemporary Implications, *Survival: Global Politics and Strategy*, 62(3), pp. 131–158.

Hughes, G. A. (2014), Syria and the Perils of Proxy Warfare, *Small Wars & Insurgencies*, 25(3), pp. 522–538.

Inspector General US Department of Defence (13 February 2020), *Audit of the DoD's Accountability of Counter-Islamic State of Iraq and Syria Train and Equip Fund Equipment Designated for Syria*, Report No. DODIG-2020-061, Washington DC: Department of Defence.

Kaldor, M. (2006), *New and Old Wars: Organised Violence in a Global Era*, Second Edition, Cambridge: Polity Press.

Kassim, Y. R. (2014), *The Geopolitics of Intervention: Asia and the Responsibility to Protect*, Singapore: Springer.

Kassimeris, G. & Buckley, J. (Eds.) (2010), *The Ashgate Research Companion to Modern Warfare*, Farnham: Ashgate.

Kaufman, S. J. (2001), *Modern Hatreds: The Symbolic Politics of Ethnic Wars*, Ithaca: Cornell University Press.

Kellner, D. (2004), Media Propaganda and Spectacle in the War on Iraq: A Critique of US Broadcasting Networks, *Cultural Studies – Critical Methodologies*, 4(3), pp. 329–338.

Krieg, A. & Rickli, J.-M. (2018), Surrogate Warfare: The Art of War in the 21st Century?, *Defence Studies*, 18(2), pp. 113–130.

Kumar, D. (2006), Media, War, and Propaganda: Strategies of Information Management During the 2003 Iraq War, *Communication and Critical/ Cultural Studies*, 3(1), pp. 48–69.

Linebarger, P. M. A. (2009), *Psychological Warfare*, Nevada City: Gateways Books and Tapes.

Loveman, C. (2002), Assessing the Phenomenon of Proxy Intervention, *Conflict, Security & Development*, 2(03), pp. 29–48.

Luttwak, E. N. (2001), *Strategy: The Logic of War and Peace*, Revised and Enlarged Edition, Cambridge: The Belknap Press of Harvard University Press.

Manoilo, A. V. & Strigunov, K. S. (2020), *Technologies of Non-Classical War: Genesis, Evolution, Practice*, Moscow: Hotline – Telekom.

Marks, T. A. & Bateman, K. J. (2019), Perspectives on the American Way of War: The US Experience in Irregular Conflict, *Small Wars & Insurgencies*, 30 (1), pp. 1–13.

Marshall, A. (2016), From Civil War to Proxy War: Past History and Current Dilemmas, *Small Wars & Insurgencies*, 27(2), pp. 183–195.

Miska, S. & Romano, S. (2019), Too Little, Too Late: Protecting American Soft Networks in COIN/CT, *Small Wars & Insurgencies*, 30(1), pp. 200–222.

Moaz, Z. & Gat, A. (Eds.) (2001), *War in a Changing World*, Ann Arbor: University of Michigan Press.

Mumford, A. (2013), Proxy Warfare and the Future of Conflict, *The RUSI Journal*, 158(2), pp. 40–46.

Nagl, J. A. (2005), *Learning to Eat Soup with a Knife: Counterinsurgency Lessons From Malaya and Vietnam*, Chicago: University of Chicago Press.

Narula, S. (2004), Psychological Operations (PSYOPS): A Conceptual Overview, *Strategic Analysis*, 28(1), pp. 177–192.

Oakley, D. P. (2019), Organising for the 'Grey Zone' Fight: Early Cold War Realities and the CIA's Directorate of Operations, *Small Wars & Insurgencies*, 30(1), pp. 62–80.

Obama, B. (14 March 2016), *Developing an Integrated Global Engagement Centre to Support Government-wide Counter-Terrorism Communications Activities Directed Abroad and Revoking Executive Order 13584*, Executive Order, Office of the Press Secretary, The White House.

Paterson, T. & Hanley, L. (2020), Political Warfare in the Digital Age: Cyber Subversion, Information Operations and 'Deep Fakes', *Australian Journal of International Affairs*, 74(4), pp. 439–454.

Paul, C. (2017), *Assessing and Evaluating Department of Defence Efforts to Inform, Influence and Persuade: Worked Example*, Santa Monica: RAND Corporation.

Paul, C., Clarke, C. P., Schwille, M., Hlávka, J. P., Brown, M. A., Davenport, S. S., Porche III, I. R. & Harding J. (2018), *Lessons from Others for Future US Army Operations in and through the Information Environment*, Santa Monica: RAND Corporation.

Payne, K. (Spring 2005), The Media as an Instrument of War, *Parameters*, pp. 81–93.

Perot, E. (2019), The Blurring of War and Peace, *Survival: Global Politics and Strategy*, 61(2), pp. 101–110.

Phillips, C. & Valbjörn, M. (2018), 'What Is in a Name?': The Role of (Different) Identities in the Multiple Proxy Wars in Syria, *Small Wars & Insurgencies*, 29 (3), pp. 414–433.

Phillips, R. (2019), Counterinsurgency in Vietnam – Schizophrenia Until Too Late, *Small Wars & Insurgencies*, 30(1), pp. 81–100.

Pillar, P., Bacevich, A., Sheline, A. & Parsi, T. (July 2020), *A New US Paradigm for the Middle East: Ending America's Misguided Policy of Domination*, Quincy Paper No. 2, Quincy Institute for Responsible Statecraft.

Poole, H. J. (2004), *Tactics of the Crescent Moon: Militant Muslim Combat Methods*, Emerald Isle: Posterity Press.

Powers, S. & O'Loughlin, B. (2015), The Syrian Data Glut: Rethinking the Role of Information in Conflict, *Media, War & Conflict*, 8(2), pp. 172–180.

Reeve, R. (February 2016), *Intervention in Libya: Why Here? Why Now? What Next?*, ORG Briefing, Oxford Research Group.

Robinson, P. (2017), Learning from the Chilcot Report: Propaganda, Deception and the 'War on Terror', *International Journal of Contemporary Iraqi Studies*, 11(1&2), pp. 47–73.

Rochlin, J. (2020), Columbia and the Transformation of Warfare: The End of Ideology?, *Global Change, Peace & Security*, 32(1), pp. 9–101.

Ross, R. T. (1998), *Truth and Deceit: Media Relations and Military Deception*, Newport: Naval War College.

Ryan, M. & Switzer, L. (2009), Propaganda and the Subversion of Objectivity: Media Coverage of the War on Terrorism in Iraq, *Critical Studies on Terrorism*, 2(1), pp. 45–64.

Samsap, J., Sung, W. S. & Jaemin, J. (2008), Propaganda or Public Relations Campaign? International Communication on the War Against Iraq, *Public Relations Review*, 34(1), pp. 63–65.

Sauliuc, A., Calopăreanu, G. & Bălășoiu, N. (2014), *Security Interests of Romania in the Wider Black Sea Area*, Cambridge: International Workshop National Strategies Observer.

Scahill, J. (2013), *Dirty Wars: The World Is a Battlefield*, London: Serpent's Tail.

Simons, G. (2015), Russian Media and Censorship: A Means or an End?, *Russian Journal of Communication*, 7(3), pp. 300–312.

Snow, N. (2003), *Information War: American Propaganda, Free Speech and Opinion Control since 9-11*, New York: Seven Stories Press.

Taylor, P. M. (2003), *Munitions of the Mind: A History of Propaganda from the Ancient World to the Present Day*, Third Edition, Manchester: Manchester University Press.

The US Army/Marine Corps (2007), *Counterinsurgency Field Manual*, Chicago: University of Chicago Press.

Ucko, D. H. (2019), Systems Failure: The US Way of Irregular Warfare, *Small Wars & Insurgencies*, 30(1), pp. 223–254.

Ucko, D. H. & Marks, T. A. (2018), Violence in Context: Mapping the Strategies and Operational Art of Irregular Warfare, *Contemporary Security Policy*, 39 (2), pp. 206–233.

USAF College of Aerospace Doctrine, Research and Education (1997), Three Levels of War, *Air and Space Power Mentoring Guide*, 1, Maxwell AFB: Air University Press.

Votel, J. L., Cleveland, C. T., Connett, C. T. & Irwin, W. (2016), Unconventional Warfare in the Grey Zone, *JFQ*, 80, First Quarter, pp. 101–109.

Waldman, T. (2019), Strategic Narratives and US Surrogate Warfare, *Survival: Global Politics and Strategy*, 61(1), pp. 161–178.

(2017), Vicarious Warfare: The Counterproductive Consequences of Modern American Military Practice, *Contemporary Security Policy*, 39(2), pp. 181–205.

Wall, T. (2010), US Psychological Warfare and Civilian Targeting, *Peace Review: A Journal of Social Justice*, 22(3), pp. 288–294.

Wang, V.-C. & Stamper, G. (2002), Asymmetric War? Implications for China's Information War Strategies, *American Asian Review*, XX(4), pp. 167–207.

Westing, A. H. (1990), Towards Eliminating War as an Instrument of Foreign Policy, *Bulletin of Peace Proposals*, 21(1), pp. 29–35.

Xie, S. & Boyd-Barrett, O. (2015), External-National TV News Networks' Way to America: Is the United States Losing the Global 'Information War'?, *International Journal of Communication*, 9, pp. 66–83.

Zollmann, F. (2017), *Media, Propaganda and the Politics of Intervention*, New York: Peter Lang.

7 The Effects of Technological Evolution and Social Media on the Individual, Society and Politics

Iulian Chifu

Introduction

The technological evolution, especially that of social media, has come with tremendous advantages for citizens: Internet connectivity allowing access to infinite amounts of information at practically no cost, possibilities of connecting and finding people we never knew but with whom we share common problems and common aspirations or beliefs. At the same time, even though we begin to use technology and social media extensively from a very early age, nobody has undertaken the much-needed social and psychological studies linked with the consequences of this tremendous evolution (Chifu & Savu, 2020).

However, this evolution comes with a lot of added value for our day-to-day lives, but also has its shortcomings; at the very least it can influence our preferences, behaviour and choices. It has made the world look differently in only a matter of a few years, and that has created difficulties for our institutions, leadership and politicians to adapt to. It promotes charismatic leaders and those with extreme ideas who are made visible and popular, dismissing the more serious experts and political figures who were much less appealing to the mass following. It has shifted the public space and changed what has been common knowledge, reality and fact, into a plethora of divided public spaces, bubbles where debate is excluded and only echo chambers exist that multiply the stringent ideas of the administrators.

This book addresses contemporary research into these fields and is trying to present the impact and changes (both positive and negative) of technology and social media in a pragmatic way, particularly on the human being, behaviour and psyche, how social psychology and sociology are changing in the current era, as society and political life change as well. We explore this in four chapters which deal with: the new generation of threats – the effects of technological evolution; technology and its impact on contemporary societies; social media and its impact on contemporary societies; and characteristics of society hypothesized and amplified by social media.

Thorough research has been conducted by the Conflict Prevention and Early Warning Centre, which has tried to cope with the ethics of the acceptance of changes and improvements in day-to-day life and avoid seeing only the negative effects of change, even though the impacts identified are huge and shake up the known rules of sociology and human life.

The New Generation of Threats to National Security

The threats and risks to national security have evolved in ways that are putting question marks against the very change in the nature of the new generation of threats. Technology, social media and their impact on society, as well as new instruments of a hybrid nature, which provoke huge rifts in our liberal democratic societies, are creating a new environment and new categories of threats that have never previously existed. Moreover, these types of changes have a tremendous impact on the resilience of democratic societies, on public support for spending in the security field and on the perception of threats, including those from conventional sources – military operations, espionage and subversion.

There is a new generation of threats to national security; internal societal and political security are at stake. Security as a concept has evolved, as well as the substance of the threats to national security. At the beginning, everything was about conventional military threats. Then we embraced the classical definition of the European Copenhagen School[1] (Romaniuk, 2018), with its five dimensions of security – military, political, social, economic, environmental (Buzan, 2012) – that became NATO's approach to security with the implementation of the Strategic Concept in Rome (NATO Strategic Concept Rome, 1991). Buzan added the need to consider three objects to be "securitized" – state, society and individual (Chifu, 2013; Buzan, 2014).

Beginning in 2010, concerns at the NATO level evolved from "emerging security challenges" to "non-traditional threats", including terrorism, the proliferation of weapons of mass destruction, cyber-attacks, disruption of energy supplies, and even extended to climate change and migration (ISIS Europe Briefing Note, 2010). That approach

[1] The Copenhagen School of security studies is an academic school that employs a critical approach to security studies. It is part of the post-positivist movement in the field of international relations (IR), which became a salient part of post-Cold War scholarship. See Romaniuk (2018).

led to the new "Emerging Security Challenges Division" (ISIS Europe Briefing Note, 2010). Then hybrid threats emerged and joined the evolution of complexity in security matters. The EU and NATO formally established the European Centre of Excellence for Countering Hybrid Threats, in Helsinki in April 2017 (NATO, 2017).

The new generation of threats to national security has a very specific format; this is why we tend to assume that we can talk about a change in the very nature of the threats to national security. The unconventional threats have already been studied (Chifu, 2020a), and are a hybrid typology of threats (Chifu, 2018a: 23–30), developed on internal vulnerabilities that become threats. It is the case that all of the characteristics of a liberal democracy, its values and principles, are considered to be vulnerabilities by some players (specifically the Russian Federation) (Chifu & Țuțuianu, 2017; Simons & Chifu, 2017); instruments are constructed to take advantage of those characteristics, becoming real threats to our societies from external sources (Chifu, 2013) arising from the assessment of the values and principles of our democratic systems (Muller, 2017; Richards, 2017).

Another source of those unconventional threats is generated by the assessment of the internal vulnerabilities of our liberal democratic societies (Kagan, 2018) arising from the evolution of technology and its impact on societies (Chifu, 2019a: 11–23). Social media and informational warfare, drones and hypersonic weapons are shaping the security environment of the future, having an impact on the approach and the way of thinking about security-related matters. And specifically, the most profound change comes from the impact of technology on democracy itself (Chifu, 2019a: 11–23).

Technology has a tremendous impact on democracy and society (Gavriluță, 2018) through direct influence on the criteria and behaviour related to the societal cohesion – solidarity, the sense of community, alienation and fragmentation (Appadurai et al., 2016; Rice & Zegart, 2018; Kirchick, 2018). On another point, the sense of lack of privacy, altered identity (individual and collective) and the need for dignity and respect (Fukuyama, 2018), are all new effects of technology on our daily lives that have an impact on our societies and participation in democratic processes, as well as on the life of the community, and on the very substance of our democracy (Gurri, 2018).

These specific processes have been underlined by a number of studies, and for others study is only just beginning. In some cases, the psychological processes linked to our reaction as humans to the speed of change (Rosling, 2018) are working against us. In those cases, more in-depth study is required and we have to consider that those evolutions themselves are creating vulnerabilities to our society (Ishinger, 2019) that we need to cope with. The bases of these reactions should be building resilience in our societies. However, this is not so easy, since resilience building needs the involvement of the state, society and individuals alike, and that cohesion can only be forged if there is a reasonable level of trust in the authorities (Chifu, 2018a: 23–30; Chifu, 2020a).

Social Media and Its Impact on Democratic Societies

We have all witnessed the evolution and use of social media in shaping society. From the 'Twitter Revolution' in the Republic of Moldova in April 2009, to the Arab Spring, the capacity for mobilization proved enormous, when the stakes and impact cross a certain threshold of emotional impulse. The use of social media in informational warfare has been extensively studied in this area (Chifu & Ivan, 2013: 221–239; Chifu, 2015: 200–209; Chifu, 2016a: 132–133; Chifu, 2016b; Chifu & Nantoi, 2016a; Chifu & Nantoi, 2016b; Chifu, 2017: 20–28; Simons & Chifu, 2017; Chifu, 2018b).

At least as important as the informational warfare is the impact of social media as an instrument that affects political and societal security. The extensive use of social media has been proven to determine important changes in individual human behaviour, on societal development and the evolution of community spirit, as well as its impact on liberal democracy as a political system.

Once again, it is not the instrument which is bad; on the contrary, it is of added value to democracy, encompassing a free flow of ideas, empowerment of people and having opinions heard. But the way this instrument is used can create side effects to a society that we, at least, must know and research in depth. The Internet itself has created a side effects by transforming those people who access the virtual reality into a far more contemplative and passive group, reacting with a click while sitting in front of a computer rather than taking to the streets or protesting and claiming their rights in the physical reality.

Once social media takes advantage of its ability to reach targeted audiences, people placed in determined geographic locations and diverse individuals from all over the world with specific political or social preferences, opinions or religious beliefs, then it constitutes an instrument that has a significant influence on changes and individual preferences in society. Moreover, it is a tool that lets everybody reach anyone across borders and barriers of any type.

However, consequences flow on different levels: societal; psychological; political impact on democracy; or political operationalization of social media. There are several ways that social media is influencing democratic societies. It has an impact on societal security through its abilities and side effects in the general fragmentation of society, in affecting participation in numerous events, in its capacity to mobilize, in modifying the human and social behaviour in several ways. With a dual effect, it also affects different types of people – active or passive, contemplative or those inclined to action.

It then has an enormous psychological impact on individuals and, as a consequence, on the society through the impactive spread of collective emotions, addressing pragmatism, accentuating emotions and senses and, at the same time, depriving some sensibilities and humanity of some events and factual realities. It also creates a deep separation between perceptions and realities, with bubbles promoting uncritical opinions and isolation from other ideas and the real debate. It creates very friendly environments for each of these opinions, but that approach creates stronger and unchallenged perceptions of truth and reality, not confronting other arguments, never verifying those perceptions with the reality and truth around us.

Finally, it is about the important impact of social media on political security, favouring strong statements and giving equal say to all participants in a conversation, despite different credentials and legitimacy. Thus, anybody can draw attention to and win a debate by popular vote as opposed to listening to experts or specialists, rejecting arguments less adapted to communication at a more general level of education in the real world.

It favours ideas that are attracting traffic that privilege nationalist and extreme statements, anti-system believers, and popular thoughts and opinions that are spread via conspiracy theories (Rice & Zegart, 2018). Social media is promoting, at high speed, extreme and shocking ideas, even those without a shred of truth or reality, sensations, emotions and personal opinions, instead of verified information, expert analysis, and traditional journalism with its customary rules of fact-checking.

Therefore, social media favours populism, nationalism, extremism of any kind, including extreme progressive ideas; it puts a huge emphasis on emotions and perceptions instead of factual realities. Identities – both classical and new – are also at the forefront of the impact of social media, linked both to the fragmentation of society, but first and foremost to the emergence of new types of political ideas, interest groups and the fight for new types of rights.

All these clash with more traditional political ideas and change majority support from mainstream politics into the promotion of extreme ideas and political parties, or populist figures; this replaces strict, conservative, decent politicians who have practically no chance of being heard. The personalization of politics is also a side effect of the impact of social media.

Social Media and the Impact on Societal Security

Social media has the side effect of the fragmentation of society. These types of isolationist bubbles are creating strong new types of identity, but are also isolated from the real debates of ideas or confrontation with other people's arguments (Appadurai et al., 2016). Fragmenting society creates a real threat to solidarity, generosity, communitarian spirit, participation in common goals of local communities, and causes rifts in society as a whole. It creates problems of cohesion on common national policies and strategic options.

As we have seen before, social media has an awkward and biased impact on participation, keeping people away from reality and protest through likes and dislikes in a contemplative virtual space; but on the other hand, it is able to mobilize huge numbers of strangers, once emotions become involved and the level of excitement reaches and touches an important number of followers. Both attributes can harm or mislead concrete ideas and initiatives for protest or the defence of rights; alternatively, they can mobilize on the basis of very awkward and non-representative ideas against a mainstream agenda with their well-organized supporters and leaders of minority groups, in order to take over the public space through marginal and controversial ideas that are tainted and presented as the important aims of a society.

Social networks operate in a "bubble" of our own opinions and beliefs like an echo chamber, a bubble that filters what we read, limiting our vision of the world. 'Groupthink' is also a side effect (Stern, 2003). The impact is the creation of strong identities, fragmentation of the social spectrum and lack of dialogue and debate, which is an integral part of the formation of opinions and development of a democratic society. Instead, we just have individuals with strong opinions on different subjects without hearing the arguments of others. This is a threat to societal security.

On another point, social media provides short, obvious, harsh and strict answers to any type of problem. In this environment, we do not have time for nuances, elaborated answers or complex evaluations, with open questions and unsolved dilemmas. Social media is simplifying reality to the level of caricaturing it, because we find it hard to understand it in its complexity. The effect on a society is the same clear and strict perception, the lack of questions on a subject and the split of the society on each question. Cohesion is hard to achieve, and any decision is criticized or supported on an emotional and not on a rational basis. Accepting the other's opinion and weighing its arguments, are the bases of liberal democracy, which disappears on such platforms.

Social media also has an impact on the behaviour of individuals linked to socialization and a humanitarian approach, as well as on community involvement and participation. It creates new types of behaviour in different individuals, fuelling fears and alienation (Simons & Chifu, 2017), or, on the contrary, forcing pragmatism and efficiency without compassion or taking advantage of emotions; such feelings are then amplified to become passionate approaches to themes promoted in a controversial manner at a national level (nuclear energy, presence of foreign troops on the national territory etc.) (Chifu & Ivan, 2013: 221–239).

However, the most important part is that linked to post-truth, perceptions and realities and the psychological impact of the social media instrument, used by specialists with bad intentions that could harm individual conscience, will or social behaviour of people (Gurri, 2018). Here the society is directly targeted by means that use social media extensively.

Democratization brought about by the technical build-up of social media has led to an explosion with enormous consequences by promoting a liberal ideal that every truth, every idea has equal value. Therefore, the Internet and social media have made each sentence and every opinion equal. We have come to an absolute democratization of the Truth and those ideas are confronted in a competition for an audience in the public space where popularity is the driving factor for the spread of the idea.

Any group or theme is an image of the people using social media and addressing a particular subject. It reflects their traits and their knowledge when they become public in such a unique media environment, if we consider social media as being also a source of news and a mass media space. It reveals level of education, habits, traditions and characteristic features (and also of society itself in terms of relationships, community spirit and its natural hierarchy). But it also alters these same characteristics via the approaches that it establishes and the multiplication of some models and leading ideas. But this situation occurs once the only criterion is the idea of attraction is established, a dramatic change in the societal hierarchies.

Social media and political security can be a detrimental mix when sentiments affect the cognitive realm of the audience, such as the spread of populism, nationalism, identity and progressive ideas. Social media introduces some tremendous changes when anybody is offered free access to infinite information, at practically no cost. And those changes are clearly in favour of a democracy. The low levels of self-censored people that do not participate in elections due to a lack of understanding of the impact and relevance of the vote, of the low civic spirit and to the lack of information, now take part in polls all over the world (Chifu, 2019a: 11–23). This has an impact on the political capacity of the society.

The second phenomenon is the creation of numerous groups of like-minded persons, with very concrete and difficult problems, knowledgeable and with nuances about their own troubles, who find each other and get together via social media. This creates problems for any administration or party that wants to solve concrete problems. This is because we are talking about tens of thousands of such concrete problems (Chifu & Ivan, 2013: 221–239) and there is no administrative capacity to analyse and respond to all of them, or to have a solution for each one.

On another note, we have studies (Maldonado, 2017) that present the "increasing sentimentalization" of democracies, through social media and their dominant presence in the public sphere. This supports and develops populism, since the sentimentalization is facilitated by social media and expresses itself in populism: "digitalization shows an important expressive and formative dimension and paves the way for a populist way of communication" (Maldonado, 2017). Populism has been described as an ideology that shapes an opposition between people, and the elite and populists claim that they are the only true representatives of the people, making use of both verbal and non-verbal strategies of representation (Maldonado, 2017).

Social media supports the development of populism, on different levels. It is not only about arousing emotions and harsh speech in order to attract numerous followers via social media (Muller, 2017; Fukuyama, 2018; Ishinger, 2019). It is also about challenging existing elites and promoting "normal people" at the forefront of the public and political institutions, giving satisfaction to large numbers of people, without any genuine ambition of ever taking public office. Therefore, traditional parties cannot hope to compete with populist movements; using balanced language, inclined to present details and nuances in speeches and trying to solve real problems, are issues that are far more complicated to communicate to large groups of followers than those populist solutions that look simple, unsophisticated, presented in clear and harsh direct messages (Pehe, 2007).

There are five key elements of populism that are enforced through the use of social media: emphasizing the sovereignty of the people; advocating for the people; attacking the elite; ostracizing others; and invoking the "heartland" (Engesser, et al., 2017: 1109–1126). Sovereignty of the people versus bureaucrats, technocrats, experts and the elite is easy to frame in the use of conspiracy theories spread by populists. Claims that populists are the real people and their representatives – and existing institutions, parties, leaders are fake – are easy to promote in closed groups on social media and in an environment driven by groupthink. Attacking the elite is an easy task on social media where emotions and harsh words have a greater reach than balanced language and rational arguments or facts; populists are very good in finding and defining enemies, and social media allows easier creation of clear-cut groups, us and them. Invoking traditions, the good old days, as ways to do things in the present is easier in an emotional environment like the one favoured by social media.

We could also add into the mix the extensive use of mass microtargeting campaigns to spread inflammatory messages among susceptible voter groups (Dans, 2018), such as during populist elections like Brexit or the American presidential elections in 2016, with Donald Trump. This tactic would have been impossible before the existence of social networks. Populism is the first political approach to profit from social media. Nationalism is another very well-defined aspect in social media, with clear ideas and coherent statements, often in harsh terms, which create a high number of followers.

Social media also has a direct impact on identities, and this also alters the political landscape, affecting political security. Social media is an easy-to-use instrument in the creation of new types of identities and groups, with different agendas, building new sets of demands for new categories of rights or advantages. The support for populism can also come through the enforcement, via social media, of new types of identities (Fukuyama, 2018). Social media is an effective way to forge identities; in some cases, the existing or classic ones – ethnic, religious and professional preferences, ideologies or attachment to different ideas or opinions – in others, the new types of identity completely linked to preferences, background or psychological structure of the components of one group or another. Creating new groups can be a way of dissolving the old traditional ones and, as in some other cases, social media and excessive Internet use could harm societal cohesion and alienate the person from the community and society, therefore harming the classical identities that are supportive of populism (Chifu, 2019a: 11–23).

The effects of social media in the support and development of populism, and in favouring the spread of populist ideas, also come from challenging the global leadership, fighting the elite, contesting traditional parties and institutions (Muller, 2017). They spread ideas of change and revolution of global security and international relations (Emmott, 2018; Rice & Zegart, 2018; Levitsky & Ziblatt, 2019), rather than altering, adapting and reforming them.

As we know, the pressure of populism has a malign effect on the liberal world order and is very real. Populism means, generally, governing with one eye on the media and the other on the polls. And it leads to Mediocracy and Mediacracy (Pehe, 2007). The influence of social media on traditional politics, ideologies and parties also has an important impact on party hierarchies, including in the mainstream parties, where populist leaders are more favoured by social media. They favour personalities with less responsibility and who are able to communicate emotions, in strong words and extreme forms that are attracting the public, instead of people with effective arguments and knowledge.

On another note, social media helps to promote extreme progressive approaches that harm individual identities and social, group identities (Gavriluţă, 2018; Kagan, 2018). Once their manifestations create the perception of a threat to a number of individuals who feel pressured, threatened and prevented from their freedom to assume and reveal their identity, the reaction of those individuals turning to the extreme and populist ideas is supported via the activities developed and favoured by social media.

Political correctness, progressive approaches, excesses of a left-wing liberal democracy, secularism, trans-humanism and support for the new technologies instead of humans (Gavriluţă, 2018), all are approaches that are favoured by social media. First, it is because social media does not make any distinction between the rational and the passionate or emotional. They are all equal as statements and posts. Second, it is a democratic approach, where all those persons who are accessing/using social media are equal, in spite of their CVs and credentials or their lack thereof. And third, through the action of those social media platforms that are taking positive action to eliminate hate speech or extreme right-wing approaches as well as other rules of the communities, the space given to those ideas has become an important aspect within the whole virtual space.

Conservative political figures as well as Christian thinkers are looking critically to those type of ideas claiming that political correctness, for instance, has become a real social phenomenon that involves social control; it is an instrument of public intimidation, introducing an excessive culpability sentiment, a social instrument of manipulation and cohesion, blame and shame, where the first target is the white, Christian, heterosexual majority. Moreover, the positive discrimination attached to political correctness is presenting the characteristics of altering the competition and sacrificing meritocracy[2] (Gavriluţă, 2018). The side effect is

[2] See Horia-Roman et al. (2018).

the perception of an attack on individual or group identity and the response is to migrate towards populist, nationalist and extreme views of those that perceive themselves under attack.

There have been changes to national security that come from social media as an instrument (Chifu, 2020a). All these societal and political security threats have a huge impact on the new threat environment, on any current threat assessment. And all come amplified in any current and future attempt to cope with the security and defence of a nation. That is why all those non-conventional threats that are harming the internal immutable realities we used to consider as fixed, constant and solid, need to be explored, known and approached in a scientific and comprehensive manner in order to prevent strategic surprise in those areas and the emergence of new types of conflicts.

It should be easy to realize that those gaps in perception and the surprises at the level of common knowledge, the impact of those technological developments on individual and societal threat perceptions are influencing internal security at the highest level. The alteration of societal security, fragmentation and alienation of the individual, losing the sense of community spirit or societal cohesion are directly influencing the security of a nation. Moreover, the new vulnerabilities could be used by external actors to take advantage of those evolutions and lack of prevention.

Support for decision makers and political strategy demands trust and understanding of the real evolutions of the threats and security concerns communicated by experts to the whole population.

Financing security needs is also linked to a clear perception of those evolutions and a general understanding where we have all gone wrong as humans and how it happens that we are all subject to those collective mistakes or misleading actions (Chifu, 2019b).

Assuming that some aspects of general information are linked to security and are not in the public domain and that there are things we cannot know, and we are not supposed to know, is another point that needs to be taught and understood at a societal level, in order to avoid mistakes and misunderstandings, as well as to accentuate vulnerabilities due to the lack of public trust.

Non-conventional Threats and the New Types of Hybrid Conflict in the Twenty-First Century

The evolution of international relations and the turbulence and acceleration of both integration and fragmentation of the International World Order have lead to unexpected and unconventional forms and types of threats to national and international security. Some come from technological development, some from the impact of those technologies on our societies, others from the rise of populism and identity, and, last but not least, from our own minds and perceptions that are influenced dramatically by our preconceptions and ease of rational thinking. All these have a huge impact on the new threat assessment, on security and the defence of a nation. That is why those facts need to be explored, known and approached in a scientific and comprehensive manner in order to prevent strategic surprise in those areas and the emergence of new types of conflicts.

There has been an evolution of security over the last thirty years, which is seen in The European School of Security. Security has proved to be much more than a collection of conventional military threats. Moreover, things have evolved much more than the classical definition of the Copenhagen School[3] (Romaniuk, 2018), with the five dimensions of security that arose with the Strategic Concept in Rome (NATO Strategic Concept Rome, Article 24, 1991), the basics of the NATO approach to security – military, political, social, economic, and environmental (Buzan, 2012). We have added three objects to the first approach to be "securitized" – state, society and individual (Chifu, 2013; Buzan, 2014).

The broader approach to looking at security has developed into a new type of methodology to examine "non-traditional threats" rather than the prior phrase "emerging security challenges", since NATO is entering new and uncertain territory in its assessment of other new threats. That area usually includes terrorism, the proliferation of weapons of mass destruction, cyber-attacks, disruption of energy supplies and even extends to climate change and migration (ISIS Europe Briefing Note, 2010). That approach was seen as so vital that in August 2010, around fifty staff members from NATO's International Staff and International Military Staff were moved into the new "Emerging Security Challenges Division" (ISIS Europe Briefing Note, 2010).

The EU and NATO established common cooperation because they are both surrounded by the same security environment. The first area of that cooperation is addressing hybrid threats. According to the strategy adopted at the Foreign Ministerial meeting, the EU and NATO established how the two organizations are going to fight hybrid threats (Press statements by the NATO Secretary General Jens Stoltenberg and the EU High Representative for Foreign Affairs and Security Policy, Federica

[3] The Copenhagen School of security studies is an academic school that employs a critical approach to security studies. It is part of the post-positivist movement in the field of international relations (IR), which became a salient part of post-Cold War scholarship. See Romaniuk (2018).

Mogherini, 2015). According to this strategy, the first responder is the state involved, but EU and NATO assistance follows directly after in support of the targeted country. This cooperation has been translated in the fact that in Helsinki in April 2017 several NATO Allies and EU members decided to formally establish the European Centre of Excellence for Countering Hybrid Threats (NATO welcomes opening of European Centre for Countering Hybrid Threats, 2017).

As a result of the above, there has been an impact of technology on community and democratic societies. But despite those developments and the speed of the evolution of our society, international relations and security are introducing new categories of non-conventional or unconventional threats. All at once, the external threats are doubled as a result of internal vulnerabilities becoming threats. That is because they belong to the hybrid typology of threats (Chifu, 2018a: 23–30), but have external sources. It is the case with all the characteristics of a liberal democracy; those values and principles that are respected because they represent our way of living, but which are considered to be vulnerabilities by some players (specifically the Russian Federation) (Chifu & Ţuţuianu, 2017; Simons & Chifu, 2017) who construct instruments in order to take advantage of those characteristics. In this way, we can very easily see them as threats from external sources against our societies (Chifu, 2013).

These threats not only come from speculation over the values and principles of our democratic systems (Muller, 2017; Richards, 2017), but also from the exploitation of those internal vulnerabilities of our liberal democratic societies (Kagan, 2018) by the evolution of technology and its impact on those societies (Chifu, 2019a: 11–23). Social media and informational warfare, drones and hypersonic weapons are shaping the security environment of the future, with an impact on the approach and the way of thinking of security-related matters. And specifically, the most profound change comes from the impact of technology on democracy itself (Chifu, 2019a: 11–23).

Technology has a tremendous impact on democracy and society (Gavriluţă, 2018) through the direct influence on the criteria for and behaviour related to societal cohesion – solidarity, the sense of community, alienation and fragmentation (Appadurai et al., 2016; Kirchick, 2018; Rice & Zegart, 2018). On another point, the sense of lack of privacy, altered identity (individual and collective) and the need for dignity and respect (Fukuyama, 2018), are all new effects of technology on our day-to-day lives that have an impact on our societies and participation in democratic processes, as well as on the life of the community, the very substance of our democracy (Gurri, 2018).

These specific processes have been underlined in several studies; in other cases, study is only just at the beginning. In some instances, the psychological processes linked to our reaction as humans to the speed of change (Rosling, 2018) are working against us. In all those cases, in-depth study is required and we have to consider that those evolutions themselves are creating vulnerabilities to our society (Ishinger, 2019) that we have to cope with. And the bases of this reaction should be building resilience within our societies. This is no easy task, since such resilience needs the involvement of the state, society and individuals alike, and that cohesion can be forged only if there is a reasonable level of trust in the authorities (Chifu, 2018a: 23–30).

Drones, Technologies and Impact on Societal Security

The unmanned aerial vehicle (UAV) or drone technology has changed the shape of warfare and made a significant impact on the psychology of democratic societies. Air vehicles and sea drones are used to hit targets and the development of this capability has a high profile today. It all began in 2002 with US surveillance drones, then came armed drones used to kill terrorist operatives in Yemen on sight[4] (List of drone strikes in Yemen), then again in Syria and Iraq. Terrorists then took advantage of the technology and used it in attacks, including in Syria against the Russian Heymim military base on 25 October 2018 (TASS Russian News Agency, 2018).

The use of drones has proved to be far more extensive, with a tremendous impact on the everyday lives of normal people. They are used in surveillance, in crowded environments, and the lack of regulation in Western cities has led to intrusion on the private lives of individuals, for example paparazzi photography and private surveillance. Drones have also been used to disrupt and block airports – for example, one Christmas Eve (see the Gatwick case in London (Taylor, 2018; Young & Holton, 2018)), and threats by ecologists (Calder, 2019).

Attacks on transportation networks and on the critical infrastructure, as well as the major attacks on production of oil occurred during 2019. First, the attacks in May, June and July on oil tankers in the Gulf of Oman and on the UAE exclusive economic zone have shown vulnerabilities of these lines of transportation – be it with air drones, unmanned

[4] On November 5, 2002, Al-Qaeda operatives in a car traveling through Yemen were killed in a targeted killing by a missile launched from a CIA-controlled Predator drone, first public known use of a drone attack, see List of drone strikes in Yemen, https://en.wikipedia.org/wiki/List_of_drone_strikes_in_Yemen#cite_note-Addicott-11.

water vessels or mines (BBC News, 2019a; Raf, 2019). Most recently, a drone attack targeted the production and supplies of oil at the global level by targeting Saudi Arabian oil facilities at Abqaiq and Khurais. The result brought the US and Iran to the brink of war but attacks were avoided at the last moment (BBC News, 2019b; LeBlanc et al., 2019; Ramkumar & Wallace, 2019). There were then new concerns after it was discovered that China was selling armed drones to Serbia (Yan, 2019) in the heart of the Western Balkans, on the borders of Romania.

We can therefore see that these types of capabilities are clearly a threat to military security if used in the theatre of operations. They could have terrorist uses or disturbing indirect effects on critical infrastructure such as airports. Drones could target economic security, and safety of supply lines, including energy security. They could be used in surveillance and on psychological operations – for example, when targeting a specific individual, harassing or observing even their most minor movements.

Drones have a societal impact, as well as individual psychological effects that harm societal security: pressurizing individuals, imposing a sense of continuous surveillance and the feeling of being constantly hunted due to the weapons capabilities installed on drones. All this has an important impact on the behaviour of individuals in a society and on the society itself – being paranoid about general surveillance and its invasion of privacy on daily life.

Digitalization: Social Media Impact on Society and Democracy

The second layer of the impact of technology on society and democracy, with consequences for political security and societal security alike, comes from digitalization. It is a natural evolution but with a tremendous speed that prevents a society, its political class and the authorities from studying its consequences to prevent side effects and negative impact. Its evolution is natural, but consequence management is very hard to put into place.

First, in the case of social media, the impact of propaganda, disinformation and informational warfare is very well known (Chifu & Ivan, 2013: 221–239; Chifu, 2015: 200–209; Chifu, 2016a: 132–133; Chifu, 2016b; Chifu & Nantoi, 2016a; Chifu & Nantoi, 2016b; Chifu, 2017: 20–28; Simons & Chifu, 2017; Chifu, 2018b). Less well known and studied are the indirect threats to security, namely the impact on political security and societal security of social media. This is done via the changes in individual human behaviour, through the impact of technology on societal development and community spirit, as well as the impact on democracy, and liberal democracy specifically.

There are several ways that social media can influence society, and any negative side effects should not detract from the huge positive effects of this tremendous technical achievement. The Internet itself has originally created a side effect in transforming the population into a far more contemplative and passive one, becoming used to reacting in front of a computer screen rather than taking to the streets to protest in the real world. Social media has an important capacity of finding people with the same concerns and way of living, thinking alike, and that leads to building uncritical and isolated bubbles from critical arguments or opposing positions. It creates very friendly environments for each and every opinion, but that approach creates stronger and unchallenged perceptions of the truth and reality that do not confront other arguments and never verify those perceptions with the reality and truth around us.

Social media has a side effect of fragmenting society. Types of bubbles that cause self-isolation are creating strong new types of identity but are also isolated from the real debate of ideas and confrontation with other people's arguments (Appadurai et al., 2016). Social media supports the development of populism, favouring strong statements and giving equal say to all participants in a conversation despite their credentials and legitimacy, so that anybody can attract the traffic and win a debate of the popular vote, turning aside arguments less adapted to communication at the general level of education and in the real world. This shows that populism wins by taking over the numbers of followers in social media (Muller, 2017; Fukuyama, 2018; Ishinger, 2019).

Nationalism is another very well-defined aspect within social media, with very clear ideas and coherent statements, often harsh ones, that attract traffic and that favour nationalist and extremist viewpoints, anti-system believers, and popular thoughts and opinions spread via conspiracy theories (Rice & Zegart, 2018). As we have seen before, social media has an awkward and biased impact on participation, keeping people away from reality and protests through likes and dislikes in a contemplative virtual space, but with the ability to be able to mobilize huge numbers of strangers, once the emotions involved and the level of excitement reaches and touches a significant number of followers.

Social media has also an impact on the behaviour of individuals linked to socialization, through a humanitarian approach as well as community involvement and participation. It creates new types of behaviour by different individuals. But the most important part is linked to post-truth, perceptions and realities and the psychological impact of this instrument when employed by specialized users with bad intentions that could harm individual conscience, will or social behaviour of people (Gurri, 2018).

At stake through this easy-to-use instrument is the creation of new types of identities and groups, making new demands and requests for specific new categories of rights, which have an impact on political security. Support for nationalist, populist and extremist discourses – already overseen by automatic mechanisms – is harming mainstream parties and elites, technical experts and specialists.

On another note, social media helps to promote extreme progressive approaches that harm individual identities and social, group identities (Gavriluţă, 2018; Kagan, 2018). Once their manifestations create the sense of threat to a number of individuals who feel pressurized, threatened and prevented from their liberty to assume and display their identity, the reaction of those individuals moving to the extreme and populist ideas is supported via the activity developed and favoured by social media.

It is not about the instrument, but about the way this instrument is used, once it has the advantage of reaching targeted audiences, geographic locations and preferences in political or social, beliefs or ways of living through religious beliefs. Moreover, being a tool that reaches all of us across borders and barriers of any type, social media has an important share of being able to have an impact on changes in society and individual preferences. Artificial intelligence has impacted upon numerous aspects of the organized human world, such as the labour market, creating a 'useless' class, becoming a digital dictator and accused of hacking the human mind. It is a tremendous tool that could help us and assist humanity on a day-to-day basis, and its positive effects are far more important. Moreover, AI has an enormous advantage over human beings in that it can utilize huge databases at tremendous speed in reacting to different stimuli and in crisis situations, making automated decisions in no time at all. The impact of AI on the individual, the society and our daily lives is even greater and less studied than social media, not to mention the possible unknown side effects.

What we can see is the impact on societal cohesion – creating differences and divisions, accelerating the fragmentation of society, community and societal will – and that it supports all the ideas that derail, criticize and break all types of cohesion or projected action; ultimately, it alters freedom of choice and behaviour at the individual and societal level.

In the medium to long term, the effects are even more complicated. AI alters the labour market, replacing humans with robots doing the repetitive and difficult work, while new jobs are developing all the time whereby humans then find it difficult to adapt to performing these new roles. This leads to people unable to adapt and creates the 'useless' class, replacing today's working class (West & Allen, 2018). Last but not least, new threats are coming from the military use of AI. Continuing its stance on informational warfare, the Russian Federation has moved to create capabilities to alter the behaviour of human beings through the use of AI (Michie et. al., 2017; Wright, 2019).

In some cases, under extreme scenarios, AI could lead to the technological destruction of the world and of humankind, as some thinkers are warning (Harari, 2018a; Harari, 2018b). Bioengineering provides divine abilities like that of creating life or selecting traits of the living creature, and we are predicting the creation of bodies, brains and minds as replacement spare parts for human beings. How can we deal with that prospect without committing enormous mistakes that could extinguish the human race? Manipulating emotions, sensations and behaviours is as risky as altering the attributes and evolution of our own bodies.

Technology is a benefit, and it will help us unless we are lacking excessive aspirations to reach the capacities of the machines, for the future humans to be created. Intelligence and discipline that help in the work and are required from the robots have an absence of compassion, neither sensibility, and have limited creativity. The prospects of a useless class of humans alters the society and the fight against exploitation could be replaced with a different type of struggle in order to fight with their irrelevance. And this changes society completely (Harari, 2018a; Harari, 2018b).

At the political level, we already have the use of AI in avoiding the circulation of unwanted ideas in the virtual space of some states – beginning in China or Russia, and arriving in Turkmenistan or North Korea – with these latter two possessing more physical barriers than the more nuanced filters of "acceptable" ideas allowed in those societies. The perspective of a general population survey that is using personal data and our behaviour, potentially leads to both digital dictator perspectives as well as a will to hack humans, such as today where some people are hacking machines, internet, computers and so forth. (Harari, 2019). AI knows and understands us better than we do, and so can learn what we do not know about ourselves, including how to modify our behaviour. That is the path that President Putin has given to its AI use in building capabilities to alter the behaviour of human beings (Daws, 2019).

The Capacity to Adapt

Human minds are very adaptable and ready to face new challenges in life. However, the inertia of our way of thinking and when looking at the speed of current evolution is taking us by surprise. The moment we learn some solid truth for the basis of our arguments, we tend to give them continuity and stability and that is why we evolve in our mindsets more slowly than the real evolution of the world. Therefore, we are subject to misleading trends due to the lack of capacity of our minds to adapt, and by our perceptions that are influenced dramatically by our preconceptions. We tend to favour the easiest ways and models in our rational thinking, and so we move further away from the truth in our judgments, reality or seeing the real evolutionary trends.

There are many studies of those differences between natural perceptions and ways of thinking developed by humans and realities which appear in several books, but the most relevant is Hans Rosling's *Factfulness* (Rosling, 2018). It presents ten human instincts – in reasoning there are some common weaknesses in evaluating the rapid evolution of today's trends. With a solid experimental basis at the highest level of knowledge, it proves that we tènd to overlook some aspects of evolution; we are inclined to find the easier categories and splits in the reality among us, and we are led by our common knowledge to fear, to be pessimistic in appreciating our own evolution, to assume constant trends and a straight-line evolution.

There is an immense amount of data and statistics that sit at the base of this book showing us that the very solid things that we think we know are obsolete and they are subject to the psychology that defines our life, the easiness of thinking, but also the reflection of basic emotions and approaches of multiple simplified models that we tend to use in our assessments, argumentation and thinking. The combination of mathematics, logics and psychology reveals to us how badly equipped we are in facing trends that change dramatically by accelerating after long periods of time, with changes in a very strict and yet simple manner of evolution.

The fact is that this is what we are facing today: huge and rapid changes, uncertainties, alteration of known models and the relativization of very clear, certain and solid grounds. Gap is a dividing model of opposite categories, negativity, constant and steady evolution, fear instinct, psychology of the size. Where big numbers, generalization approach, destiny instinct or existence of a single perspective constitute a unique way of seeing things, blame and urgency sense, which are all proving the limits of our approach to grounding in reality.

All those models which develop from education and experience, that we use to learn from past realities of the world and our conservatism, our psychological reactions to speedy change, are limiting our ability to realize the dramatic evolutions we are witnessing. And it is worth studying and knowing this reality since it is the "normal" way used by our co-citizens and the whole of humanity to perceive and react to the large-scale impact, as the one we are facing nowadays. And the reaction of the general public is subject to that limitation of our minds.

Societal and political security have a direct impact on hard security. All societal and political security threats have a huge impact on the new threat environment, and on any current threat assessment. Moreover, all become amplified in any current and future attempt to cope with the security and defence of a nation. That is why all those non-conventional and unconventional threats that are harming the internal immutable realities we used to consider as fixed, constant and solid need to be explored, known and approached in a scientific and comprehensive manner in order to prevent strategic surprise in those areas and the emergence of new types of conflict.

It should be easy to realize that those gaps in perception and the surprises at the level of common knowledge, the impact of those technological developments on individual and societal threat perceptions are of influence at the highest level of internal security. The alteration of societal security, the fragmentation and alienation of the individual, losing the sense of community spirit or societal cohesion are directly influencing the security of a nation. Moreover, the new vulnerabilities could be used by external actors who could take advantage of those evolutions and lack of prevention.

Financing security need is also linked to a clear perception of those evolutions and a general understanding where we are all wrong as humans and how it happens that we are all subject to those collective mistakes or misrepresentation (Chifu, 2019b). Assuming that a part of the general information linked to security is not a subject of the public area and that there are things we cannot know, and we are not supposed to know – classified information, intelligence – is another point that needs to be taught and realized at a societal level, in order to avoid mistakes and misunderstandings, as well as to accentuate vulnerabilities due to the lack of public trust.

Technology and Democracy: The Impact on the Evolution of Security and International Relations

Technology has evolved in multiple spheres so much that it now impacts us in day-to-day life. Social media and informational warfare, drones and hypersonic weapons are all shaping the security environment of the future, with an impact on the approach and the way of thinking towards security-related matters, on the perception about those fields as well as on the way security and international relations are adapting to those changes. But the most profound change comes from the impact of technology on democracy itself and the changes in the democratic system, at the internal level, inside states and groups and in the international arena. And the trends are frightening, if we are to talk about the speed of the evolution and the difficulties that mankind and the international establishment are facing when trying to adapt.

In the methodology, we are not trying to cover such deep and extensive aspects as that of the whole influences of technology on the evolution of democracy, neither to look further at the impact of those changes on the security establishment or the international relations framework as a whole. But coming back to prospective studies (Chifu, 2014), there is one case, one methodology, the Global Trends (National Intelligence Council, 2012) of the National Intelligence Committee, which introduces in its long-term evaluation the technological breakthroughs, such as energy development, capacity and efficiency of storage, or even the change of the generation of energy sources, such as drones, hypersonic missiles etc.

We have learnt from our previous exercises on prospective studies related to specific events – the annexation of Crimea and Russian-backed aggression in Eastern Ukraine (Chifu, 2014) – or on countries and regions – Ukraine (Chifu, 2014; Chifu et al., 2015), the wider Black Sea Region (Chifu & Bălăşoiu, 2018) – that it is not just about the direct impact. The secondary effects and influences are side effects; via third indicators – indicators from processes outside of the direct interference of the actors in the game, like it is the case for instances of complex global interdependencies. This is what happens with social media; there is an assessment of the direct impact, but never of the resulting side effects on third types of evolution, who also contribute to the changes in international relations and the security establishment, the object of our study today. Changes are direct – consequential. They could be secondary effects(side effects) and third level categories of influences(influences of the secondary(side) effects of an intended action. This complicates the anticipation and are still relevant for the evolution of international relations.

On social media, there is a lot of information on its impact in the Arab Spring evolutions (Howard & Muzammil, 2011; Khamis & Vaughn, 2011; Aday et al., 2012; Lim, 2012), on the new social movements (Castells 2012; Gerbaudo 2012; Mason, 2013), on mobilization (Langman, 2013: 510–524; Tejerina et al., 2013: 377–392; Diani & Kousis, 2014: 387–404; Simiti, 2014; Mavrommatis, 2015: 432–449; Aslanidis & Marantzidis, 2016: 125–157). But nobody assessed the full and deep effects that social media has had on our societies. Technology has had a direct impact on our way of living, and at a secondary level through the changes it drives inside society which also alters and changes our way of living. All this certainly impacts democracy and how we view it.

We select from those changes just two case studies. First, the direct impact of social media on democracy via the informational warfare and cyber-attacks on electoral systems, and indirectly via the new types of politics that it enshrines. The second case covers the impact of social media on the rise of populism, adhocracy and anti-system parties, with the secondary effect of those changes on the security and international relations framework.

Ultimately, the impact of social media as a technological product on the evolution of security and international relations is the cumulated product of the direct impact as well as of the influence of social media on democracy, the society, the ways of doing politics and, as a consequence, on security doctrines, strategies and arrangements, as well as international relations, institutions, framework and developments.

Social Media and Its Impact on Liberal Democracy in Western Countries

Social media has introduced some tremendous changes as it offers free access to infinite information to anyone, at practically no cost. And those changes are clearly in favour of democracy. The self-censored people who do not participate in elections due to a lack of understanding of the impact and relevance of the vote, of the low civic spirit and to the lack of information now move to participation in worldwide polls. Having access to information has made the self-censorship disappear; everybody thinks they have enough information in order to make a choice, support a position and a favourite candidate and to vote for them.

Social media has empowered and activated deep layers of society and attracted them into democratic activity, first and foremost to vote in elections other than local ones. The increase in participation is a way of enforcing the democratic process. Unfortunately, this happens alongside another creeping phenomenon: the absence from voting of the middle to upper wealthy and educated class. Also, corporate employees appear less interested in the politics of their country as long as any thresholds that may cause alarm are not reached. That part of society is less interested because its livelihood does not depend on respective government policies and has little to gain from national strategies, as they are highly competitive and able to adapt to working in any part of the world. Thus, their absence is registered and documented.

The fresh voters are those coming from the younger generation and from the less-educated public, and are the most exposed to malign influences through social media (Chifu & Nantoi, 2016: 1). They have the lowest individual resilience and means to react to propaganda and informational warfare (Simons & Chifu, 2017). They are also exposed to manipulation through fake news and propaganda, and also to the more sophisticated means of some international actors and local competitors (Chifu & Ţuţuianu, 2017).

The second phenomenon has been the creation of numerous groups of like-minded persons, with concrete and difficult problems. Those virtual groups have begun to play a more and more important role; they can find themselves on social media, exchanging ideas and solutions, experiences and problems that they face, in very concrete and narrow fields. Those specialized and niche areas are producing important and very tough questions, creating problems for the administration and the conventional parties, interested in solving problems.

This is not a problem when it comes to a person with specific or multiple administrative issues and diverse concerns of numerous citizens, but it becomes a challenge with structured groups of numerous citizens facing more or less the same very complicated problems. However, when this multiplies to the level of thousands or tens of thousands of groups with respective unsolved and very complicated problems, technical ones specifically, it begins to be an issue.

Finally, social media has its original role in activism, mobilization, or finding people with the same problem (Chifu & Ivan, 2013: 221–239.); it has also been concerned with creating anonymous groups or non-assumed mobilization groups, or even revolutions without assigned leaders, reaching an objectification of social constructs in the abstract space. The impact on contemporary administrations, as well as on traditional parties and politicians, is a pressing one that enshrines changes in the political spectrum.

The refined and nuanced mode of addressing those issues is important since every step further towards a solution means hard choices, often divisive for society, meaningful on ideological grounds, or costly for the parties of political figures that want to address them for fear of excluding half of the voters unhappy with final policy decisions. The multiplication of such groups and issues create the sense of a failed administration with large deficiencies or of an unprepared political class, unwilling or completely unsuited to solving the real and complicated issues of its citizens. The situation is made more complicated to navigate using informational warfare and deep fake tactics.

We now arrive at the impact that social media has had on technology, humankind and democracy. Through multiplication of fears and alienation by malign influences, informational warfare has a channel for an immediate and powerful impact, since everybody can access a large number of people in a particular area from anywhere in the world (except from those countries with filters and censorship) (Simons & Chifu, 2017). We are approaching the fourth generation of informational warfare (Chifu, 2020b: 67–105).

Bernard-Henri Lévy was someone who tried to explain the paradigm of post-truth, with all its components. The French philosopher speaks about the three components of this development of human reality that we will have to adapt to; there is "a philosophical suspect, a technological culprit and a very responsible politician" (Levy, 2018) in present-day life and this shocking post-truth. All those who question the Truth have their origins in Friedrich Nietzsche's philosophical thinking, which questions the Truth as an absolute, considers it relative, and furthermore, places it as an equal to the Lie. Nietzsche wondered why Truth had greater value than Lie (Nietzsche). It is the renaissance of the oldest idea in philosophy – the school of thinking of the Sophists (Braunstein & Pepin, 1997) – and the perpetuation of the struggle of ideas between the Philosophers and the Sophists. The Sophists questioned the owner, the absolute Truth holder.

Technological interference, with the advent of the Internet and social networks, has led to absolute democratization in a virtual space but it did this by directly attacking the mind and perception of the real man connected to this space. Democratization has led to an explosion with enormous consequences by promoting a generous idea, that every truth, every idea, has equal value. It has confronted the competition of other ideas, to find what is most listened to, most prized. It is not a question of verification of human nature, it is suggested, an absolute check of the way in which humankind really looks on average, and which is the nature of the human spirit, how individuality and the spirit of the flock matter. It includes education, habits, traditions, characteristic features, and society, relationships, community spirit and its natural hierarchy.

Therefore, the Internet and social media have made each sentence and every opinion equal. We have come to an absolute democratization of the Truth, which has pushed far beyond Nietzsche's relativization of the Truth. It is no longer about giving an ethical value superior to the Truth in relation to the Lie, here it is the equation of all nuances – which have similar value in the competition of sentences on the Internet – between the extreme Lie and the Absolute Truth. Moreover, it equalizes the source of these defended ideas, the idiot put together with the genius, the knowledgeable, the expert, the professional.

Finally, populist politicians are the third source guilty of the problem being discussed. Here we are already entering concepts that remind one of Kafka or Eugen Ionescu, a theatre of the absurd. Populists resort to emotions, but their main idea is that "Truth is not the Reality, Truth is what I say is true" (Goebbels). This is how Nazism, Modern Absolutism, and Authoritarianism emerged. It is the new idea that pushes, through consequences, to the absurd and to the destruction of freedom, and to the absolute control from the centre, whatever it may be, of peoples, societies, nations, or the world.

While post-truth means the value of Truth is relative, questioning the "owner" of the truth (Sophists), equalizing the value of the Lie with Truth (Nietzsche), democratizing Truth (social media, all sentences are equal, regardless of the author and the connection with the Truth), it is about the connection of human perception, human understanding with the Truth, with the reality (Chifu, 2015, 2016b). We talk about interpretations, emotions that alter the understanding of Truth, or obstruct and guide our individual Truth. But it does not affect the facts.

Post-humanity is even more tragic, and it links directly to another human creation, artificial intelligence (AI). An automatic, independent machine-learning mechanism capable of infinitely manipulating more data than the human being, with infinitely more efficiency, within fractions of a second and capable of reproducing the reality of any human sense. It can alter the human mind in a programmed manner.

Everything started from the over-use of AI in detecting *fake news*. It taught cars to corroborate language elements, taught them to translate, to read, to write lyrics (not poems!!!), to understand grammars and syntaxes, to create correct sentences in any language, more accurate than real people's speech. Moreover, it taught them to recognize faces, to interpret gestures, to create grimaces, and today the program with which a text and an image of anyone, together with his voice, can create "authentic" factual statements that will take up to two more years to perfect (Buterin, 2018; David). Then we should watch out for informational warfare!

What will it be like to wake up to an AI engine that creates from algorithms and mega-data, the life-like image of a leader making a speech, starting from only the text? Looking, speaking and behaving more authentically than the original. Forecasters are predicting this technological achievement within two years. We have here a clear case of a mechanism that is superior to humans in terms of memory, speed and access to information that affects the personal internal universe of each individual and psychologically alters us. This is post-humanity.

This is not an isolated case; it is about a whole human population that will need to confront and adapt to the stress of this post-human reality, which affects us deeply on an individual level. Here, human resilience and the possibility of existence and consolidation of adaptability to these developments of a near future matters. It will also show the content and capacity of humanity in its entirety to adapt, we will know how the average individual, humanity in its entirety, struggles with these real mental illnesses of the future of concurrently living in parallel virtual spaces, a type of schizophrenia caused by the constant shift between computer-generated realities and everyday life, factuality, truth and humanity.

Of course, there are forms of reacting to post-reality, deep fake, now and in the future. Every human construct also has its antidote, even if it is already used as a weapon against humanity itself. We see today the reactions to generations of past informational warfare; we see today the constant antidote accompanying the responsible construction of new technical discoveries, including AI in action.

Niall Ferguson, the well-known historian at the Hoover Institution, Stanford University, argues that "Technology is important, but less important, in fact, in the future, it changes very quickly and creates frequent ruptures, so the life of a technology is much smaller than it used to be. Computers and social media have a major impact, but they are eliminated more quickly than in the 50s" (Ferguson, 2018). The difference is not only access to virtually infinite databases, at practically no cost to anyone in the world who is connected to the Internet, but also means anyone can access large masses of people within any society, controllable, influential, known via their personal data and intimate experiences reflected by social networks.

Ferguson argues that the impact of decentralization of information and the strengthening of capacity and impact on people's opinions, however, has led to the recent upgrading of power. Even AI can become, over time, an instrument that leads to an increased state control over the individual. That could be true, as excess of work with social media and alternative avatars for profiles could lead to detachment from the reality, alienation, schizophrenia or even retreat or isolation from a world where some of our co-citizens can no longer see the difference between true and false and are no longer able to deal with this.

A question to pose at this stage: are Populism, Adhocracy, or anti-system parties more attractive than the conventional ideologies? If so, why? As we have seen above, there are important changes that social media is fuelling inside a society in terms of virtual specialized groups with niche problems, in terms of mobilization, getting people together would otherwise never known about each other. However, it also influences informational warfare, propaganda and fake news. In political terms the democratization by bringing to the vote and to political debates the less-educated and poorer representatives from the lower classes is also a phenomenon linked to social media and information or pseudo-information sharing via those networks.

The big unknown is the level of influence of social media in the development of populism. There are not many studies looking at the impact of social media in extending the support for populism. Certainly, the emergence of populism, populist leaders and parties are not directly and completely linked to social media, but some processes are quite clear. So, if our thesis is "Social media drives changes in society and the applied democracy which support the spread of populism, adhocracy and anti-system parties", it would be difficult to prove, beyond the reasonable scientific doubt and in the finest detail required by a scientific analysis. What we can say for sure is that there is an impact and direct influence, but many other phenomena are also undergoing such development, which is not linked to social media.

What we have demonstrated up to now is the fact that traditional parties find it difficult to confront populist movements due to, firstly, their will to solve issues and govern, and not only to win votes. Second, we can support the statement that populist leaders and movements are favoured by social media because of its simple, unsophisticated, clear and tough direct messages (Pehe, 2007) that they are communicating, in contrast with conventional parties that need nuances, elaborate projects, more complicated ways of addressing things that are certainly less easy to communicate or be understood by large groups of people.

Politicians use Facebook and Twitter for populist purposes. Five key elements of populism are derived from the literature: emphasizing the sovereignty of the people; advocating for the people; attacking the elite; ostracising others; and invoking the 'heartland' (Engesser et al., 2017: 1109–1126). Social media gives to the populist actors the freedom to articulate their ideology and spread their messages. That aspect is proven. But social media helps also democratization and mobilization for different purposes.

On another note, we have studies (Maldonado, 2017) that present links between the rise of populism, the "increasing sentimentalization" of democracies, and the digitalization of the public sphere. We have seen that "populism is strengthened by digitalization and affectively charged; sentimentalization is facilitated by digitalization and expresses itself in populism; digitalization shows an important expressive-cum-formative dimension and paves the way for a populist way of communication" (Maldonado, 2017). Populism has been described as both a thin ideology that rests upon the opposition between the people and the elite and as a political style where the populist's performance plays a key role in creating the people itself, making use of both verbal and non-verbal strategies of representation (Maldonado, 2017).

In the analysis, we have discussed a lot about the role of social media in spreading populism and amplifying it, but the level of methodology used as ideas developed is far from being acknowledged in the field. It is true that we see an extensive use of mass microtargeting campaigns to spread inflammatory messages among susceptible voter groups (Dans, 2018). This tactic would have been impossible before social networks.

But social networks reaffirm in a "bubble" our own opinions and beliefs like an echo chamber, a bubble that filters what we read, limiting our vision of the world and simplifying the reality to the level of caricaturizing it, because we find it hard to understand it in its complexity. Groupthink is also a side effect of it (Stern, 2003). This does not mean direct support for populism, or even that social media necessarily favours populism or populist ideas. Additionally, arguments like "historical coincidences" (Stern, 2003) are never acceptable for a deterministic relation between social networks and the rise of populism.

It is also true that it is easier to spread misinformation on social media than it is to correct it, and it is easier to exacerbate social divisions than to build societal cohesion using them (Beauchamp, 2019). But it is questionable if "the very nature of how we engage with Facebook and the rest now helps far-right, authoritarian factions weaken the foundations of democratic systems – and even give themselves an easier pathway to seizing power" (Beauchamp, 2019). It is more controversial to not assess that "social media, in the way that it's used now, is an authoritarian medium" (Beauchamp, 2019).

Therefore, we can conclude, at most, that social media is an instrument that amplifies populist ideas and adhocratic leadership. We have presented some arguments and links to support this. Social media helps to move those types of ideas – populist, adhocratic, anti-system – into places where they do not usually feature, presenting them to people more open to the consumption and influence of those sorts of concepts. It is simplistic but pervasive, and with measurable effects.

Conclusion

Another important point widely discussed and related to the debate about the support for populism is that of identity (Fukuyama, 2018). One explanation for the populist reaction is the fact that liberal democracy in its progressive forms, imposes a type of political correctness together with evolutions that harm personal identity and the collective identities of humans, so the reaction to such pressures can take the form of an exacerbation of nationalism, divisive identities and support for populist ideas and for populism (Muller, 2017).

In that respect, there are more studies that link identity and populism. If we return to the effects of social media, it is its capacity to find like-minded people and to gather them in groups according to their preferences that is one of its leading effects; a way for forging identities. In some cases, the existing and more classical identities – based on ethnic, religious, professional ideologies or attachment to different ideas or opinions – are in some ways linked to the new types of identity, completely linked to preferences, background or psychological structure of the components of one group or another. This drive to create new groups could be a way of dissolving the more traditional ones, but, as in other cases, social media and excessive Internet use could harm societal cohesion and alienate the person from the community and society, therefore harming the classical identities that are supportive of populism.

If we look at the effects of social media on democracy, by supporting and developing populism or at least favouring the spread of populist ideas to target audiences via social media, we can foresee the impact that it has on the security framework and the nature of international relations. First and foremost, there is a global, unbinding group of supporters on social media that challenge the global leadership and the elite, contesting traditional parties and institutions (Muller, 2017). Other phenomena are also adopting the same ideas of revision and revolution of global security and international relations (Emmott, 2018; Rice & Zegart, 2018; Levitsky & Ziblatt, 2019), rather than simply changing, adapting and reforming them, the pressure on the liberal world order is very real and is supported by social media.

The second impact comes from the attempt to win votes rather than governing. In some cases, populist parties that gain power, facing the burden of governing and the limits of the administration and real economics, transform into traditional parties. This is the case of Syriza in Greece, but less of Podemos in Spain. In other cases, populist parties win elections and want to govern under the same ideas and promises as those used in the electoral campaign, trying to implement the platform that won them the elections as is the case with Italy's Lega and to some respect Cinque Stelle.

Populism means, generally, governing with one eye on the media and the other on the polls, but this leads to Mediocracy and Mediacracy (Pehe, 2007). Its influence on traditional politics, ideologies and parties also has an important impact in party hierarchies; and chosen and designated politicians for several different positions are selected more or less on the same basis. It offers the same level of responsibility or lack of responsibility, and those phenomena touch upon the very substance of international institutions and security arrangements.

Another important point in the international arena and global security is the fact that populist agendas do not have an alternative strategy (Lamy et al., 2017; Kirchick, 2018). It is not about changing or reforming a system. On the contrary, what we can see in those cases is just promoting change for change's sake. Without a proper project, they are more inclined to deconstruction and elimination of institutions or attributes of those institutions. Populist agendas diminish the boundaries enshrined in commitments in such organizations, or push for the relativization of international rules, or even challenge previous agreements, without a very clear picture of the future of the security framework.

So, technological achievements and evolution, specifically social media, are creating new threats, challenges and vulnerabilities in the field of security, directly linked to the lack of rules and the number of possibilities and liberties present in this virtual space. They also generate direct threats via the support for populism, adhocracy and anti-system parties, populist ideas and changes to the existing system, without a proper program, design or strategy for the one that will replace it. This frequently leads to turbulence in the financial system, as well as attacks on the prestige of institutions in the fields of intelligence, security, defence and international affairs, and challenges to their continuity.

References

Aday, S., Farrell, H., Lynch, M., Sides, J. & Deen, F. (2012), *Blogs and Bullets II: New Media and Conflict After the Arab Spring*, Peaceworks: United States Institute for Peace.

Appadurai, A., Porta, D. della, Fraser, N. & Geiselberger, H. (coord.) (2016), *Marea Regresie: De ce trăim un moment istoric*, Bucharest: Art Bucharest.

Aslanidis, P. & Marantzidis, N. (2016), The Impact of the Greek Indignados on Greek Politics, *Southeastern Europe*, 40(2), pp. 125–157.

Beauchamp, Z. (22 January 2019), *Social Media Is Rotting Democracy from Within*, Vox, www.vox.com/policy-and-politics/2019/1/22/18177076/social-media-facebook-far-right-authoritarian-populism.

Braunstein, F. & Pepin, J. F. (1997), *Marile Doctrine. Politice. Economice, Filosofice*, Bucharest: Antet.

Buterin, V. (13–15 September 2018), *Will Blockchain Change the World?*, at 15th Yalta European Strategy, Annual Meeting, Kiev, "The Next Generation of Everything", available at https://yes-ukraine.org/en/photo-and-video/video/15-a-shchorichna-zustrich-yes/chi-zminit-blokcheyn-svit.

Buzan, B. (2014), *Popoarele, Statele şi Frica*, Chişinău: Cartier.

(2012), *Popoarele, Statele şi Teama*, Chişinău: Cartier.

Calder, S. (11 September 2019), Heathrow Drone Protest: When Are Climate Change Activists Going to Shut Down Airport and Why?, *Independent*, at www.independent.co.uk/travel/news-and-advice/heathrow-drone-protest-pause-climate-change-airport-extinction-rebellion-nats-a9101346.html.

Castells, M. (2012), *Networks of Outrage and Hope: Social Movements in the Internet Age*, Cambridge: Polity Press.

Chifu, I. (2020a), *Non-Conventional Threats and the New Types of Conflicts of Hybrid Nature in the 21st Century*, Conferinţa Securitatea şi Apărarea între istorie, teorie şi politici publice, 6–7 November, to be published in the Magazine of the National Defense Staff, *Gândirea Militară*.

(2020b), The Fourth Generation of Informational Warfare, *Romanian Intelligence Studies Review* (Revista Română de Studii de Intelligence - RRSI), RRSI nr.22/2019, Bucharest, pp. 67–105.

(2019a), Technology and Democracy. The Impact of the Evolution of Security and International Relations, in proceedings, 15th International Scientific Conference "Strategies XXI" Strategic Changes and International Relations, 11–12 April, Bucharest, National Defence University, pp. 11–23.

(2019b), Intelligence and Crisis Decision-making: A Bridge Too Far? in IAFIE – EUROPE 2019, "Mihai Viteazul" National Intelligence Academy, Bucharest, 14–17 April, to be published, in proceedings.

(2018a), *Războiul hibrid şi rezilienţa societală. Planificarea apărării hibride*, Infosfera Magazine, February, pp. 23–30.

(2018b), *Cum alterăm realitatea în războiul informaţional: Post-adevăr, post-factual, post-umanitate*, Infosfera Magazine, December.

(September 2017), *Trei generaţii de război informaţional. Nivelul dezvoltării domeniului în partea sa publică*, Infosfera Magazine.

(2016a), "Informational warfare, from STRATCOM to Psyops", Iaşi: Working Papers of the "World LUMEN Congress Logos Universality Mentality Education Novelty", 12–17 April, International conference, Lumen Media Publishing UK, 2016, (ISI Thomson), pp. 132–133.

(2016b), *Războiul informaţional: componentă a războiului hibrid şi instrument de acţiune agresivă cu relevanţă strategică*, Infosfera, no. 3.

(2015), *Război hibrid, lawfare, război informaţional. Războaiele viitorului,* in Universitatea Naţională de Apărare, Strategii XXI. Complexitatea şi dinamismul mediului internaţional de securitate, Publishing House of UNAp "Carol I", pp. 200–209.

(2014), *Perspectives on Ukraine Crisis. Scenarios for a Mid–Long- term Evolution*, Bucharest: Publishing House of the Institute of Political Sciences and International Relations "Ion I. C. Brătianu" of the Romanian Academy.

(2013), *Gândire Strategică*, Bucharest: Publishing House of the Institute of Political Sciences and International Relations "Ion I. C. Brătianu" of the Romanian Academy.

Chifu, I. & Bălăşoiu, N. (2018), *Prospective Studies of the Wider Black Sea Region. Scenarios For Its Future in Times of High International Turbulence*, Bucharest: Publishing House of the Institute of Political Sciences and International Relations "Ion I. C. Brătianu" of the Romanian Academy.

Chifu, I. & Ivan, C. (2013), *Rolul democrației participative, al meta-narațiunilor și autorului colectiv în secolul XXI. Mesaj din viitor*, Bucharest: Romanian Intelligence Studies Magazine, RRSI no.10/December, pp. 221–239.

Chifu, I., Nantoi, O. & Getmanchuk, A. (2015), *Prospective on Ukraine Crisis. A Trilateral Approach*, Bucharest: Publishing House of the Institute of Political Sciences and International Relations "Ion I. C. Brătianu" of the Romanian Academy.

Chifu, I. & Nantoi, O. (2016b), *Information Warfare. The Pattern of Aggression*, Bucharest: Publishing House of the Institute of Political Sciences and International Relations "Ion I. C. Brătianu" of the Romanian Academy.

(2016a), *Război informațional: tipizarea agresiunii informaționale a Federației Ruse*, Bucharest: Publishing House of the Institute of Political Sciences and International Relations "Ion I. C. Brătianu" of the Romanian Academy, p. 1.

Chifu, I. & Savu, L. (2020), *The Impact of Technology on Human Beings, Society and Politics*, Bucharest: Romanian Institute for Political Science and International Relations of the Romanian Academy.

Chifu, I. & Țuțuianu, S. (2017), *Torn Between East and West: Europe's Border States*, London and New York: Routledge.

Dans, E. (3 December 2018), Did Social Networks Cause Populism?, Forbes, www.forbes.com/sites/enriquedans/2018/12/03/social-networks-and-populism-cause-and-effect/#19a17d475276.

Darrell, M. W. & John, R. A. (24 April 2018), How Artificial Intelligence is Transforming the World, Brookings, at www.brookings.edu/research/how-artificial-intelligence-is-transforming-the-world/.

David, M., Artificial Intelligence and Deep Fake, available at https://yes-ukraine.org/en/photo-and-video/video/15-a-shchorichna-zustrich-yes/shtuchniy-intelekt-ta-deep-fake.

Daws, R. (31 May 2019), Putin Outlines Russia's National AI Strategy Priorities, available at https://artificialintelligence-news.com/2019/05/31/putin-russia-national-ai-strategy-priorities/.

Diani, M. & Kousis, M. (2014), The Duality of Claims and Events: The Greek Campaign Against the Troika's Memoranda and Austerity, 2010–2012, *Mobilization, An International Journal*, 19(4), pp. 387–404.

Emmott, B. (2018), *The Fate of the West. The Battle to Save the World's Most Successful Political Idea*, London: Profile Books Ltd.

Engesser, S., Ernst, N., Esser, F. & Büchel, F. (2017), Populism and Social Media: How Politicians Spread a Fragmented Ideology, *Information, Communication & Society*, 20(8), pp. 1109–1126. DOI: 10.1080/1369118X.2016.1207697.

Ferguson, N. (13–15 September 2018), The Next Generation of Everything. A Historian's View, at 15th Yalta European Strategy, Annual Meeting, Kiev, "The Next Generation of Everything", available at https://yes-ukraine.org/en/photo-and-video/video/15-a-shchorichna-zustrich-yes/maybutnye-pokolinnya-vsogo-poglyadi-istorika-ta-investora.

Fukuyama, F. (2018), *Identity. Contemporary Identity Politics and the Struggle for Recognition*, London: Profile books Ltd.

Gavriluță, N. (2018), *Noile religii seculare. Corectitudinea politică, tehnologiile viitorului și trans-umanismul*, Iași: Polirom.

Gerbaudo, P. (2012), *Tweets and the Streets: Social Media and Contemporary Activism*, London: Pluto Press.

Global Trends 2030. Alternative Worlds. A publication of the National Intelligence Council (December 2012), available at https://globaltrends2030.files.wordpress.com/2012/11/global-trends-2030-november2012.pdf.

Gulf of Oman Tanker Attacks: What We Know (2019a), BBC News, 18 June, available at www.bbc.com/news/world-middle-east-48627014.

Gurri, M. (2018), *The Revolt of the Public and the Crisis of Authority in the New Millennium*, Stripe Press, Stripe Matter Inc.

Harari, Y. N. (12–14 September 2019), New Challenges of the 21st Century, 16th YES Annual Meeting "Happiness Now. New Approaches for a World in Crisis", at www.YouTube.com/watch?v=WA4t9uK2ZX8, 5.26.00–6.20.00.

(2018a), *21 de lecții pentru secolul XXI*, Iași: Polirom.

(2018b), *Homo Deus. Scurtă istorie a viitorului*, Iași: Polirom.

Howard, P. & Muzammil M. H. (July 2011), The Role of Digital Media, *Journal of Democracy*.

Ishinger, W. (2019), *Lumea în Pericol. Germania și Europa în vremuri nesigure*, Bucharest: Curtea Veche.

Kagan, R. (2018), *The Jungle Grows Back. America and Our Imperilled World*, New York: Alfred A. Knopf.

Khamis, S. & Vaughn, K. (2011), Cyberactivism in the Egyptian Revolution: How Civic Engagement and Citizen Journalism Tilted the Balance, *Arab Media and Society*, 14(3), 1–25.

Kirchick, J. (2018), *Sfârșitul Europei. Dictatori, demagogi și Noul Ev Întunecat*, Iași: Polirom.

Lamy, P., Gnesotto, N. & Baer, J. M. (2017), *Încotro se îndreaptă lumea?*, Bucharest: Niculescu.

Langman, L. (2013), Occupy: A New Social Movement, *Current Sociology*, 64(4), pp. 510–524.

LeBlanc, P., Atwood, K., Diamond J. & Westwood, S. (16 September 2019), Trump: US 'locked and loaded depending on verification' of attack on Saudi oil field, CNN, available at https://edition.cnn.com/2019/09/15/politics/trump-us-saudi-arabia-attack-iran-iraq/index.html.

Levitsky, S. & Ziblatt, D. (2019), *How Democracies Die*, New York: Broadway Books.

Levy, B. H. (13–15 September 2018), Who Killed Ms. Truth?, at 15th Yalta European Strategy, Annual Meeting, Kiev, "The Next Generation of Everything", available at https://yes-ukraine.org/en/photo-and-video/video/15-a-shchorichna-zustrich-yes/hto-vbiv-mis-istinu.

Lim, M. (2021), Clicks, Cabs, and Coffee Houses: Social Media and Oppositional Movements in Egypt, 2004–2011, *Journal of Communication*, 62(2), April, pp. 231–248, https://doi.org/10.1111/j.1460-2466.2012.01628.x.

List of drone strikes in Yemen, https://en.wikipedia.org/wiki/List_of_drone_strikes_in_Yemen#cite_note-Addicott-11.

Maldonado, M. A. (20–22 September 2017), Rethinking Populism in the Digital Age: Social Networks, Political Affects and Post-Truth Democracies, XIII AECPA Congress, Santiago de Compostela, GT 1.8 The Contemporary Public Sphere – Protest Movements and Populist Actors, available at https://riuma.uma.es/xmlui/bitstream/handle/10630/14500/Rethinking%20Populism%20in%20the%20Digital%20Age%20-Manuel%20Arias%20Maldonado%2C%20GT%201.8%2C%20XIII%20Congreso%20Aecpa.pdf?sequence=1.

Mason, P. (2013), *Why It's Still Kicking Off Everywhere: The New Global Revolution* (Revised and updated second edition), London: Verso Books.

Mavrommatis, G. (2015), Hannah Arendt in the Streets of Athens, *Current Sociology*, 63(3), pp. 432–449.

Merlyna, L. (2012), Clicks, Cabs, and Coffee Houses: Social Media and Oppositional Movements in Egypt, 2004–2011, *Journal of Communication*, 62(2), pp. 231–248, https://doi.org/10.1111/j.1460-2466.2012.01628.x.

Michie, S., Thomas, J., Johnston, M., Aonghusa, P. M., Shawe-Taylor, J., Kelly M. P., Deleris, L. A., Finnerty, A. N., Marques, M. M., Norris, E., O'Mara-Eves, A. & West, R. (18 October 2017), The Human Behaviour-Change Project: harnessing the power of artificial intelligence and machine learning for evidence synthesis and interpretation, available at www.ncbi.nlm.nih.gov/pmc/articles/PMC5648456/.

Muller, J. W. (2017), *Ce este populismul*, Iaşi: Polirom.

NATO Strategic Concept Rome, Article 24, in "The Alliance's New Strategic Concept agreed by the Heads of State and Government participating in the Meeting of the North Atlantic Council" (7 November 1991), available at www.nato.int/cps/en/natolive/official_texts_23847.htm.

NATO Welcomes Opening of European Centre for Countering Hybrid Threats (11 April 2017), available at www.nato.int/cps/en/natohq/news_143143.htm.

NATO's New Division: A Serious Look at 'Emerging Security Challenges' or an Attempt at Shoring Up Relevance and Credibility? (September 2010), ISIS Europe Briefing Note, No. 51, available at www.natowatch.org/sites/default/files/NATOs_New_Division_0.pdf.

Niezsche, F., *On Truth and Lie in an Extra-Moral Sense*, available at www.kth.se/social/files/5804ca7ff276547f5c83a592/On%20truth%20and%20lie%20in%20an%20extra-moral%20sense.pdf.

Patapievici, H. R., Liiceanu, G., Kagan, R., Ferguson, N., Gavriluţă, N. (2018), *Noile religii seculare. Corectitudinea politică, tehnologiile viitorului şi trans-umanismul*, Iaşi: Editura Polirom, ISBN 978-973-46-7203-5, p. 221.

Pehe, J. (2007). Populism in the Age of Mediocracy and Mediacracy. *Populism in Central Europe*, available at www.pehe.cz/Members/redaktor/populism-in-the-age-of-mediocracy-and-mediacracy.

Press statements by the NATO Secretary General Jens Stoltenberg and the EU High Representative for Foreign Affairs and Security Policy, Federica Mogherini (1–2 December 2015), available at www.nato.int/cps/en/natohq/opinions_125361.htm.

Raf, S. (13 June 2019), US Blames Iran for Oil Tanker Attack in Gulf of Oman Amid Rising Tensions, available at www.telegraph.co.uk/news/2019/06/13/oil-tankers-hit-fresh-attack-gulf-oman/.

Ramkumar, A. & Wallace, W. (16 September 2019), Crude Prices Soar After Attack on Saudi Facilities, Threatening Global Growth, The Wall Street Journal, available at www.wsj.com/articles/oil-prices-soar-after-saudi-attack-11568585160.

Rice, C. & Zegart, A. (2018), *Political Risk. Facing the Threat of Global Insecurity in the Twenty First Century*, London: Weidenfeld & Nicolson.

Richards, S. (2017), *The Rise of the Outsiders. How Mainstream Politics Lost Its Way*, London: Atlantic Books.

Romaniuk, S. N. (2018), Copenhagen School in Arrigo, B. A. (Ed.), *The SAGE Encyclopaedia of Surveillance, Security, and Privacy*, Thousand Oaks: SAGE Publications, Inc.

Rosling, H. (2018), *Factfulness. Ten Reasons We're Wrong About the World – and Why Things Are Better Than You Think*, London: Sceptre Publishing.

Sahar, K. & Vaughn, K. (Summer 2011), Cyberactivism in the Egyptian Revolution: How Civic Engagement and Citizen Journalism, *Arab Media and Society*.

Saudi Oil Attacks: US Says Intelligence Shows Iran Involved (2019b), BBC News, 16 September, available at www.bbc.com/news/world-middle-east-49712417.

Simiti, M., *Rage and Protest: The Case of the Greek Indignant Movement*, Hellenic Observatory Papers on Greece and Southeast Europe, GreeSE paper no. 82, available at eprints.lse.ac.uk/56229/.

Simons, G. & Chifu, I. (2017), *The Changing Face of Warfare in the 21st Century*, London and New York: Routledge.

Stern, E. (2003), *Crisis Decision Making. A Cognitive Institutional Approach*, Stockholm: Swedish National Defence College, Elanders Gotab.

TASS Russian News Agency (25 October, 2018), Drone Attack on Russia's Syrian Airbase Was Elaborate Pentagon Operation, Says Expert, available at https://tass.com/defense/1027834.

Taylor, J., Gatwick Drone Attack Theories – Who Is Behind the Airport Sabotage?, The Mirror, available at www.mirror.co.uk/news/uk-news/gatwick-airport-drone-attack-theories-13771643.

Tejerina, B., Perugorría, I., Benski, T. & Langman, L. (2013), From Indignation to Occupation: A New Wave of Global Mobilization, *Current Sociology*, 61 (4), pp. 377–392, available at https://doi.org/10.1177/0011392113479738.

West, D. M., & Allen, J. R. (2018). How Artificial Intelligence Is Transforming the World. Report, April, 24, 2018.

Wright, N. (Ed.) (January 2019), AI, China, Russia, and the Global Order: Technological, Political, Global, and Creative Perspectives, available at https://nsiteam.com/ai-china-russia-and-the-global-order-technological-political-global-and-creative-perspectives/.

Yan, S. (11 September 2019), China Sells Armed Drones to Serbia Amid Concerns Arms Deal Could Destabilise Region, The Telegraph, available at www.telegraph.co.uk/news/2019/09/11/china-sells-armed-drones-serbia-amid-concerns-arms-deal-could/.

Young, S. & Holton, K. (20 December 2018), Drones Paralyze British Airport, Grounding Christmas Travellers, Reuters, available at www.reuters.com/article/us-britain-drones-gatwick/drones-paralyze-british-airport-grounding-christmas-travelers-idUSKCN1OI2TN.

8 Western Hybrid Warfare: Crisis and Subversion in Regime Change

Greg Simons

Introduction

Western hybrid warfare is a broad concept that uses the perception of projected binary 'realities' to allow, through public opinion, the commission of acts of inhumanity in the name of humanity. It is used to attack the intangibles of an opponent, which are used as a means to build a case for the context and eventual use of tangible assets in an open kinetic war. The intangibles refer to the reputation, credibility and brand of an opponent/targeted country or entity. If these are negative, and opinion is turned against the political and military leadership of a target, then the likelihood of being able to choose the military option (by the communicator) is increased. These 'humanitarian' wars are enshrined in the notion and practice of communication in terms of diametrically opposed sets of norms and values. In spite of these tactics being fairly standard, and used on a number of previous occasions, they are still somewhat successful.

In order to create and project the roles played by norms and values, which open the path to the military option (often the only considered choice from the beginning of a 'crisis'), there needs to be a means of operationalising the projection of perception in order to manipulate public opinion and sentiment. The creation and manufacture of a crisis is the first step, because in addition to the tangible properties of any crisis there are also intangible aspects. A physical crisis also involves the associated information flows that are related to it, the actors that lose control of those informational flows also have their operational choices limited by their opponent. This comes at a time when a crisis demands a political mobilisation and call for action to 'resolve' it. This highlights the importance of mass media and NGOs in carrying the message in order to influence audiences in the predetermined and desired manner of the communicator.

Brecher (1996: 127) argues that "crisis, conflict and war are intricately interrelated, both conceptually and empirically. All are characterised by mutual mistrust between adversaries, turmoil, tension and hostility." The purpose of this chapter is to examine the phenomena of crisis and subversion for the purposes of implementing a regime change operation from a more conceptual and theoretical perspective. In a perhaps 'controversial' approach, the focus will not be upon the usual 'suspects' of the practice that many have come to refer to as hybrid warfare, but rather the attention is upon the Western conceptualisation and practice of this covert form of warfare. One of the motivating factors for this choice is that the state of the practice is more developed and better supported in the West than in other countries, such as Russia, and a more widely accepted form of international relations policy by the Western political and mass media elite. The question is posed in this chapter is, what are the necessary information and cognitive domain environmental conditions that need to be prepared and set for attempting the act of regime change within the context of hybrid warfare?

For the purposes of preparing an answer to this research question, it is necessary to evaluate and analyse the informational and cognitive domains of the constructed human world in order to make sense of the actions conducted in the physical domain. This chapter begins with a brief overview of what is meant by a crisis and the implications and opportunities that stem from this labelling. The following section involves the process of constructing an iconic crisis event as a means of attempting to facilitate a more overt act of aggressive subversion in order to finalise the regime change. Following from this section is an overview of propaganda, and specifically its role in contributing to the fog of war. The fourth section concerns the concept of hybrid warfare and its use as a tool of foreign policy. The final section details the act of subversion leading to regime change that is facilitated by the creation of a war of ideas.

Manufacturing a Crisis and Its Implications

As an initial point of departure, it is necessary to ask and define, what is a crisis? A crisis can mean and represent different things to different individuals and groups, where perceptions and interests tend to diverge. In addition, there is the false perception that a crisis only represents a threat, whereas, in fact, a crisis can provide an opportunity to select goals and interests. A crisis is generally considered as being an extraordinary event, a situation and condition that breaks and threatens the normal routine and functioning of society (Boin et al., 2005). The term "crisis" is often invoked in the public sphere; often, it is not clearly defined what exactly is meant and what it entails. From a theoretical perspective, a crisis can be defined as existing when three simultaneously present aspects can be found: 1) A *threat to values*, which can involve a threat to human life, property, the economy, state integrity and many other examples. 2) The element of *uncertainty* when the shortage of reliable information and the often-rapid pace of events make the ability to predict what is going to happen next difficult, which can raise the perceived levels of risk and threat being faced. 3) *Time constraint* is the final part of the puzzle, which is connected to the simple proposition that the longer a crisis event continues the more physical and psychological damage is done, which necessitates a rapid resolution in order to ensure a speedier and effective recovery process on the road back to 'normality' (Boin et al., 2005: 2–4).

When a crisis situation arises, the lead actor involved already has some form of predetermined communicative stance. This influences how an organisation responds to a crisis with stakeholders. "Stances are placed on a continuum that has advocacy and accommodation as its anchor points. Advocacy involves an organisation arguing for its interests, while accommodation involves the organisation making concessions to the other parties. The stance an organisation should take depends on the nature of various factors that make up the conflict situation" (Coombs, 2014: 18). This situation demonstrates the relevance and role of information and communication in a crisis; the information domain can be subjected to 'cultivation' in order to project certain factors that enable the manufacturing of consent in the cognitive domain.

When the physical crisis strikes there is a corresponding crisis of information (in terms of the quantity and quality of information and communication). As noted by Joseph Scanlon in 1975, "every crisis is also a crisis of information [...] Failure to control this crisis of information results in a failure to control the crisis, including its directly operational aspects" (Ogrizek & Guillery, 1999: xi). The nature and scale of crisis communication is determined by the type of break-out event that marks the beginning of a particular crisis. Communication on and around a crisis may also be capable of prolonging a crisis or even triggering a relapse (Ogrizek & Guillery, 1999: xii). When it comes to understanding what societal values are at stake, these become readily apparent in the communication around a crisis. "These values tend to be broadly defined societal values such as security, safety, economic development, democratic legitimacy, environmental protection, and so on" (Nohrstedt, 2011: 201). Value expectations are based upon notions and understandings of socially imposed rules or norms, interpretations of law and regulations, national and international mandates, and public expectations/opinion. If the right tone is achieved at the right time, the target audience can be successfully emotionally primed and mobilised to create opportunities for some actors and reduce opportunities for others.

Crises can be selectively used to shape instruments to enable greater leeway for an actor in the event of a crisis. "Crisis evaluations might contribute to institutionalise moral standards into formal rules or codes of conduct" (Nohrstedt, 2011: 201). This is owed, at least in part, to the fact that a crisis event precipitates a political call for action that is nominally and rhetorically meant to resolve the crisis at hand. It is intended to offer a specific situational logic, sometimes a false one, which is meant to prime and mobilise the target audience to consent to the suggestion; this is in order to alleviate the psychological discomfort that it may experience within the communicated framework of humanitarian dilemmas.

Within the context of a crisis and the communication flows that occur, new challenges for the management of an event are occurring. One of these is the technical change that has shortened the news and information cycle to no more than a few minutes at the most, which gives little time to officially and adequately respond. The second challenge is a massive proliferation of 'information outlets' that are radically altered and have broadened the assumed role of media in a crisis. "As a result, the new reality repeatedly challenges, then threatens, political and institutional power. It also distorts the credibility of leaderships and those that serve them" (Gowing, 2009: 10). War is framed within the fog as a crisis, given the tangible and intangible threats to values and norms as well as human life and property; the logic is the need to urgently resolve it, even if the premises of the logic are in fact false, misleading or unclear. Within the context of armed conflict, an international crisis is initially precipitated by a foreign policy crisis.

> The trigger to a foreign policy crisis derives from three interrelated perceptions generated by a hostile act, disruptive event or environmental change: perceptions of (1) a threat to one or more basic values; (2) a finite time for response; and (3) a heightened probability of involvement in military hostilities before the challenge is overcome. The third condition is the most crucial: a higher than normal perception of likelihood of war necessarily includes a perceived higher than normal threat to values; but a threat can exist without the expectation of violence. (Brecher, 1996: 127)

As noted above, much depends upon the aspect of perception, which may or may not reflect the physical reality. To operationalise militarised foreign policy goals within the context of a projected and assumed crisis, a norm-based approach has been used where the notion of Responsibility to Protect (R2P) is being represented as a global norm. The basic premise and assumptions of R2P are that the age of sovereignty was gradually passing and that there was an urgent need to enforce and demonstrate that states could no longer act with impunity and treat their citizens in any manner that they saw fit (Badescu & Weiss, 2010; McCormack, 2010; Chesterman, 2011). McCormack (2010: 74) noted that there were significant shifts in the way in which powerful states and institutions comprehended sovereignty and acted. In addition, the timing of when this debate began is highly significant, beginning almost immediately after the end of the Cold War. However, R2P is vague and therefore open to wide interpretation and operational application. It does not clearly define the responsibility of who it defends, from what particular risk of threat and by what exact means. Therefore, a politically subjective and convenient interpretation can be used to advance specific geopolitical interests in a powerful actor's foreign policy on the international stage under the guise and façade of benevolent 'humanitarian' values and norms in order to make it more acceptable through manipulating perception and interpretation of the act. The twenty-first century has marked the shift in how people are increasingly perceiving and interpreting the constitution of the global geopolitical order, from unipolar towards a more multipolar configuration.

This marked the end of the bipolar world marked by the Cold War and the emergence of the US as the world's sole superpower. There were immediate implications for how wars and military interventions were narrated. "Certainly for Western states, major military interventions were justified in highly moral and altruistic terms, of being fought on the behalf of others" (McCormack, 2010: 74). R2P is, at a rhetorical level, aimed at addressing a crisis where the loss of life is the primary value at stake. It is a political call that is designed to prime and mobilise publics. Moreover, R2P, at a policy level, is intended to gain political consensus in order to engage in possibly contentious foreign policy practice.

The first live practical test and application of R2P came with the so-called Libyan crisis, which was framed as the Libyan government randomly targeting its own civilians and in 'need' of international intervention that would ultimately lay the foundations for spreading the model of Western liberal democracy. In some quarters, this event marked a 'coming of age' of the R2P norm (Morris, 2013; Thakur, 2013). However, as noted by Thakur (2013: 61), "the use of force – no matter how benevolent, enlightened, or impartial in intent – has dramatic consequences. It shapes the struggle for power and helps to determine the outcome of political contests, which is why it is inherently controversial." However, the excesses of Libya, namely the exceeding of the UNSC mandate, has made R2P much more difficult to apply to Syria

(Morris, 2013: 1274–1277; Thakur, 2013: 70–72). As a result, R2P as a norm and an operational tool has been compromised. This has been further confirmed in a British Parliamentary report that stated the Government had gone to war on the grounds of faulty and misleading information flows concerning the crisis in Libya.[1] This demonstrates the significance of the role and power that propaganda and the fog of war has on the decision-making process by influencing opinion and perception.

The Iconic 'Crisis' Event

Iconic events are commonly seen as snapshots of popular representations of key and symbolic moments in politics and international relations. They shape and persuade audience perceptions with resonating values that are associated with and communicated to them, stirring the public's imagination and social memory. Leavy (2007: 4) defines an iconic event as "an event that undergoes intense interpretive practices but also becomes mythic within the culture through its appropriation into other political or social discourses and its eventual use within commercial culture." There are numerous examples of iconic events within the field of armed conflict, such as: the sinking of the armoured cruiser USS Maine in Havana Harbour in 1898; sea evacuations of encircled British, French and Belgian troops from Dunkirk beach in 1940; the Japanese naval air attack on Pearl Harbour in 1941; the Al Qaeda terror attacks of September 11 on symbolic landmarks on the US mainland; and the toppling of the Saddam Hussein statue in Baghdad's Fidas Square during the US-led invasion. These manipulated events can also be used as a pretext for a demand for military operations to 'solve' a humanitarian crisis by engineering public consent through the power of symbolism, memory and emotions. This was certainly the case for the sinking of the USS Maine in Havana Harbour that was used as a pretext for the US declaring war on Spain in 1898, Pearl Harbour that quickly united Japan and its entry into Second World War (departing from its earlier isolationist public sentiment), the 9/11 attacks quickly evolved into the current Global War on Terrorism and the cycle of endless wars, and Dunkirk was presented as a "miracle" and a means to boost military and public morale for continuing Britain's fight with Nazi Germany.

[1] House of Commons Foreign Affairs Committee, *Libya: Examination of Intervention and Collapse and the UK's Future Policy Options*, HC 119, Third Report of the Session 2016–2017, https://publications.parliament.uk/pa/cm201617/cmselect/cmfaff/119/119.pdf, September 2016.

An iconic crisis event performs the role of emotionally priming and mobilising the intended target audience by engineering perceptions and opinions of symbolically interpreted events in the physical domain via the information domain. These are invariably framed within Orwellian sets of projected values and a 'defensive', yet 'unavoidable' humanitarian posture of the need for good to triumph over evil. This use of frames and narratives was on display in the Middle East North African region during the Arab Spring. An example of this has been the various uses of alleged chemical weapons attacks that were immediately attributed to the Syrian government, in spite of the less than convincing nature of the argumentation to justify overt military aggression in a conflict that has been driven for many years at various levels of covert subversion and hybrid warfare (Simons, 2019). This attack comes after a number of previous attacks, which made use of different chemical weapons against their opponents or the civilian population. Often these occasions rapidly became highly politicised in nature and are used as a means to gain political advantage over their enemy, involving a call for military intervention as a 'humanitarian' gesture against an inhuman act. However, past events have shown these highly mediatised events to be an attempt at lobbying and influencing the public audience into accepting an otherwise unacceptable means – military intervention. This is often done within a false logic promoted within a frame of a humanitarian crisis, where the political calls for action are superficially intended to alleviate the situation, but operationally they serve another purpose (sometimes regime change).

Looking back at the events of the recent past, in Libya and also in Syria, it is an often-repeated script. There is a lot familiar with the current mainstream media framing of the chemical weapon event that took place outside of Damascus in 2013 (Simons, 2016).

From this point onwards, the chapter will try to map the mainstream media (MSM) narratives, especially those that assign guilt and suggest courses of action. Then the focus will shift to the counterarguments and information that will detail the weaknesses in the MSM arguments. As a logical next step, the nature and timing of the reporting of such stories implies that there is some practical reason for trying to influence and persuade public opinion on the issue at this particular point in time. Finally, a conclusion in the way of a summary and the main points to learn, in order to provide the reader with some concrete knowledge and understanding of how such mediatised events work. This is all part of projecting the propaganda of the act as a means of enabling aggressive foreign policy through the creation of popular geopolitical narratives that are manipulative in nature.

Propaganda and Creating the Fog of War

Marlin (2002: 13) remarked that "out of the last two centuries a globally interconnected, mass mediated society has emerged. The events of the 20th century in particular have shown the enormous power, for good or evil, possessed by those who directly control or know how to manipulate the mass media." Mass media and propaganda in the twenty-first century are collaborating in such a manner which results in an increased fog of war. Media attempt to shape public opinion and to reinforce certain themes and ideologies, downplaying other opposing or contrasting versions. In this regard media institutions can be critical in the process of the formation of public opinion, and in the restriction of access to alternative forms of critical information and news, especially in the context of manufacturing public consent for war (Western, 2005; DiMaggio, 2009). Some authors have argued that, without the technological developments that have permitted the creation and use of means of mass communication, it would be more difficult for propaganda to exist.

> The development of a mass society, informed by a cohesive national press, has obvious ramifications for the development of national public opinion. Mass society provides a dual influence upon the concept of propaganda; increasing exposure to coherent and managed information while at the same time providing the illusion of greater choice and a variety of analysis. (Willcox, 2005: 25–26)

The above text fits well with the definition of propaganda by Marlin (2002: 22) as "the organised attempt through communication to affect belief or action or inculcate attitudes in a large audience in ways that circumvent or suppress an individual's adequately informed, rational, reflective judgement." Instead of serving a notion of public interest through a fourth estate function, the indication is that mass media permits and facilitates mass propaganda. This is achieved by manipulating public perception and opinion through the selective release of information. Media and journalism, in effect become the "instrument of war" that was detailed by Payne, owing to the fact that modern wars' outcomes are gradually being determined more and more by political and not military factors (Payne, 2005). Thus, there is a tendency by governments to try and use mass media and mass communication to support military operations by potentially bestowing it with a façade of legitimacy (Simons & Chifu, 2017). There have been academic observations that indicate a certain consistency in the use and function of propaganda in international relations.

> The pervasiveness of international propaganda has increased not only as a result of the availability of new channels, but also because of the recognition of

propaganda as a regular, permanent function of national governments, in peace as well as in war, and the expansion of the audience for international communication to include whole populations rather than only leaders, fighting men, or the educated classes. (Davidson, 1971: 5)

As noted above, propaganda and the role of organised persuasive communication are being used by governments as a means to engineer public consent through manipulating the information domain with subjective interpretations of reality. The façade of legitimacy has been recently seen in the US approach to regime change operations across different presidencies and political party lines (namely George Bush and Barak Obama). Both of the presidents concerned understood the normative significance of an appeal to domestic and international law in the use of military force, and both sought to gain the presumed sense of legitimacy that comes from a United Nations Security Council (UNSC) resolution. However, UNSC approval was by no means a necessary precondition for initiating the regime change (Ettinger, 2017: 10). Given the use of the term propaganda, there is an urgent need to define exactly what is meant by it, and what the possible implications may be.

Propaganda can be seen and communicated in many different forms. Some of the aspects and markers in common that can distinguish propaganda are: the use of persuasive function; a sizeable target audience sought; the representation of a specific agenda; and the use of faulty reasoning and/or emotional appeals. In spite of its notoriety though, propaganda does not necessarily have to be always harmful, but by its nature it is always manipulative given its intention to gain control over other people's thoughts and actions (Shabo, 2008: 5). Given the negative associations with the term propaganda, there is a tendency to try and project propaganda as something the enemy does whereas 'we' do information (Paul, 2008: 9). The fog of war is used to create a projected reality that is composed of starkly contrasting diametrically opposed forces of good and evil in a manner that is used to not only brand the enemy, but also the war itself so that automatic assumptions create a sense of recognition, understanding and acceptance among the target audience (Waller, 2007: 84). This can be clearly seen in the US 'democracy assistance' programmes around the world, and how these are marketed and spun to create a more benevolent and worthy cause. "The imperial methods behind National Endowment for Democracy (NED) and US 'democracy assistance' programmes conflate democracy with neo-liberal expansionism" (Sussman, 2007: 191). By engaging in a sustained propaganda campaign and creating a fog of war, the intention is to lay the conditions in the political and informational environment that would permit the use of subversion against a chosen target.

An example that best illustrates the Western communicative approach to driving regime change and subversion against a target country is found in Syria. There has been a lot of effort and expense that has been 'invested' in regime change in Syria at a multiple number of different levels. In a recent exposé of the Syria propaganda operation, the nature of the breadth and depth became more readily apparent. A significant part of the operational aspects are conducted by Western government contractors and 'NGOs', which attempt to influence the informational flows and narratives on the conflict in Arabic and English language media coverage. The aim of these organisations is to create a steady stream of pro-regime change coverage and to dominate the information domain. This is done through a number of means by US and UK government programmes to train ("opposition") media activists through the Basma programme, to produce their own pseudo-news for media outlets, the re-branding of Salafi-jihadist groups to soften their image to the global public (for example the work of Dubai-based Analysis Research Knowledge (ARK) 'humanitarian' NGO), creating parallel state institutions, linkages between political movements and media outlets, training communicators, creating newsworthy events and personalities for media coverage, reinforcing the core narratives of the regime change logic, having lost the military aspects of the war try to push for a lawfare (political) continuation of the regime change operation.[2] This demonstrates the role and importance of the façade of reality in the information streams that surround and shape public perception and opinion of highly mediated hybrid warfare events.

Investigative journalist John Pilger stated that "the information age is actually a media age. We have war by media; censorship by media; demonology by media; retribution by media; diversion by media – a surreal assembly line of obedient clichés and false assumptions."[3] By nurturing the information domain in this manner, it is one of the necessary steps and preconditions for creating the façade of consensus to permit acts that may, if truthfully represented, could cause a sense of discomfort in the public and therefore may in fact prove an obstacle in permitting more aggressive foreign policy options.

[2] B. Norton, *Leaked Documents Expose Massive Syria Propaganda Operation Waged by Western Government Contractors and Media*, The Gray Zone, https://thegrayzone.com/2020/09/23/syria-leaks-uk-contractors-opposition-media/, 23 September 2020 (accessed 25 September 2020).

[3] J. Pilger, *War by Media and the Triumph of Propaganda*, Johnpilger.com, https://johnpilger.com/articles/war-by-media-and-the-triumph-of-propaganda, 5 December 2014 (accessed 6 February 2015).

Hybrid Warfare as a Foreign Policy Tool

The idea of US exceptionalism in international relations is a highly hot and controversial topic in international relations. This creates an impact upon the conceptual notion and role it plays in global politics and geopolitics, which is reflected in the nature of US foreign policy. It is used in invoking 'legitimacy' in subverting other countries and in regime change attempts, which other countries cannot or would not use to justify aggressive foreign policy (Hughes, 2015). Exceptionalism in US foreign policy takes the form of three categories: 1) supporting multilateral regimes only if they allow exceptions for US citizens or practices; 2) signing human rights conventions with reservations; and 3) refusing to ratify international treaties or only doing so after lengthy delays (Hughes, 2015: 529). It is possible to add a fourth category to this list, which is *regime change by semi-covert means in the name of human values and norms.* The point where US exceptionalism meets geopolitics and foreign policy is found in the "empirical law" that is enshrined in the orthodoxy of knowledge notion that democracies do not fight each other, which has been used as an excuse by American statesmen for launching a foreign policy of promoting "democratization"[4] in an Orwellian rhetorical means of achieving the false logic of global peace. Grigoryan (2020: 158) notes that the US-led West applies the Wilsonian principles ("democracy support") on a basis of geopolitical convenience – "contrary to the popular narrative, the West has supported democracy only when that support has been reinforced by material interests, and rarely, if ever, when it posed a threat to such interests." This behaviour cuts across modern political divides and different presidential administrations.

In the twenty-first century there are a number of state-based and non-state actors that are seen as challenging or competing with what are viewed as US foreign policy interests, within the framework of exceptionalism, and that seek to engage five tiers of actors in global geopolitics. The first level involves major powers, the second level regional powers, and the third, fourth and fifth are lesser local powers (Cohen, 2003: 3–4). A noticeable increase in the number of regional and major powers in the international system was noted early into the twenty-first century (Cohen, 2003: 30), which has resulted in a more complex geopolitical world with increased global competition for influence. As noted by Mastanduno (1997: 85–88), "US strategy has been responsive to the

[4] E. Mansfield & J. Snyder, *Democratization and War*, Foreign Affairs, www.foreignaffairs.org/articles/china/1995-05-01/democratization-and-war, May/June 1995 Issue (accessed 24 June 2015).

constraints and opportunities of the international structure and to the US position within it. [...] Unipolarity imposes less of a constraint and affords the United States more room to manoeuvre." The lack of geopolitical balancing in the immediate post-Cold War world enabled unilateral military actions leading to regime change against various lower tier countries undertaken within the framework of exceptionalism foreign policy. The relative weakening of US global hegemony in the twenty-first century and the beginnings of the emergence of a multipolar global order have created an increasingly unstable international relations environment, where the US seeks to retain its position against various first and second tier state-based challengers and non-state actors (various terrorist and insurgent forces) (Bennis, 2003; Trautsch, 2015; Simons & Glaser, 2019).[5] These building geopolitical tensions have resulted in the increasing traction and public declaration of the new popular geopolitical narrative, Cold War II or New Cold War.[6] This creeping crisis situation has led to the search for a viable coercive foreign policy mechanism that is indirect and covert in nature to subvert targets in the international system.

When the concept "hybrid warfare" is mentioned or invoked, it is most often associated as being used by Russia against the West and Western interests. The concept of the "Gerasimov Doctrine" is given as being 'proof' of the existence and practice of Russian hybrid warfare. However, when reading the article by Gerasimov (2013) in the original text and not the translations and interpretations of the original text, the content and meaning is something quite different. This is a warning concerning the Western (primarily American) foreign policy conceptualisation and practical use of hybrid warfare against countries targeted for regime change and the call for a viable scientific approach to counteract it.

The term hybrid warfare is an imprecise term that has become a buzzword in security studies and international relations. As such, there are various terms and definitions offered that tend to add further confusion and a distinct lack of clarity (Reichborn-Kjennerud & Cullen, 2016). For the purposes of this chapter, the following description of the activities is offered.

[5] J. Von Mittelstadt & E. Follath, *Interview with Henry Kissinger: 'Do We Achieve World Order Through Chaos or Insight?'* Spiegel Online, www.spiegel.de/international/world/interview-with-henry-kissinger-on-state-of-global-politics-a-1002073-druck.html, 13 November 2014 (accessed 17 November 2014).

[6] W. R. Mead, *Mike Pence Announces Cold War II*, Opinion, The Wall Street Journal, https://wsj.com/articles/mike-pence-announces-cold-war-ii-1539039480?mod=e2fb, 8 October 2018 (accessed 10 October 2018).

> Fomenting armed rebellion by covert (or at least officially denied) arms deliveries, troop contributions and military operations, propping up friendly local leaders, propaganda, promises of economic benefits and threats of economic reprisals. The aim can be regime change or succession of part of the territory (which can then become a puppet state). (Biscop, 2015: 1)

The definition hints at the fact that the information battlefield is prepared in advance of the physical battlefield. This is a form of subversive warfare that falls short of war, and therefore often creates a grey zone threshold for counter-measures by the target and its allies in terms of retaliation against the aggressor. These descriptions are in keeping with other attempts to bring clarity to the hybrid warfare concept (Reichborn-Kjennerud & Cullen, 2016; Wither, 2016: 76–77).[7] It is far from being a new and innovative form of warfare as multiple regime changes in Africa, Asia, Latin America and the Middle East during the Cold War do testify. It is also relevant in the contemporary field of geopolitical practice in global international relations where the US-led West has been engaging in *non-classical war*[8] with target countries around the globe in order to compel these countries to do their will by creating instability and regime change (Manoilo & Strigunov, 2020) through the mechanisms identified by Biscop. The result has been that countries that are targeted (such as China, Iran, Venezuela and Russia), or likely to be targeted by Western hybrid warfare, begin the process of trying to insulate against this indirect and covert form of warfare, which can be narrated as being "democracy resistance". One of the ways of achieving this by the state elites has been to reduce the vulnerabilities from indirect communication into their internal affairs from NGOs through increasingly regulating the NGO legal environment and foreign financing (Gilbert & Mohseni, 2018).[9] Gudrun Persson argues that the high-level Russian thinking on Western hybrid warfare as posing the state a significant threat should not be dismissed as 'conspiracy theories' and should be taken seriously.[10]

[7] J. Ball, *What Is Hybrid Warfare?* Global Security Review, https://globalsecurityreview.com/hybrid-and-non-linear-warfare-systematically-erases-the-divide-between-war-peace/, 10 June 2019 (accessed 22 June 2020).

[8] This is contemporary terminology used to denote "hybrid warfare" in the Russian information warfare community parlance, as the term of hybrid warfare is not a 'native' term but is used to denote the actions of the West against other countries.

[9] P. Felgenhauer, *Fears of Western 'Hybrid Warfare' and Suppression of the Russian Opposition*, The Jamestown Foundation, Eurasian Daily Monitor, 16(115), https://jamestown.org/program/fears-of-western-hybrid-warfare-and-suppression-of-the-russian-opposition/, 8 August 2019 (accessed 22 June 2020).

[10] G. Persson, *Book Review on Russian Thoughts on Hybrid War and Colour Revolutions*, Russian Studies Series 1/20, NATO Defence College, www.ndc.nato.int/research/research.php?icode=625, 16 January 2020 (accessed 22 June 2020).

There is a parallel with the US approach on the eve of the outbreak of the Second World War when there was an attempt to defend national interest and security against any hybrid intrusions by Nazi Germany through the creation of the Foreign Agents Registration Act (FARA), which was at least partly prompted by and based on the successful and influential British propaganda campaigns in the United States during the First World War. Thus, the practical role and application of state-sponsored subversion through a war of ideas is not specific or new to the twenty-first century.

The Act of Subversion and Regime Change within a War of Ideas

Subversion is a powerful tool, which has been used through centuries of human history and endeavour and is a conscious act to overthrow an existing political regime by use of psychological coercion and/or physical force by a 'domestic' actor (even if supported by an external power). Propaganda is used to influence and support subversion, where a proxy force and its foreign backer can cooperate together, in addition to the physical support rendered to a proxy (such as arms, money, intelligence, military support and other supplies) (Beilenson, 1972: v–ix). The US path to understanding and practising subversion was born in the late Cold War at a period when some Neo-Conservative thinkers considered that they were on the losing side, and the reason for this as they saw it was the success of the subversive triumphs of Marxist Revolutionary Warfare. This led to the policy pathway where Marxist Revolutionary Warfare was co-opted and engineered by the US for its own foreign policy and geopolitical purposes, which, for example, is visualised through the use and role played by the Contras in Nicaragua in the 1980s.

The question came to be, how does one turn the tool of subversion upon its heirs? One of the suggestions from the 1970s was that "the United States should give dissidents against all Communist governments protracted sustained aid – initially money for propaganda-agitation, with supplies and arms added where feasible and warranted by a developing situation" (Beilenson, 1972: 241). Ideological and geopolitical interests are a significant motivating factor for an external act to engage in and support the subversion of a chosen foreign government. Beilenson (1972: 252) noted one of the shortcomings of the Cold War approach by the US: "we are carrying on our present political warfare with armed might, diplomacy, treaties, and traditional subversion, three and a half tools against four on the other side." He went further and noted that "in terms of general strategy, political warfare is analogous to war, and passive defence wins neither." Thereby the situation is transformed into advocating for an offensive application of subversion in pursuit of foreign policy and security objectives.

Various considerations detailed above create a situation where subversion and a war of ideas can be tied together for the purpose of initiating regime change against a selected country and a particular government. The framing of this is critical in order to enable a greater chance of success, which is often hidden within the fog of war that has been created by the informational flows concerning the crisis event.

> Winning the "war of ideas" isn't about the triumph of ideas, per se. "Shaping" is a much better metaphor for US goals in the war of ideas. A well-orchestrated shaping campaign ties information efforts at every level and military action to operational or policy goals. Getting people in other nations to "like" or "agree with" US ideas is not a good policy goal. Reducing support for anti-US activities and increasing support for US policies is. (Paul, 2008: 120)

The war of ideas is intended to benefit the communicating party at the expense of the target in a manner where informational flows surrounding the crisis event in the media and public sphere are used to simultaneously increase the operational and policy choices of one actor at the expense of the other. This is achieved through acquiring information dominance in the segments of the information domain that are responsible for shaping opinion and perception of a specific 'crisis' event in focus. It concerns the realisation of information dominance in order to maximise strengths and opportunities whilst minimising threats and weaknesses. One of the key aspects in shaping the war of ideas is the universal applicability of ideology.

> Ideology provides people with a unifying identity and sense of community. It gives them a cause they identify with. It provides a sense of purpose, meaning and shape to their lives. Ideology also provides someone else to blame for a people's misfortunes, and building up an image of an enemy to fight. Perhaps most importantly, ideology offers hope that direct action will make for a better future, either in this life or the next. (Waller, 2007: 21)

Even though this tract of text was intended as the basis for understanding how to counteract political Islam, and the Soviet Union before it, it also provides an interesting framework to understand the Western means of subversion in the framework of hybrid warfare. For example, the assumptions of democratic peace, encompassing the notion that 'democratic' countries do not go to war with each other, denies the fact that democratic countries are much more likely to militarily attack a 'non-democratic' country (Rosato, 2003). This is also infused with the liberal notion and assumptions of having a superior political and economic system to which others wish to aspire. It also assumes a higher ethical and moral purpose, which leads to such conceptual constructs as R2P. This means that "words and images are the most powerful weapons in a war of ideas. Used skilfully, they can serve the cause well" (Waller, 2007: 38). The above-mentioned situation falls within the context and practice of political warfare, which involves the forceful political expression of policy (Codevilla, 1989: 77). These ideas have been seen in practice in the twenty-first century, in terms of revolutionary waves.

Branded revolutions are a form of political warfare. A brand is infused within an ideological component, where the ideology constitutes a cognitive primer and mobiliser by activating a sense of one of Maslow's higher order needs (such as self-actualisation of the active participants); the brand acts as a means of reputation and expectation management of the 'in-group' as well as creating a sense of community. The Colour Revolutions and Arab Spring contain these elements, where the ideology is represented by projected norms and values, such as the illusion and promise of democracy and freedom to participants. It is also meant as a constraint on the targeted government, as logic implies that to resist these 'aspirations' is to go against the will and approval of the rhetorical in-group construct of the *international community* (a form of bandwaggoning and psychological pressure on the target intended to coerce conformity against their interests and even against their survival). The brand is used to call to arms, prime and mobilise a collective mob with the promise of a reward for their active participation and delineates those who are the in-group and 'righteous' through the use of slogans, colours and forms of visual and rhetorical symbolism. Branding is also used in the management of audience expectations, for example the use of "Spring" in "Arab Spring" that promises an 'inevitable' better and more prosperous time ahead (Spring as a metaphor for a period of growth and positive development) for the people with a successful regime change where Western norms and values of individualism and freedom will triumph over collectivism and oppression in a binary depiction of ethical and moral-based 'realities'. This was also seen in the reaction to the results of the Arab Spring, which were predictably disastrous for the region, with the contra invocation of the Arab Winter.

In spite of the façade narrative that the US is a force for good in the world and supports 'democracy' wherever that may be, its record reveals another story altogether. The regular accusations of other international actors (mostly enemy or competing countries) mask the level of US covert and overt action of interference with foreign elections, contrary to its attempted international branding. While admitting to some 'limited' use of covert operations (involving the CIA) to influence another country's elections, the reaction and excuse is that they have moved away from such covert practices "unlike Moscow". Denials of CIA involvement include its current counter-terrorism focus and that the end of the Cold War "robbed the CIA of its long-running purpose: to counter the Soviet Union."[11] One can always be a little suspicious why current and former CIA officials would talk frankly with a researcher. However, other agencies can be involved in overt and covert foreign election meddling, such as the State Department. The reasoned logic used does not have to be true, and when put under scrutiny the logic is not solid. One can argue that the so-called New Cold War has re-invigorated the motive to subvert other countries' elections in an increasingly competitive global environment of geopolitics; the only question is finding the right opportunity to successfully manage an operation. One could point to the coup in Bolivia that saw the overthrow of the elected leader and the establishment of a pro-US regime at a time when the interest in terrorism was declining and a renewed emphasis on getting back to the basics of the Munroe Doctrine in Latin America came into focus. This was also evident in the attempts at political regime change in Venezuela that occurred in the same time period. How is political warfare defined?

"Political warfare is the marshalling of human support, or opposition, in order to achieve victory in war or in unbloody conflicts as serious as war" (Codevilla, 1989: 77). Therefore, subversion can be seen as being a constituent part of political warfare; it is a coordinated and calculated pursuit that seeks to gain political power through the use of coercion and force that is applied at the right place and point in time. Furthermore, "political warfare, then, is a broad concept involving acts both overt and covert" (Codevilla, 1989: 79). This can involve the use of such assets and tools as grey and black propaganda, support for foreign groups and the agents of influence as well as a wide array of different non-state and state-based actors shaping the information domain in order to influence the cognitive domain of the target audience.

The 'glue' that holds any given society together operates at two different and distinct levels: mass and elite. Mass society (those who are governed), the public, is held together with common emotionally laden symbols of meaning – flags, national songs, national sports, coats of arms, ideology, religion and so forth. At the elite level (those who govern), this glue is found in the perception of common mutual and pragmatic interests, such

[11] D. Shimer, *When the CIA Interferes in Foreign Elections: A Modern Day History of Covert Action*, Foreign Affairs, www.foreignaffairs.com/articles/united-states/2020-06-21/cia-interferes-foreign-elections, 21 June 2020 (accessed 25 September 2020).

as political, social or economic power. In order to increase the chances of success of subversion, a targeted society needs to be split: mass society is separated emotionally from elite society; and the elite society divided along the lines of breaking a sense of pragmatic mutual interests. This has been clearly seen in the Colour Revolutions, for example, where mass society was primed and mobilised through common colour symbols (including flags) and common simple emotional slogans ("Pora" meaning "it is time" – in the Orange Revolution in Ukraine). The powerful foreign broker(s) also attempt to split the elite and therefore the decision-making capability and capacity to successfully resist subversion by use of carrot and stick. The carrot strategy, for example, could be by the use of financial or other incentives to change sides or use of bandwaggoning logic to join the 'righteous' or 'winning' side. And the stick strategy, for example, by the selective use of targeted economic sanctions or travel bans against key members of the ruling political and economic elite of the target country under the guise (false logic argument) of humanitarianism by punishing the 'bad' elite because they are denying the 'good' revolutionaries from achieving self-actualisation.

One of the recent concepts that is used to provoke a rethinking in the way international relations are practiced is the idea of *guerrilla diplomacy*. The genesis of the idea as Copeland (2004; 2009) has argued is that diplomats and diplomacy are being increasingly marginalised in an era of increasing conflict, and when they are needed the most. The drivers behind this are globalisation and the increasing militarisation of foreign policy. A result of this situation is that there is a "dominance for a few and a dependence for many" (Copeland, 2009: 50). The role of guerrilla diplomacy as envisaged by Copeland (2009: 161; Copeland, 2018: 376–377) was a new-style Foreign Service officer that was capable of autonomous action and flexible in deployment, engaging in networking and advocacy with the local population, which places the practice as a subset of public diplomacy. Copeland (2004; 2009; 2018; Copeland & Potter, 2008) argued for the role of diplomats to be to bring or maintain peace, rather than to foster conflict, although it does involve a foreign country striving to secure its goals and interests in a foreign country. However, the idea is not without its criticism.

> Insurgency and public relations can be used for ill as well as good; guerrilla or public diplomats with bad intentions might well have similarly dramatic effects as those envisaged by unconventional diplomats with good ones. We lack in the contemporary international system any kind of agreement as to why, when and how we ought to speak to foreign publics and try to shape their views. Without that, the guerrilla risks being treated as a saboteur and the public diplomat a propagandist. (Hall, 2010: 256)

These reservations are justified given the current fog of communication in foreign policy and international relations that is used to shape the cognitive domain. There are stark contrasts and contradictions seen in how subversive events are depicted and the different sides involved in the conflict. "In the majority of countries in the internal affairs in which the United States intervenes as "peacemaker", they bet and cooperate with the very political forces and regimes that the world calls "terrorist" and "extremist"" (Karpovich & Manoilo, 2015: 112). A problem that was verified in the Podesta emails that were released, when one of those emails to Hillary Clinton noted that Al Qaeda and the US were in effect allies and working for the same goals in Syria. The email dated 12 February 2012 from Jacob Sullivan of the State Department to Hillary Clinton stated that "AQ [Al Qaeda] is on our side in Syria."[12] Thus there are contradictory and nonsense slogans and catch phrases such as "the enforcement of democracy", that is, that 'democracy' is something that needs to be imposed on a country and its people by a more militarily powerful actor. Not to mention that rhetorically the US was fighting terrorism but working for the same purpose in Syria.

The US and its allies champion themselves as being brokers in the conflicts of the Arab Spring, where individual events in the chain were characterised and narrated as being a crisis of humanitarian values, norms and practice that require an urgent political solution along the lines outlined in the crisis section of this chapter. It is readily apparent though that they are not an honest broker, but a powerful one. A political entity is favoured over the others and is assisted to power. In spite of the attempts to try to label the Syrian conflict as a 'civil war', the August 2012 Defence Intelligence Agency report clearly labelled it as a proxy war (DIA, 2012), which clearly places it within the framework and understanding of this section. This statement by the Defence Intelligence Agency is further substantiated in the research of Oleg Karpovich and Andrei Manoilo (2015: 111–130). Interestingly, studies have shown that foreign imposed regime change generally does not improve relations between the states involved (Downes & O'Rourke, 2016). Even if the policy of regime change does not actually bring the desired and expected results, it is about attempting to leverage advantages and opportunities and minimise those for the opponent.

The act of successful regime change is a symbolic event that at the moment of its achievement can be constructed into an iconic event to capture this moment and to inspire future events. Use was made of

[12] US Department of State Case No. F-2014-20439, Doc No. C05789138, Date: 10/30/2015.

subversion during the Cold War in different manners by different actors, to halt the progress of Communist influence, such as in Guatemala in 1954 (Gordon, 1971), 1954 in Vietnam, 1957 in Indonesia or 1962 in Burma (Kahin & Kahin, 1995), and in Southeast Asia where attempts were made to counter Communist subversion (McKnight, 2005). All these actions resulted in disastrous consequences for the citizens of these countries. During the Cold War (between the years 1947–1989), the US tried to change other countries' governments on some seventy-two different occasions. Moreover, sixty-six of these were covert operations and six involved overt regime change operations. Of these covert operations, twenty-six of them were successful in bringing a US-backed government to power and the remaining forty failed. Tactics varied along with the results; supporting militant groups to topple a foreign government resulted in five successful operations and thirty-one failed attempts. However, sponsoring coups proved to be more successful with nine out of fourteen attempted coups bringing a US-backed leader to power. There was also the method of influencing foreign elections (covert funding, advising and spreading propaganda for preferred candidates), sometimes beyond a single election cycle, on sixteen identified occasions where US-backed candidates won their elections 75 per cent of the time.[13] As noted by Luttwak (1979), who confirms the above-mentioned research review, the coup is the most frequently attempted method of changing government and the most successful means of achieving it. Luttwak (1979: 27) defines a coup as consisting "of the infiltration of a small but critical segment of the state apparatus, which is then used to displace the government from its control of the remainder." This is not without risk of course, even if regime change is successfully achieved.

However, a great deal of uncertainty and unpredictability can result from such acts, even when successful; a US-backed foreign leader can face increasing pressure from domestic opposition if they do the bidding of the US at the expense of the local population. Furthermore, successful regime changes can lead to severe instability and conflict in the target country, such as Iraq and Libya. But there is also a long history of it.

> The so-called "Bruce-Lovett Report", to which he (Ambassador Joseph P. Kennedy) was a signatory, described CIA coup plots in Jordan, Syria, Iran,

[13] L. A. O'Rourke, *The US Tried to Change Other Countries' Government 72 Times during the Cold War*, Monkey Cage, The Washington Post, www.washingtonpost.com/gdpr-consent/?next_url=https%3a%2f%2fwww.washingtonpost.com%2fnews%2fmonkey-cage%2fwp%2f2016%2f12%2f23%2fthe-cia-says-russia-hacked-the-u-s-election-here-are-6-things-to-learn-from-cold-war-attempts-to-change-regimes%2f, 23 December 2016 (accessed 24 December 2016).

> Iraq and Egypt, all common knowledge on the Arab street, but virtually unknown to the American people who believed, at face value, their government's denials. The report blamed the CIA for the rampant anti-Americanism that was mysteriously taking root "in many countries in the world today." The Bruce-Lovett Report pointed out that such interventions were antithetical to American values and had compromised America's international leadership and moral authority without the knowledge of the American people. The report also said that the CIA never considered how we would treat such interventions if some foreign government were to engineer them in our country.[14]

However, this policy has continued and even intensified with the Global War on Terrorism, the Colour Revolutions, Arab Spring and so forth. For example, the iconic media event of the Libyan woman that entered the hotel occupied by foreign journalists in Benghazi to tell them of the systematic mass rapes committed by Libyan government forces; the allegation was the basis for formulating the crisis of humanitarian values at stake that required an urgent political solution that led to a further chain of events. This began with the support of 'rebels' and attempts to split the Libyan elite; when this proved insufficient to complete the regime change, NATO airpower was overtly used for the purpose under the guise of R2P, which earned it the reputation in some quarters as "Al Qaeda's Airforce". The result has been an increasingly dysfunctional and hostile environment of international relations and politics. There are an increasing number of dire predictions concerning the re-emergence of great power conflict within the context of the New Cold War that has seen tensions between the US-led Western global hegemony and non-Western powers, such as China and Russia. However, some argue that if or when the conflict comes it will not take the traditional form of previous conflicts.

> Should great-power conflict come, however, it will bear little resemblance to the traditional interstate wars analysts study, that academics teach, and for which militaries train. Those wars rarely occur anymore, and that is a good thing for humanity. Instead conflict plays out indirectly, through a kind of proxy warfare called "foreign subversion." Foreign subversion is a covert, indirect form of modern statecraft. It involves empowering illicit and armed non-state groups that act as extensions of a sponsor state.[15]

[14] R. F. Kennedy (Jr.), *Why the Arabs Don't Want Us in Syria*, Politico, www.politico.com/magazine/story/2016/02/rfk-jr-why-arabs-dont-trust-america-213601, 23 February 2016 (accessed 29 February 2016).

[15] M. M. Lee, *Subversive Statecraft: The Changing Face of Great-Power Conflict*, Foreign Affairs, www.foreignaffairs.com/articles/2019-12-04/subversive-statecraft, 4 December 2019 (accessed 19 June 2020).

This form of unpredictable and extremely volatile type of tension and conflict can already be seen in contemporary international relations and politics: in Latin America, Africa, Eastern Europe, Asia and the Middle East. The US-led West's attempts to retain its position of global economic and military dominance is likely working against its own interests and security via heavy-handed tactics and destructive results on the populations in the areas of geopolitical contestation. The resulting negative attitudes are often dismissed deceptively and wrongly as hating 'our freedoms' or of anti-Americanism. This situation leads to unstable domestic situations in the West, especially given the physical and emotional connections between migrant communities and their sense of home region and people that can be inflamed through subversive foreign policy. The problem is illustrated in the environment of subversion and counter-subversion in Muslim migrant communities in Europe in an effort to stoke terrorism or insurgency and on the other hand to prevent it (Kilcullen, 2007). However, the long-term consequences of subversive foreign policy can be overlooked or ignored in favour of myopic short-term objectives.

In order to gain a freer hand in exercising foreign policy and pursuing those objectives, which include practices such as regime change, the US needed to change the manner in which the rules and boundaries of international relations were conceived and practised. To meet this situational objective, the idea of the centuries old practice of the Westphalian system needed to be broken down as it presented a hindrance to the selective application of political, economic, diplomatic and military power against a selected government that had been designated for regime change. Hence the first section of this paper considers the issue of invoking a crisis that in turn leads to a political mobilisation for an apparent 'resolution' to the situation that quickly and logically follows. One of the most recent and useful tools for the purposes of skirting the legal and ethical issues of regime change within the Westphalian system is to invoke the notion of R2P, which is a highly fluid concept (bordering upon slogan) that superficially appears to be benevolent. In fact, it is so vague that it can be moulded to almost any purpose if the event is 'correctly' narrated and the reality is sufficiently surrounded by the fog of war created by a sustained effort of propaganda and subversion.

Conclusion

This chapter's research question is: what are the necessary information and cognitive domain environmental conditions that need to be prepared and set for attempting the act of regime change within the context of hybrid warfare? In order to answer this question, it is necessary to contextualise and describe the interactions and effects of the information and cognitive domain on the perceptions and understandings of the physical domain.

The aim of hybrid warfare is, in this instance, regime change in the physical domain. It is a clear-cut foreign policy goal, but one that does not come without tangible and intangible risk. Often the aim of regime change is very pragmatic and tangible in nature too, as a means of installing a leader who will do the bidding of the sponsoring state that put them into power. Therefore, the short-term physical domain goals are often logical, even if they are flawed. To commit an act of subversion that leads to regime change requires a perceived legitimate cover to enact, and this requires an extraordinary situation rather than an ordinary one. As the logic can be stretched that extraordinary times require extraordinary measures, a crisis in its definition is an extraordinary event and one that must be urgently resolved. This needs to be communicated through the information domain.

On the surface, there is the oft-held excuse of journalistic good practice of short-cuts owing to the highly competitive 24/7 news cycle. Mass communication possibilities give rise to the prospects of mass influence results, in order to control various informational streams that give life and meaning to physical events through symbolic representation and interpretation. A result of this is to narrow the permissible breadths of describing a particular event as there is a consensus that forms around some central ideas and narratives. In other words, information is transformed as orthodoxy of knowledge on the given topic, which has the effect of excluding alternatives, even if the knowledge is based on faulty and/or inaccurate information. Subversion often begins in the intangible realm first (to influence perception, opinion and decision making), which assists in the tangible realm by making political and military opposition to the attempts to take power less effective. Mainstream liberal mass media tend to rather uncritically carry the popular governmental geopolitical narratives that are not only stereotypes but provide the façade of legitimacy that glosses over legal and ethical contradictions to the intended act. One example is the notion of American exceptionalism in the global system, where other countries committing the policy acts would face heavy criticism and sanctions.

Media are being used to try and shape public perception and consciousness with an emotional logic that is designed to prime and mobilise them politically into accepting foreign policy that is not only greatly flawed, but morally and ethically completely bankrupt. If emotions and sentiment are ignited there is a greater likelihood that the public will behave in a manner that does not benefit their own interests. What is required in this day and age is a clear and consistent use of rational logic to debunk these informational programmes that are intent on getting us to accept and believe in courses of action that are not just or acceptable in any way, shape or form to us and to the targets of the foreign policy. It is time to take a pause, remembering back to those narratives of justification and the promises of a better future for the people that were used in justifying the overthrow of Gadaffi in Libya by use of military force; the moment of his capture, torture and execution forms a rather gruesome crisis iconic event moment. There is nothing humanitarian about the process or the final result, rather it is a cynical and calculated act of inhumanity dressed as a 'good' cause.

Hybrid warfare is a covert and indirect means of waging war on an opponent, it can come in different forms that include subversion and is not always successful. There are a number of different physical, informational and cognitive elements that increase or decrease the chances of conducting a successful regime change operation. This includes physical elements, such as the amount of time and resources expended, assuming that they are used in the most effective manner (directed at the weakest points in the target), and the ability of the target to be able to effectively deploy sufficient resources to counter subversion attempts where it will be the most effective. However, the information domain can have greater effects on the physical and psychological environments of the contested space. Given the covert nature of the actual subversive acts in hybrid warfare, but the overt marketing and publicising of the crisis aspects of the event, it creates a dualistic and contradictory information environment of uncertainty and risk. This space can shape the cognitive domain of the various actors on all sides – decision makers, practitioners and the public. This is the key, where the control, or lack of control, of the information flows around the crisis event in international relations yields or cedes the struggle for influencing perception and influencing their opinions and actions that may either win or lose the cognitive battle that will decide the outcome of the politicised physical war. The information domain forms the mechanism that connects the physical and cognitive domains across the political and battle space as well as the public global information space.

References

Badescu, C. G. & Weiss, T. G. (2010), Misrepresenting R2P and Advancing Norms: An Alternative Spiral?, *International Studies Perspectives*, 11(4), pp. 354–374.

Beilenson, L. W. (1972), *Power Through Subversion*, Washington DC: Public Affairs Press.

Bennis, P. (2003), *Before and After: US Foreign Policy and the War on Terrorism*, New York: Olive Branch Press.

Biscop, S. (June 2015), *Hybrid Hysteria*, Egmont Security Brief No. 64, Egmont Royal Institute for International Relations.

Boin, A., t'Hart, P., Stern, E. & Sundelius, B. (2005), *The Politics of Crisis Management: Public Leadership Under Pressure*, New York: Cambridge University Press.

Brecher, M. (1996), Introduction: Crisis, Conflict, War – State of the Discipline, *International Political Science Review*, 17(2), pp. 127–139.

Chesterman, S. (2011), "Leading from Behind": The Responsibility to Protect, the Obama Doctrine, and Humanitarian Intervention After Libya, *Ethics & International Affairs*, 25(3), pp. 279–285.

Codevilla, A. M. (1989), Political Warfare in Lord, C. & Barnett, F. R. (Eds.), *Political Warfare and Psychological Operations: Rethinking the US Approach*, Washington DC: National Defence University Press Publications.

Cohen, S. B. (2003), Geopolitical Realities and United States Foreign Policy, *Political Geography*, 22(1), pp. 1–33.

Coombs, W. T. (2014), *Applied Crisis Communication and Crisis Management: Cases and Exercises*, Thousand Oaks: Sage.

Copeland, D. (2018), Innovation, Adaptation and Foreign Policy in the Age of Globalisation: Is Global Affairs Canada Fit for Purpose? *Canadian Foreign Policy Journal*, 24(3), pp. 372–377.

(2009), *Guerrilla Diplomacy: Rethinking International Relations*, Boulder: Lynne Rienner Publishing.

(2004), Guerrilla Diplomacy: Delivering International Policy in a Digital World, *Canadian Foreign Policy Journal*, 11(2), pp. 165–175.

Copeland, D. & Potter, E. V. (2008), Public Diplomacy in Conflict Zones: Military Information Operations Meet Political Counter-insurgency, *The Hague Journal of Diplomacy*, 3(3), pp. 277–297.

Davidson, W. P. (1971), Some Trends in International Propaganda, *Annals of the American Academy of Political and Social Sciences*, 398(1), pp. 1–13.

Defence Intelligence Agency (DIA) (August 2012), Information report, 14-L-0552/DIA/287.

DiMaggio, A. R. (2009), *Mass Media, Mass Propaganda: Examining American News in the "War on Terror"*, Lanham: Lexington Books.

Downes, A. B. & O'Rourke, L. A. (2016), You Can't Always Get What You Want: Why Foreign Imposed Regime Change Seldom Improves Interstate Relations, *International Security*, 41(2), pp. 43–89

Ettinger, A. (2017), The Patterns and Implications of American Military Interventions in the Post-September 11 Era, *Canadian Foreign Policy Journal*, 23(3), pp. 205–217. https://doi.org/10.1080/11926422.2017.1341842

Gerasimov, V. (27 February 2013), Ценность науки в предвидении: Новые вызовы требуют переосмыслить формы и способы ведения боевых действий (The Value of Science in Anticipation: New Challenges Need to Rethink the Forms and Methods of Warfare), Военно-промышленный курьер (*Military-Industrial Courier*), www.vpk-news.ru/articles/14632.

Gilbert, L. & Mohseni, P. (2018), Disabling Dissent: The Colour Revolutions, Autocratic Linkages, and Civil Society Regulations in Hybrid Regimes, *Contemporary Politics*, 24(4), pp. 454–480, DOI: 10.1080/13569775.2018.1471645.

Gordon, M. (1971), A Case History of US Subversion: Guatemala, 1954, *Science & Society*, 35(2), pp. 129–155.

Gowing, N. (2009), *'Skyful of Lies' and Black Swans: The New Tyranny of Shifting Information Power in Crises*, Oxford: Reuters Institute for the Study of Journalism.

Grigoryan, A. (2020), Selective Wilsonianism: Material Interests and the West's Support for Democracy, *International Security*, 44(4), Spring, pp. 158–200.

Hall, I. (2010), The Transformation of Diplomacy: Mysteries, Insurgencies and Public Relations, *International Affairs*, 86(1), pp. 247–256.

Hughes, D. (2015), Unmasking an Exception: A Critical Genealogy of US Exceptionalism, *Review of International Studies*, 41(3), pp. 527–551.

Kahin, A. R. & Kahin, G. M. (1995), *Subversion As Foreign Policy: The Secret Eisenhower and Dulles Debacle in Indonesia*, New York: New Press.

Karpovich, O. & Manoilo, A. (2015), *Colour Revolutions: Techniques in Breaking Down Modern Political Regimes*, Bloomington: Author House.

Kilcullen, D. J. (2007), Subversion and Countersubversion in the Campaign Against Terrorism in Europe, *Studies in Conflict & Terrorism*, 30(8), pp. 647–666.

Leavy, P. (2007), *Iconic Events: Media, Politics, and Power in Retelling History*, Lanham (MD): Lexington Books.

Luttwak, E. (1979), *Coup d' État: A Practical Handbook*, Cambridge: Harvard University Press.

Manoilo, A. V. & Strigunov, K. S. (2020), *Technologies of Non-Classical War: Genesis, Evolution, Practice*, Moscow: Goryachaya Liniya – Telekom (in Russian).

Marlin, R. (2002), *Propaganda and the Ethics of Persuasion*, Ontario: Broadway Press.

Mastanduno, M. (1997), Preserving the Unipolar Moment: Realist Theories and US Grand Strategy After the Cold War, *International Security*, 21(4), pp. 49–88.

McCormack, T. (2010), The Responsibility to Protect and the End of the Western Century, *Journal of Intervention and Statebuilding*, 4(1), pp. 69–82.

McKnight, D. (2005), Western Intelligence and SEATO's War on Subversion, 1956–63, *Intelligence and National Security*, 20(2), pp. 288–303.

Morris, J. (2013), Libya and Syria: R2P and the Spectre of the Swinging Pendulum, *International Affairs*, 89(5), pp. 1265–1283.

Nohrstedt, D. (2011), Uncertainty, Accountability, and the Conduct of Post-Crisis Inquiries in Svedin, L. (Ed.), *Ethics and Crisis Management*, Charlotte: Information Age Publishing.

Ogrizek, M. & Guillery, J.-M. (1999), *Communicating in Crisis: A Theoretical and Practical Guide to Crisis Management*, New York: Aldine De Gruyter.

Paul, C. (2008), *Information Operations Doctrine and Practice: A Reference Handbook*, Westport: Praeger Security International.

Payne, K. (Spring 2005), The Media as an Instrument of War, *Parameters*, pp. 81–93.

Reichborn-Kjennerud, E. & Cullen, P. (2016), *What Is Hybrid Warfare?* Policy Brief #1, Oslo: Norwegian Institute of International Affairs.

Regan, P. M. & Aydin, A. (2006), Diplomacy and Other Forms of Intervention in Civil Wars, *Journal of Conflict Resolution*, 50(5), pp. 736–756.

Rosato, S. (2003), The Flawed Logic of Democratic Peace Theory, *American Political Science Review,* 97(4), pp. 585–602.

Shabo, M. E. (2008), *Techniques of Propaganda and Persuasion*, Clayton: Prestwick House Inc.

Simons, G. (2019), Syria: Propaganda As a Tool in the Arsenal of Information Warfare in Baines, P., O'Shaughnessy, N. & Snow, N. (Eds.), *The Sage Handbook of Propaganda*, London: Sage, pp. 443–460.

(2016), News and Syria: Creating Key Media Moments in the Conflict, *Cogent Social Sciences,* 2(1), DOI: 10.1080/23311886.2016.1170583.

Simons, G. & Chifu, I. (2017), *The Changing Face of Warfare in the 21st Century*, London: Taylor and Francis.

Simons, G. & Glaser, M. (2019), New Cold War and the Crisis of the Liberal Global Order, *Outlines of Global Transformations: Politics, Economics, Law*, 12 (3), pp. 61–77.

Sussman, G. (2007), Globalising Politics: Spinning US 'Democracy Assistance' Programmes in Dinan, W. & Miller, D. (Eds.), *Thinker, Faker, Spinner, Spy: Corporate PR and the Assault on Democracy*, London: Pluto Press.

Thakur, R. (2013), R2P After Libya and Syria: Engaging Emerging Powers, *The Washington Quarterly,* 36(2), pp. 61–76.

Trautsch, J. M. (2015), Who's Afraid of China? Neo-Conservative, Realist and Liberal-Internationalist Assessments of American Power, the Future of "the West" and the Coming New World Order, *Global Affairs*, 1(3), pp. 235–245.

Waller, J. M. (2007), *Fighting the War of Ideas Like a Real War: Messages to Defeat the Terrorists*, Washington DC: The Institute of World Politics Press.

Western, J. (2005), *Selling Intervention and War: The Presidency, the Media, and the American Public*, Baltimore: Johns Hopkins University Press.

Willcox, D. R. (2005), *Propaganda, the Press and Conflict: The Gulf War and Kosovo*, London: Routledge.

Wither, J. K. (2016), Making Sense of Hybrid Warfare, *Connections: The Quarterly Journal*, 15(2), pp. 73–87.

9 Geopolitics in the Age of Social Media: The Struggle for Influence on Ukraine

Greg Simons

Introduction

In the last decades geopolitics has been experiencing a revival of interest, both as a political pursuit and as an academic interest (Hepple, 1986). The end of the Cold War and the breakup of the Soviet Union were two events that significantly changed the global geopolitical landscape. This created a situation where the one remaining superpower was able to further increase its influence in territory that had once been the preserve of its Cold War rival. This situation caused a growing resentment in the new Russia (Suslov, 2014), which asserted itself in accordance with the traditional means of classical hard power geopolitics in the 2008 Georgian-Russian War. Another conflict emanating from geopolitical competition has emerged in Ukraine with the events of Euromaidan (civil unrest) from late 2013, the overthrow of President Yanukovych of Ukraine and the ensuing violence that followed. However, this time, geopolitics has been waged according to the rules of the new social media-based information age.

Therefore, it is critical to understand and address the issue of how information, and especially new media, connects with geopolitics and conflict. The nature of, and relationships between, the mass media, information and the public have changed. "First and foremost, public relations practitioners need to realise that managing the media is no longer an option. The media is now anyone with a cell phone, a laptop, a digital camera, or a tape recorder. With the advent of Wikipedia, You Tube, and, of course, blogs, the concept of the media as a gatekeeper or content provider is laughable" (Duhé, 2007: xiv). A situation is created where it is difficult to uniformly control what people think, owing to the nature of the information environment, but it is possible though to influence what people think about.

One of the ways to influence what people think about is to create relationships that are based on shared values and norms, and using selective emotion, which can unite and/or divide communities. The emotions that most affect social mobilisation and political behaviour are fear and enthusiasm, which are linked to the motivations of avoidance and approach (Castells, 2012: 13–14). The focus of this chapter is to understand and analyse the use of new media assets in the Ukraine conflict, with attention paid to the events of the Euromaidan in Ukraine and how this was characterised by selected pro-Euromaidan[1] and anti-Euromaidan platforms that publish material in the English language. The research question is: how do the opposing sides use the social media information domain in order to influence the audience's cognitive domain on a geopolitically contested event? It is important to emphasize that this chapter was written before the current Russia–Ukraine War that began on 24 February 2022.

The actor's reward for effective communication is the potential of being able to shape the accepted public narrative that defines the constructed reality of an event at the expense of other competing communicated realities. Given the time focus of this chapter, the period during the Western-supported regime change of Yanukovych, and before the consolidation of the new regime in Kyiv, the importance of attempting to set the stereotypical geopolitical narrative was more than a solely governmental task (among all interested parties) in order to try to create and maintain the orthodoxy of knowledge on the issue. Concepts and theories that are used in this chapter are defined, such as the concepts of persuasion and influence. The theory of geopolitics and Web 2.0 communication and how the two relate to each other is tackled next. Following this, the case studies, of pro-Euromaidan and anti-Euromaidan new media platforms engaged in political mobilisation and shaping perception and influencing different publics, are then introduced. Each is put in a background context of the politics that shaped the communication style and content of these different Web 2.0 platforms.

Conceptual Considerations

Persuasion and Influence

The representation and practice of international relations events and processes are composed of a competitive marketplace of contradictory

[1] Pro- and anti-Euromaidan refers to the geopolitical positioning of actors in this conflict, pro being those in favour of the revolution and regime change that ousted Victor Yanukovych, and anti being those against the process of regime change, which coincide with the positions of the most significant foreign actors in the conflict – the United States and Russia.

and clashing ideas, each vying for influence by persuading target audiences. Persuasion involves five different components: 1) It is a symbolic process, 2) in which people persuade themselves, 3) which involves an attempt to influence and 4) the transmission of a message and 5) which requires free choice (Perloff, 2010: 12–15). In terms of impact, persuasion can be used for three broad effects. One effect is to *shape* attitudes and opinions on something. A second use is to *reinforce* attitudes and opinions in an audience. The third effect is to *change* attitudes and opinions (Perloff, 2010: 24–25; Jowett & O'Donnell, 2012: 33–34). This is achieved through the control of a two-way flow of information. "The most successful movements simultaneously seek information about their targets while managing facts about themselves" (Bob, 2007: 52). There may in fact be different publics that need to be segmented as there may be incompatible goals between the different publics, for example an international public and a foreign government. The goal may be to garner support for regime change (although formulated in a more palatable and diplomatic form) among the international public, and to try and limit the perceived options available to a government.

There has been a refinement of tactics used in regime change that try to reduce the overall perception of deep-rooted national interests in the process of regime change, which uses a variety of different state and non-state actors as well as international and domestic (in the target country). Keck and Sikkink make an interesting observation in the opening of their book on transnational advocacy networks.

> World politics at the end of the 20th century involves, alongside states, many non-state actors that interact with each other, with states, and with international organisations. These interactions are structured in terms of networks, and transnational networks are increasingly visible in international politics. (Keck & Sikkink, 1998: 1)

International relations, including the act of regime change, in the twenty-first century are no longer the sole preserve of state-based actors and include an increasing wide variety of different non-state-based actors and interests. It is remarked that "advocacy networks are helping to transform the practice of national sovereignty" (Keck & Sikkink, 1998: 2). However, these networks can and are used for other purposes as well; by eroding the concept and strength of national sovereignty, it potentially exposes a targeted country to informational attack for the purposes of regime change.

"Despite their differences, these networks are similar in several important respects: the centrality of values or principled ideas, the belief that individuals can make a difference, the creative use of information, and the employment by non-governmental actors of sophisticated political strategies in targeting their campaigns" (Keck & Sikkink, 1998: 2). These groups often work in areas where there are high-value content and informational uncertainty. Information exchange is a central aspect in their relationships, which is used to mobilise informational strategy in order to persuade, pressure and influence their target group (Keck & Sikkink, 1998). These groups seek to influence the political environment by trying to use the implementation of certain norms and values as leverage.

Critical Geopolitics and Communication

Kelly (2006) contends that there are two distinct versions of geopolitics that contend in their academic formats as *classical* and *critical*. This chapter will employ the theoretical frame of critical geopolitics, which is defined as a "problematizing theoretical enterprise that places the existing structures of power and knowledge in question" (Ó Tuathail, 1999: 107). It consists of a broad set of literatures and tendencies. It has been argued that both of the variations of geopolitics are equally 'correct' and can complement each other in informing an academic problem (Kelly, 2006: 49–50). In the post-Cold War era of international relations and politics, critical geopolitics has been identified as being of policy relevance for a number of reasons.

> Critical geopolitics is relevant to policy making in that it can help deconstruct the persistence of such stereotypical geopolitical conceptions and notions in popular and political culture. With its sensitivity to geographical difference and its critique of ethnocentrism, it forces strategic thinking to acknowledge the power of ethnocentric cultural constructs in our perception of places and dramas occurring within them. (Ó Tuathail, 1999: 117)

The warning by Ó Tuathail of the persistence of stereotypical concepts in popular and political culture remains to the present day, for example, the New Cold War geopolitical narrative construct. These constructions create a lack of clarity in this concept, and also reveals the presence of qualities of traditional geopolitics that have come to be in popular usage (Ciuta & Klinke, 2010: 330). Although there are a number of issues and problems identified with using the New Cold War narrative, it is highly politically symbolic in signalling the re-emergence of a geopolitically divided Europe of conflicting and opposing poles (Orenstein, 2015). With the conflict in Ukraine emerging, the New Cold War geopolitical narrative became an increasingly popular trope used in mass media framing of this conflict (Ojala & Pantti, 2017). The symbolism of highly selective language and its meanings play a crucial role in priming and mobilising global audiences.

This is made easier to communicate by an increasingly diverse number of actors and interests with the development of communicational tools such as social media. Mathew Fraser speaks of *geopolitics 2.0*, and he notes three significant shifts: 1) States to individuals; 2) real world to virtual world mobilisation and power; 3) old media to new media. He also notes that states have reacted to these changes by either censoring and/or deploying web platforms to achieve their goals and assert their influence (Fraser, 2009). Maréchal (2017) has noted that the role and significance of the *geopolitics of information* is likely to play an increasingly important and significant role in shaping and influencing international relations. The idea of geopolitics 2.0 fits well with the concept of popular geopolitics, which is associated with the broader literature on critical geopolitics (Dittmer & Dodds, 2008: 440). Popular geopolitics is traditionally focused on the elite visions of media moguls, movie directors and media workers as the focus has been on media and popular culture artefacts (Dittmer & Gray, 2010: 1664). This has led to the call for scholars of popular geopolitics to "complement an interest in the discursive analysis of representations with a concern for audiences and the meanings that they construct out of popular culture and related texts" (Dittmer & Dodds, 2008: 453). Web 2.0 provides the audience (as prosumers) with a mechanism and a forum, which can be searched and analysed to engage in this form of academic enquiry.

Web 2.0

Significant differences exist in executed communication strategies that are based within traditional media (newspapers, TV and radio), and those that are run in the new media (Cunningham, 2010: 110). The failure to appreciate the differences and to adapt can put a communicator at a significant disadvantage. Social media is far from being a neutral platform and has changed the conditions and rules for social interaction. In theorising social media logic, van Dijck and Poell (2013: 2) identify four grounding principles, which are programmability, popularity, connectivity and datafication. These principles are noted as becoming increasingly entangled with mass media logic. Popularity and connectivity are the principles of greatest interest to this chapter. The term Web 2.0 is a specific one, which dates back to possibly as early as 2004 by some accounts, occupying a segment of cyberspace. It is a highly interactive form of many to many communications.

> The social web, often referred to as Web 2.0, is made up of a second generation set of software applications, enabling users to collaborate, work, and share online.

It is characterised by popular web applications such as You Tube, Facebook, Twitter, Wikis, and myriad others. (Richter, 2012: 107)

As noted above, the communication in new media is a dialogic form, which is based upon a many-to-many message flow. Simultaneously the sending and receiving of information takes place. This differs from the traditional media sphere where monologic communication takes place based upon one-to-many communication flows, where one individual or group sends a message to many different people. Traditional media content creates consumers, whereas with new media the audience are prosumers (both consumers and producers of content) (Cunningham, 2010: 111). In terms of political activity and activism in the Internet age, Castells notes several critical aspects.

First, social movements in the Information Age are essentially mobilised around cultural values. [...] In this context, communication of values, mobilisation around meaning, become fundamental. Cultural movements (in the sense of movements aimed at defending or proposing specific ways of life and meaning) are built around communication systems – essentially the Internet and the media – because they are the main way in which these movements can reach out to those who would adhere to their values, and from there to affect the consciousness of society as a whole. (Castells, 2003: 140)

Aspects such as expressed values and cultural identification of the group or network are an important foundation for establishing communication. These also form the basis for appealing to a broader audience, especially when this involves lobbying and influencing activities. A certain appeal and angle need to be presented to the desired target audience(s). Answers need to be provided to the questions: who are we and what are our values? Social movements have other factors to consider as well.

The second feature characterising social movements in the network society is that they have to fill a gap left by the crisis of vertically integrated organisations inherited from the industrial era. [...] loose coalitions, semi-spontaneous mobilisations, and ad hoc movements of the neo-anarchist brand substitute for permanent, structured, formal organisations. Emotional movements, often triggered by a media event, or by a major crisis, seem often to be more important sources of social change that the day-to-day routine of dutiful NGOs. (Castells, 2003: 140–141)

As noted above, the social movement needs to be able to position itself and give context within the frame of the society in which it is present. There should be some kind of perceived gap that has been created by the absence of other organisations (including those of the state). This answers the question, 'why are we here?'.

There is a third major factor specifying social movements in our age. Because power increasingly functions in global networks, largely by-passing the institutions of the nation-state, movements are faced with the need to match the global reach of the powers that be with their own global impact on the media, through symbolic actions. (Castells, 2003: 142)

This final factor revealed by Castells demonstrates the need to not only clearly define the earlier aspects, but the social movement needs to be noticed and its words and deeds clearly present in the information sphere. Social media provides social movements with a platform to enable the geopolitics of civic engagement and the possibility to deploy it as an instrument by a foreign power to promote its geopolitical agenda (Staeheli & Nagel, 2012). The informational dimension is critical to the life of social movements that must not only make sense of their purpose, but be able to project their purpose.

"The processes of conflictive social change in the Information Age revolve around the struggles to transform the categories of our existence, by building interactive networks as forms of organisation and mobilisation" (Castells, 2003: 143). This exerts an impact upon messages that enter the new media sphere, whereby the original meaning of messages can be entirely transformed.

Method

A research project that interviewed protestors in Kyiv found that 49 per cent of respondents learned about the protests from Facebook, 35 per cent from VKontakte and 51 per cent from different and diverse Internet news sites.[2] Facebook was the favourite form of social media, which has been confirmed in subsequent discussions with Ukrainian friends and acquaintances. There are many different platforms in social media that engage on both sides of the debate in the current Ukraine crisis. This necessitated a selection of the total number to narrow down the massive information flow on the subject. Present on Facebook, were individuals who were engaged in the informational war in Ukraine, such as Graham Phillips (www.facebook.com/grahamwilliamphillips?ref=ts&fref=ts) and Alexander Mercouris (www.facebook.com/alexander.mercouris). However, individuals were discounted, and groups or news aggregates were targeted instead. The underlying reason is that these groups are more likely to fit a role within the frame of transnational advocacy.

[2] O. Onuch, *Social Networks and Social Media in Ukrainian "Euromaidan" Protests*, Monkey Cage, The Washington Post, www.washingtonpost.com/blogs/monkey-cage/wp/2014/01/02/social-networks-and-social-media-in-ukrainian-euromaidan-protests-2/, 2 January 2014 (accessed 8 November 2014).

A number of social media platforms, for and against Euromaidan, were chosen from an extensive search of different sites shortly after Euromaidan and in the immediate aftermath. These sites were chosen on the basis of their organisational aims and objectives (please see these below), which were aimed at international audiences, and especially in those instances where they positioned themselves as a source for mainstream media. Two sites from each camp (for and against Euromaidan) were chosen, both of whom had a presence on the social media platform (as well as elsewhere in some instances; the material posted on the Facebook sites was analysed for references to values and norms, and rhetoric and storyline were critically evaluated. The relatively low number of websites chosen is intended to permit a qualitative study to understand possible indications of the role of social media in the Ukraine crisis and how this fits into the bigger picture of persuasion and influence in this geopolitical conflict. More detail is provided below on the particular pages chosen, their self-declared goals and objectives.

Following from Dittmer and Gray (2010: 1673), this study intends to use "qualitative methods to focus on the everyday intersection of the human body with places, environments, objects, and discourses linked to geopolitics." The approaches to textual analysis include content analysis (quantifications of different elements in text), argumentation analysis (the structure of argumentation used), and the qualitative analysis of ideas in the content (with a focus on people and events) (Boréus & Bergström, 2017:7–9). The combination of these approaches is expected to yield results on the ontology (what exists) and epistemology (knowledge and how we 'know' things) of reactions to social media textual depictions of Euromaidan and resulting regime change within the context of the early phases of the Ukrainian conflict. The objects of study include people, foreign power policy, oppression and freedom, war and peace and so forth (Boréus & Bergström, 2017:1–2). The four selected Facebook groups and pages are as follows, beginning with those of Pro-Euromaidan orientation.

Pro-Euromaidan

The *Ukraine Media Crisis Centre* (http://uacrisis.org/) publishes material in five different languages (English, French, German, Russian and Ukrainian), and in addition to the website has a presence on different social media platforms – Facebook, You Tube, Twitter and Google Plus. On their Facebook site (www.facebook.com/uacrisis) they state that "we provide the international community with objective information about events in Ukraine and threats to national security." They describe their mission as being "to provide the international community with objective information about events in Ukraine and threats to national security, particularly in the military, political, economic, energy and humanitarian spheres. During this crisis period, the Centre, on a 24/7 basis, will provide support to all the media who cover events in Ukraine."

Stop Fake is a site (found at the following URL – www.stopfake.org/en/) that publishes material in English and Russian. It has a social media presence on You Tube, VKontakte, Google Plus, Twitter, Pinterest and Facebook (www.facebook.com/stopfakeukraine). They describe their role (in Russian): "the website StopFake.org are working journalists who consider the wrong and extremely dangerous spread of false information concerning the events in Ukraine as a whole and in the Crimea in particular. Our task is reasoned refutation of various kinds of fakes and deception in all forms of media."

Anti-Euromaidan

The Facebook group *Repression in Ukraine* (found at the following URL – www.facebook.com/groups/436727636459512/) is a public group that has over 400 members. It situates itself as "a news aggregator for collecting information of all incidents of repression in Ukraine committed by the fascist junta in power in Kiev supported by Western regime-backed Banderite-Orangist groups." They describe their mission as being to "have a central source for distributing data to all interested parties. Help anti-fascist activists, journalists, scholars and all interested parties in their work for spreading the truth about what is happening in Ukraine."

The *Anti-Maidan + say no to Fascism Community* on Facebook (which can be found by following this link – www.facebook.com/antimaidan.eng) had some 799 'likes'; information in Russian can be found at VKontakte with a link provided on their page (http://vk.com/antimaydan). A mission statement is provided concerning their purpose: "Our community's mission is to share trustworthy information on the situation in Ukraine and highlight Russia's, EU and US positions relating to this issue." A rather long description is provided, but the positioning is clear: "'anti' means 'against' AND 'instead of'. We are not just against the #Maidan as a political manipulation which was paid and staged by some #Ukrainian oligarchs and some #US-politicians and processed as the greatest tragedy of #Ukraine taking away lives and prosperity of so many people. We stand for the opposite." There are a number of different pages on Facebook that have similar names and objectives as this group, which are given links from this page.

Background to the Early Stages of the Ukrainian Conflict

The particular spark that is associated with the protests, which quickly developed into a revolution, occurred on 21 November 2013 when President Viktor Yanukovych suspended preparation for signing an Association Agreement and Free Trade Agreement with the European Union.[3] The primary narratives that come from Western governments and mass media depict events around Euromaidan and the seizure of political power in Kyiv as being a spontaneous local grassroots action, by democratically minded forces opposing the corrupt and authoritarian, "pro-Russian" regime of President Yanukovych.[4] A value-based catch-phrase was used by those protesting – "Ukraine is Europe".[5]

There are vastly different geopolitical narratives for explaining the beginning of the conflict in Ukraine, which are used to promote ideas as moral and good, while ignoring obvious flaws and contradictions (Boyd-Barrett, 2017). These competing geopolitical narratives are intended to favour the US or Russian preferences in terms of the outcome (Ericson & Zeager, 2015). Western mainstream mass media reporting has tended to be supportive of the regime change as being "legitimate" and critical of Russia's role in the conflict, which coincides with US and EU policy on the issue (Ojala & Pantti, 2017: 51). This geopolitical 'reality' contrasts significantly to the communicated geopolitical reality from Russia. The early official narrative in Russia, especially with regard to Crimea, was a historical injustice being corrected and a modern injustice being thwarted that was precipitated by Western interference in Ukraine's domestic affairs (Suslov, 2014; 2016). As such, there are conflicting communicated popular geopolitical narratives concerning the exact nature of the events, and who is responsible.

The early events in Ukraine have been characterised as a crisis, and not a revolution or war, which is important in trying to distance public perception of something that occurs 'naturally' rather than being contrived. By focusing attention on events as they unfolded in Kyiv, and especially in Maidan (Independence Square), the information content was well-suited to an entertainment-like news format for the international public and Kyiv's Independence Square was developed into an

[3] *Ukraine Protests After Yanukovych EU Deal Rejection*, BBC News Europe, www.bbc.com/news/world-europe-25162563, 30 November 2013 (accessed 26 May 2014).

[4] J. Woodruff, *Why Did Ukraine's Yanukovych Give in to Russian Pressure on EU Deal?*, PBS News Hour, www.pbs.org/newshour/bb/world-july-dec13-ukraine2_12-02/, 2 December 2013 (accessed 26 May 2014).

[5] *Ukraine Protests After Yanukovych EU Deal Rejection*, BBC News Europe, www.bbc.com/news/world-europe-25162563, 30 November 2013 (accessed 26 May 2014).

iconic event (together with the symbolism, values and norms expressed by this act). The violent radicalisation of the initially peaceful protests was used as a means to build a more efficient leverage against the government as the peaceful protests proved a failure (Ishchenko, 2020). This made the country ungovernable.

Subversion of the Yanukovych regime has proved to be successful in this case for a variety of different reasons: 1) Ukraine has been suffering from a protracted period of economic crisis. 2) The perceived diplomatic defeat, symbolised by Yanukovych's refusal to sign the Association Agreement and Free Trade Agreement. This was used symbolically as a means to justify and try to crystallise public discontent. 3) The chronic instability of the political system in Ukraine (together with elections that were scheduled for May 2015). 4) Lack of active dialogue between the government and society. 5) Foreign assistance to those domestic actors seeking regime change, which weakened the Ukrainian government's ability to meet the internal and external challenges it faced.

Foreign intervention and assistance came in a number of different forms – tangible and intangible. The tangible forms included financial support (Under-Secretary of State for Europe and Eurasia, Victoria Nuland's claim of US$5 billion spent[6]), as well as other support activities, such as the Soros Foundation support for creating the Ukrainian Crisis Media Centre to 'inform' the international community.[7] Intangible forms included attempts at giving the coup government political legitimacy through symbolic speech acts and access to high-level meetings with Western leaders. A statement by protestors on 29 November 2013 set the tone for the ultimate aim of the Euromaidan movement. There were three primary demands from a letter that was signed by the organisers and leaders. First, to form a co-ordinating committee to communicate with the European community. Second, the statement that the president, parliament and the Cabinet of Ministers aren't capable of carrying out a geopolitically strategic course of development for the state alongside calls for Yanukovych's resignation. Third, was the demand for the cessation of political repression against Euromaidan activists, students, civic activists

[6] *American Conquest by Subversion: Victoria Nuland's Admits Washington Has Spent $5 Billion to "Subvert Ukraine"*, Global Research, www.globalresearch.ca/american-conquest-by-subversion-victoria-nulands-admits-washington-has-spent-5-billion-to-subvert-ukraine/5367782, 7 February 2014 (accessed 26 May 2014).

[7] S. Weissman, *Meet the Americans Who Put Together the Coup in Kiev*, Reader Supported News, http://readersupportednews.org/opinion2/277-75/22758-meet-the-americans-who-put-together-the-coup-in-kiev, 25 March 2014 (accessed 26 May 2014).

and opposition leaders.[8] This call was an attempt to impose a counter-political discourse to the Yanukovych regime as well as an attempt to begin the process of attacking his intangible elements of power and to open the door to external political pressure from the West. It also begins to establish the narrative of the coming crisis as a conflict between 'positive' Western and 'negative' Russian influence in the context of a geopolitical struggle.

Social media plays an important role in civic mobilisation and shaping perceptions of the conflict and events in Ukraine. The crisis in Ukraine is certainly seen in the crisis of information where different actors and interests seek to manage information and perception in order to project a 'desired' perception of events and people that is often wrapped in a specific set of competing values and norms that shape the framing and narratives used.

There are a number of narratives that have become the 'orthodoxy of knowledge' on the Ukrainian conflict via communicated stereotypes and generalisations, both in academia and mass media. For example, the majority of the protestors were young and there to protest cancellation of the association agreement with the European Union. However, some fieldwork and research demonstrate a more complex picture in terms of the demographics and motivations for the civil unrest. From 27 November 2013, some twenty interviewers were present at the Kyiv protest sites carrying out interviews amongst those participating. Five preliminary findings were announced in January 2014, some of which contradicted the popular cultural and political stereotypes of the time. "1) Protestors are older than expected. 2) Protestors are more diverse than expected. 3) Social media are important, but not only as a provider of information about the existence of protests. 4) Social networks – both within and outside of social media – seem to be highly influential in bringing people out into the streets. 5) Social media and Internet news sites seem to have been successfully used as key framing devices for protest themes."[9]

The volume of propaganda and agitation found in the mass media coverage of events and amongst people in Ukraine has already begun to generate alternative views or satire concerning the lack of objectivity – from all sides involved. This has even reached some reflective segments

[8] EuroMaidan Rallies in Ukraine – Dec. 9, Kyiv Post, www.kyivpost.com/content/ukraine/euromaidan-rallies-in-ukraine-live-updates-332341.html, 10 December 2013 (accessed 26 May 2014).

[9] O. Onuch, *Social Networks and Social Media in Ukrainian "Euromaidan" Protests*, Monkey Cage, The Washington Post, www.washingtonpost.com/blogs/monkey-cage/wp/2014/01/02/social-networks-and-social-media-in-ukrainian-euromaidan-protests-2/, 2 January 2014 (accessed 8 November 2014).

in mainstream mass media in the West. A good example was the recent German satire of the anti-Russian propaganda found in mainstream German media. The show "Die Anstalt" appeared on German TV (ZDF), which mocked the lack of objectivity and professional standards being displayed by the media in their coverage of the events in Ukraine.[10] There have been accusations and counter-accusations cast back and forth between the different information spaces, each blaming and condemning the other for making use of blatant propaganda.[11] Some of the key/iconic moments that were used to generate an emotionally charged atmosphere also came to be questioned.

Some events came under critical scrutiny by Western media outlets, which cast doubt upon the earlier assumed versions and 'truths'. For example, the Maidan sniper shootings provides one such event. In April 2014, the German TV channel ARD ran an investigation of the shootings and came to a very different conclusion than those reached by the Kyiv authorities and many Western political leaders. The TV report concluded that there were many unanswered questions, but there were indications of the involvement of elements of the protest movement.[12] Reuters also found flaws in Kyiv's court case against those accused of the massacre of civilians in Maidan. "A Reuters examination of Ukraine's probes into the Maidan shootings – based upon interviews with prosecutors, defence attorneys, protestors, police officers and legal

[10] *German Satire Show Mocks Biased Ukraine-Coverage - "Die Anstalt" with English Subtitles*, You Tube, www.youtube.com/watch?v=ovdGpOHP9Uw, 25 September 2014 (accessed 3 October 2014).

[11] A. Yuhas, *Russian Propaganda Over Crimea and Ukraine: How Does It Work?*, www.theguardian.com/world/2014/mar/17/crimea-crisis-russia-propaganda-media, The Guardian, 17 March 2014 (accessed 13 October 2014); V. Goldstein, *Western Media Coverage of the Ukraine Crisis Is as Distorted as Soviet Propaganda*, The Nation, www.thenation.com/article/179986/western-media-coverage-ukraine-crisis-distorted-soviet-propaganda, 22 May 2014 (accessed 13 October 2014); N. Abdullaev, *Is Western Media Coverage of the Ukraine Crisis anti-Russian?*, The Guardian, www.theguardian.com/world/2014/aug/04/western-media-coverage-ukraine-crisis-russia, 4 August 2014 (accessed 13 October 2014); P. Rogov, *Media Propaganda and the Ukraine Crisis*, Global Research, www.globalresearch.ca/media-propaganda-and-the-ukraine-crisis/5383859, 26 May 2014 (accessed 13 October 2014); O. Sukhov, *The Media War Behind the Ukraine Crisis*, The Moscow Times, www.themoscowtimes.com/news/article/the-media-war-behind-the-ukraine-crisis/495920.html, 11 March 2014 (accessed 13 October 2014); R. Parry, *US Media Propaganda Machine and Powerful 'Group Think' on Ukraine: Blaming Everything on Putin*, Global Research, www.globalresearch.ca/u-s-media-propaganda-machine-and-the-powerful-group-think-on-ukraine-blaming-everything-on-putin/5396426, 18 August 2014 (accessed 13 October 2014).

[12] ARD German television, Fatal Shootings in Kyiv: Who is Responsible for the Carnage from Maidan, 10 April 2014 in Johnson's Russia List, 2014-#212, 10 October 2014. For example, please see the academic works of Professor Ivan Katchanovski, www.academia.edu/8776021/The_Snipers_Massacre_on_the_Maidan_in_Ukraine.

experts – has uncovered serious flaws in the case against Sadovnyk and the other two Berkut police officers."[13] Mainstream media have been using material and identified emerging news stories from social media sources, such as Twitter, Facebook and LinkedIn. There is a rush to be the first to publish breaking stories on social media (many mainstream media have a simultaneous presence in social media too). It is only in the aftermath that some kind of reflection may occur, and more 'sober' journalism emerges, such as seen in the Maidan sniper story. Reuters has also begun to introduce new information and possible theories into the exact cause of the shooting down of MH-17. On 27 October 2014, an article opened with the sentence "Dutch prosecutors investigating the crash of Malaysian Airlines MH 17 believe the aircraft might have been shot down from the air but that a ground-to-air missile is more likely."[14]

There have been cases of mainstream mass media using social media content, such as The Guardian that published a data blog of some 13,000 Instagram photos taken over a 144-hour period of the violence in Kyiv during February 2014.[15] Although this present chapter focuses upon social media, it should not be seen as being remote from the mainstream media environment. Social media can exert an influence not only in the new media sphere, but in the traditional mass media information sphere too.

Web 2.0 and the Information War about Ukraine

Pro-Euromaidan

Protests were projected as a 'natural' and spontaneous reaction to the excesses of corruption and nepotism of the Yanukovych regime and as an expression of the desire and will of the Ukrainian people. The political crisis from the anti-Yanukovych forces was characterised as being non-violent, spontaneous and democratic in nature (to enjoy the same basic rights as enjoyed in Western countries). The violent overthrow of Yanukovych was seen as being justified as the framework of democratic

[13] S. Stecklow & O. Akymenko, *Special Report: Flaws Found in Ukraine's Probe of Maidan Massacre*, Reuters, www.reuters.com/article/us-ukraine-killings-probe-special-report-idUSKCN0HZ0UH20141010, 10 October 2014 (accessed 13 October 2014).

[14] T. Escritt, *MH 17 Prosecutor Open to Theory Another Plane Shot Down Airliner: Der Spiegel*, Reuters, www.reuters.com/article/us-ukraine-crisis-mh17-idUSKBN0IG1X420141027, 27 October 2014 (accessed 29 October 2014).

[15] *Visualising the Ukrainian Revolution Using Instagram*, The Guardian, www.theguardian.com/news/datablog/2014/oct/08/visualising-the-ukrainian-revolution-using-instagram, 8 October 2014 (accessed 20 October 2014).

means would have been unlikely to succeed (the excuse that the elections would be rigged), and it was the popular will of the Ukrainian people. The presence of radical elements in the ranks of the protests was downplayed.

The supposed point of 'no return' came when Yanukovych turned down the EU association agreement in exchange for the more lucrative offer from Russia. This was used to project the image of Ukraine being diverted from the popular desire for a Western civilisational trajectory and being held in the Russian orbit. Hence the label of Yanukovych as being simultaneously under Russian influence and initiating it, and therefore being out of touch with the will of the Ukrainian people.

Crowds increasingly gathered in Kyiv, protesting the decision and increasingly against Yanukovych. Protests began peacefully enough, but gradually degenerated into violence, which was blamed on the police and the political leadership of the country. The reaction of Western governments and organisations was to criticise the Ukrainian government on a number of fronts and to praise the activists for their bravery. Secretary-General of NATO, Anders Fogh Rasmussen, stated that "first of all, I strongly condemn the excessive use of police forces we have witnessed in Kyiv. I would expect all NATO partners, including Ukraine, to live up to fundamental democratic principles including freedom of assembly and freedom of expression."[16] The police response to the protests was predictable and probably calculated into the timeline of the growing revolution. By the protestors being there, the organisers could count on an eventual response from police, which then could be used to 'justify' increasingly violent protests, which then developed into revolution when the conditions became 'right'.

Ukrainian Crisis Media Centre takes a professional approach to content format and presentation. Material on their Facebook page appears in English, Russian and Ukrainian languages. However, in spite of the claim that they provide 'objective' information on the current events in Ukraine, the nature of the content demonstrates a clear and distinct bias in the selection of content. Information that appears clearly contains an anti-Russian bias and that supports the tenets of the Euromaidan narrative. For example, on 23 October 2014 it shared a link that quoted the Australian Foreign Affairs Minister that she had no doubt in Russia's involvement in the Malaysian Airlines MH-17 crash (www.bloomberg.com/news/articles/2018-05-25/netherlands-australia-hold-russia-liable-for-its-part-in-mh17-jhlqz5ti?). However, there was no substantial proof

[16] *Rasmussen Hopes Ukraine Lives Up to Democratic Principles*, Interfax-Ukraine, http://en.interfax.com.ua/news/general/178955.html, 3 December 2013 (accessed 26 May 2014).

to back up the claims made, and furthermore the caption of the posting was not substantiated by the interview on Bloomberg TV and is somewhat misleading.

In addition to supporting the role of reinforcing the script of Russia as the 'bad guy' in the conflict, undermining and destabilising the situation and acting contrary to international standards and norms, a counter-propaganda function is observed. One posting (also 23 October 2014 – "Wake Up Europe" by George Soros) even went as far as to claim that Russia is not only a threat to the existence of Ukraine, but Europe as a whole. This technique is meant to project the threat and give a personal perception thereby creating an atmosphere of empathy based upon the idea of a common threat. Human Rights Watch (HRW) criticised the Ukrainian military for using outlawed munitions in the conflict. A post on 23 October 2014 refuted the claims.

> The report, published by the Human Rights Watch on October 20 states that the Ukrainian government forces used cluster munitions in populated areas in Donetsk. The OSCE Special Monitoring Mission spokesman Michael Bociurkiw said that their 100 monitors saw no such evidence. Ukrainian officials have also denied the use of such weapons in the area. Making such bold statements is fraught with serious consequences for any international organization. (www.facebook.com/uacrisis)

The text and visuals set out to try and sow seeds of doubt about the veracity of the claim by HRW. Moreover, OSCE monitors are used as 'evidence', but there is no indication that they were in a position to see whether banned munitions were used or not; it is implied, but not overtly stated. In general, there is little use of derogatory labels of the other side in the conflict. On 23 August 2014 a post appeared that compared Putin and Stalin, and on 20 July a post referred to a group of "Russian militants" as a "terrorist group". There have been substantial improvements in the quality of the English language text used over time. The earlier postings were amateurish and clumsy compared to the more contemporary posts.

Stop Fake has a function that is focused upon 'debunking' or countering news from Russian information sources. The site encourages visitors to nominate possible fake stories that they can expose. The Facebook page has material that is only in the Russian language; there are less postings here than on the Ukrainian Crisis Media Centre. One of the source targets of Stop Fake is the news portal www.ukraina.ru. For example, on 20 October 2014 the posting concerning the Reuters investigation into the Maidan sniper massacre (from 13 October 2014) was picked up and published by Ukraina.ru (www.stopfake.org/en/reuters-did-not-report-that-people-on-maidan-were-killed-by-opposition-s-snipers/). The main point of the posting was to show that the original material was embellished to suit the political agenda of the Russian state. It is stressed that the Reuters article found no one was to blame for the act, but merely found 'flaws' in the investigation.

Another posting on 19 October 2014 concerned a manipulated photo being carried by Ukrainian nationalists which included a photo of Hitler on a banner beside Ukrainian Insurgent Army (UPA) leaders, and was circulated on the social networking site VKontakte (www.stopfake.org/en/fake-photo-portrait-of-hitler-on-poster-of-ukrainian-nationalists/). This fake and counter-fake story demonstrates the importance of the battle of values and norms within the context of political and armed conflicts.

However, Stop Fake goes beyond the scope of the conflict in Ukraine in attacking information that appears in Russia's information space. A posting titled *Channel One Falsely Interpreted Jen Psaki's Words in Order to Accuse the USA in Financing the Revolution in Hong Kong* appeared on 19 October 2014. There was an attempt to justify adding such information on the basis that "at first sight the revolution in Hong Kong and financing of non-governmental organizations by the USA do not concern Ukraine. But with the help of the Kremlin voicer – the Channel One – Moscow tells that the USA finances all the revolutions, including those in Hong Kong and Ukraine" (www.stopfake.org/en/channel-one-falsely-interpreted-jen-psaki-s-words-in-order-to-accuse-the-usa-in-financing-the-revolution-in-hong-kong/).

Both of the social media sites studied and analysed here displayed a reasonably professional journalistic standard in terms of the format and presentation of material, and generally the absence of derogatory or hate speech. However, there was a clear case of material bias, where information presented supported the pro-Euromaidan message and there was nothing to contradict it. So in spite of the promise of 'objective' information and coverage, there was a clear subjective element. There are many examples given of Russian news fakes, yet no such equivalents from Western or Ukrainian media sources. At times, there was misleading material, such as the 'interpretation' of what the Australian foreign minister said concerning the MH-17 disaster.

Anti-Euromaidan

The expressed values and beliefs in this category differed greatly from the pro-Euromaidan groups. One of the first points that needs to be raised is that the events of Euromaidan were characterised as being Western organised and sponsored, and not as spontaneous grassroots political protest.

Protestors were viewed as being provocative and violent in nature, which contrasts with the non-violent emphasis in Western mass media. The eventual violent overthrow of Yanukovych was denounced as being undemocratic, arguing that even if he was corrupt, he was still the democratically elected head of state. Emphasis was placed upon the radical elements of the anti-Yanukovych forces, such as the extreme right-wing organisations Right Sector and Svobodo. This led to the suggestion of a rise in fascism and of a fascist coup, bringing in the political symbolism of the Great Patriotic War (1941–1945) and the fight against Nazism.

Blame for the crisis situation was laid clearly and solely at the feet of Yanukovych and his government by Western political circles; the increasingly violent protestors absolved of any wrongdoing. However, there were some media events that projected the worst and the damaging complicity of the United States in supporting regime change in Ukraine. The most obvious example was the recorded phone conversation between Nuland and the US Ambassador to Ukraine, Geoffrey Pyatt.[17] In the recording, which was broadcast on YouTube, Nuland and Pyatt discuss who should be included in the new government to be and who should be excluded. For example, Arseniy Yatsenyuk was approved and Vitali Klitschko (leader of the UDAR party) was not. A number of articles have appeared in non-mainstream and non-corporate media that refer to the events and result of Euromaidan as an exercise in nothing less than regime change. The primary argument being, whatever his personal flaws, Yanukovych was the duly elected leader of Ukraine and he was overthrown by non-democratic means that was supported by the United States and the West at large.[18] The specific form of power that is utilised is coercive typology.

One of the 'early' controversies that emerged, and only did so thanks to an intercepted phone call between the EU foreign affairs chief,

[17] *Nuland: Fuck the EU*, YouTube, www.youtube.com/watch?v=r5n8UbJ8jsk, 6 February 2014 (accessed 26 May 2014); J. Marcus, *Ukraine Crisis: Transcript of Leaked Nuland-Pyatt Call*, BBC News Europe, www.bbc.com/news/world-europe-26079957, 7 February 2014 (accessed 28 April 2014).

[18] J. Giambrone, *Obama Off the Ukrainian Deep End*, Foreign Policy Journal, http://www.foreignpolicyjournal.com/2014/03/17/obama-off-the-ukrainian-deep-end/, 17 March 2014 (accessed 26 May 2014); R. McGovern, *Ukraine: One 'Regime Change' too Many?*, Consortium News, http://consortiumnews.com/2014/03/01/ukraine-one-regime-change-too-many/, 1 March 2014 (accessed 26 May 2014); B. O'Neil, *Ukraine: This Isn't a Revolution – Its Regime Change*, Spiked, www.spiked-online.com/spikedplus/article/ukraine-this-isnt-a-revolution-its-regime-change#.U4NRK_mSyPU, 25 February 2014 (accessed 26 May 2014); S. Weissman, *Meet the Americans Who Put Together the Coup in Kiev*, Reader Supported News, http://readersupportednews.org/opinion2/277–75/22758-meet-the-americans-who-put-together-the-coup-in-kiev, 25 March 2014 (accessed 26 May 2014).

Catherine Ashton and the Estonian foreign minister Urmas Paet, concerned the story behind the Kyiv snipers. This was assumed and blamed on Yanukovych. However, the leaked phone call[19] told another story, and indicated that it may in fact be snipers employed by someone within the Euromaidan movement.[20] The story was carried in some mainstream Western media, such as The Guardian, simply because this was too big a story to ignore and was potentially damaging. Social media contain numerous groups that take an anti-Euromaidan stance, and their approaches are wide and varied.

The group *Repression in Ukraine* is only found on Facebook, and the primary language used is English (some video material is in Russian with English subtitles). They position themselves as communicating to an international audience (hence English language) and as having a counter-propaganda function. There is no reference to objectivity in their description, with everyone permitted to post on their page (although the right to delete posts is reserved). Postings appear on a regular basis on the site. A political agenda is set with references to it being a repository for the truth against fascism and the Western-backed authorities in Kyiv. The themes of the information and material are varied and include atrocity stories, the Kyiv regime's links to fascism and Nazism, volunteers fighting for the separatists in Eastern Ukraine (especially foreign volunteers) and many others.

Specific stories surveyed on the Facebook page tend to support the key anti-Euromaidan frames and narratives. On 28 October 2014, a post was titled "West turns Eastern Ukraine into a new Gaza". This is meant to undermine the West as a benevolent actor in world politics by comparing the result of their actions with the more well-known and recent events in Gaza. "Voting to divide" was a caption in a post from 27 October 2014. The post counters the democracy narrative by saying that this is a façade that is being used to justify and bring legitimacy to the government that was brought to political power through a coup. A post on 22 October 2014 concerned an article referring to the use of cluster weapons by the Ukrainian military, stories concerning Ukraine's economic situation and 'failed state' ranking appeared on the 20th and 17th of October 2014 respectively. Putin's criticism of human rights violations in Ukraine by the authorities appeared on 14th of October, and a media report that was

[19] www.youtube.com/watch?v=ZEgJ0oo3OA8, 5 March 2014 (accessed 27 May 2014)

[20] T. Durden, *"Behind the Kiev Snipers it was Somebody from the New Coalition" – A Stunning New Leak Released*, Zero Hedge, www.zerohedge.com/news/2014-03-05/behind-kiev-snipers-it-was-somebody-new-coaltion-stunning-new-leak-reveals-truth, 5 March 2014 (accessed 28 April 2014)

critical of a human rights report by the United Nations on 13th of October. A story depicting the United States as a police state was posted on the 10th of October. These stories depict Ukraine as a country ruled by an extremist government that is supported by Western political and economic interests in order to serve a geopolitical agenda.

Anti-Maidan + say no to Fascism Community was the other anti-Euromaidan page that was analysed. It also appears on Facebook. The stated mission is to share trusted information concerning the positions of Russia, the US and EU on Ukraine and the unfolding events there. Material is mostly in the English language. Links are shared on the page, with many graphic pictures and content relating to current atrocities and those in the past (namely associated with the Holocaust).

On the 28 October 2014 a post was published that posed the header "inconvenient questions to the Maidan supporters." A total of twenty-eight points were presented that were intended to provoke reflection on the question of who is to blame for the current and unfolding situation in Ukraine. They were intended to make the reader begin to question the underlying motives and agenda of domestic and foreign supporters of the Euromaidan movement. Another post on the same day questioned why so much information concerning the downed airliner, MH-17, was still not made publicly available at this stage. The implication being that 'inconvenient' information was being suppressed in order to maintain the pro-Euromadian narrative. There are various references and stories on the undemocratic nature of the regime in Kyiv. Posts concerning the comments of US and European sources that support the general anti-Euromaidan narrative are used, such as a post from the 26th of October from a French diplomat who was quoted as saying the West should not arm Ukraine as Putin has already won. The site does not present the 'other' side of the story (EU and US perspectives) given the nature of the stories that appear, and the source of many of the articles (from Russian state media).

These two social media projects both act in accordance with supporting the anti-Euromaidan perspective. The stories serve to cast significant doubt upon the fundamental foundations of the pro-Euromaidan legitimacy, questioning the legality of the coup, stressing the violent nature of the new regime and the fact that there has been no significant change (changing the faces of power but not the nature of it), the unconstructive and dangerous actions of the West in the regime change and subsequent anti-terrorist operation (ATO), and the constructive role of Russia in comparison.

Analysis

The current situation in the information environment in and about Ukraine validates the observation by Duhé (2007) that managing the media is no longer an option owing to portal information and communication technologies potentially allowing a large pool of people to collect and disseminate information. Therefore, this becomes an exercise in not controlling what people think precisely, but rather what they think about. Information and communication are not only about 'informing' publics, but influencing opinion, perception and social mobilisation. The mechanism for achieving this is through eliciting strong emotional impulses that cause fear or enthusiasm – motivating feelings of avoidance or approach. This is aided if the messenger can get the audience to become actively involved and form a relationship, transforming them from a consumer to a prosumer. It adds a greater sense of perceived legitimacy and trust, because the interaction is based upon freedom of choice, the interaction is symbolic (especially in terms of norms and values expressed) and it is a process of self-persuasion.

The effects of these groups' communicational measure of activity are to manufacture a space in the information domain that consists of binary opposing sets of stereotypical geopolitical narratives that are intended to legitimise or de-legitimise attempts to shape audience perception and opinion on the Ukrainian conflict. These consist of information and opinion on people, the role and actions of foreign powers, visions of oppression versus freedom and war versus peace. These were linked to the human experience in the geopolitical chaos and turmoil of Euromaidan and the regime change period. The two pro-Euromaidan sites chosen, sought to position themselves as an information exchange, albeit with a clear geopolitical bias, to mass media outlets and the international publics. However, the anti-Euromaidan sites served more as a more amateur public forum for those (mostly individuals) that held a negative perception and opinion of the geopolitical events and processes occurring in Ukraine in the early post-Yanukovych era.

Web 2.0 allows advocacy networks[21] to transform the practice of national sovereignty, which bears out the relevance and significance of critical and popular geopolitics in the conflict and contestations in defining the 'realities' people, places and events in international

[21] There is some relevance in using public diplomacy, new public diplomacy in particular, as a lens to gaining further insights and understanding in the role of these actors. This is beyond the scope of this chapter.

relations practice. The transnational advocacy networks examined here display the common traits highlighted by Keck and Sikkink (1998) – the central role of values or principled ideas, the belief in individuals being able to make a difference and the creative use of information. Their role and significance are boosted in the environment of informational uncertainty that characterises the current conflict. The information flows through worldwide information networks have reshaped the perception, influence and meaning of geographic borders and time. Messages are sent instantaneously to networks that can and often do, span the globe. Communities that are remote in terms of geographical proximity and time are closely connected by shared values and norms via 'virtual' networks. This connects to the point concerning cultural aspects, where those groups or communities are connected to each other emotionally via shared values and norms, which influences how they perceive and process worldly events.

This was particularly evident in the two groupings that were assigned, pro-Euromaidan and anti-Euromaidan. These groups had central values and principles that supported their version of events and clashed with the key understandings and narratives of the 'other'. Those belonging to the group came from diverse geographical locations but were emotionally united by their shared world view. Those actively engaging in the respective groups (prosumers) were mostly in agreement with the worldview and politics expressed by the social media platforms. When disagreement and divergence emerged, there was a terse exchange of accusations and counter accusations, for example, referring to the perceived other as a CIA paid troll or a Kremlin paid troll.[22] One of the problems faced by Ukrainians caught up in the current situation is the dilemma of being forced between different poor alternatives. In a conversation with a Ukrainian friend from Kyiv, he expressed this problem well, stating that the Euromaidan brand has been 'brandjacked' (co-opted) by extremist elements; however, he is not prepared to state this publicly because this would have the effect of validating one of the key narratives expressed by Russia.[23] There is an element of uncertainty insofar as it is hard to know if these sites are 'independent' or acting on behalf of an organisation or government. Stopfake is certainly financed by organisations and governments[24] with a geopolitical stake in the Ukrainian conflict.

[22] See e.g., the viewer comments under the main article of this page, www.stopfake.org/en/how-to-identity-a-fake/.

[23] Conversation conducted via Facebook instant messaging on 21 August 2014.

[24] See www.stopfake.org/en/about-us/ for some of the funding organisations.

Epilogue: Social Media in the Russia–Ukraine War

This is a very brief summary of the continued relevance of social media's role in the Russia–Ukraine War that began in February 2022, and is not intended as a full and complete review of the conflict. Needless to say, social media continues to play a vital and important role in the armed conflict, more for Ukraine than Russia owing to the asymmetry in tangible kinetic military capability and capacity. Wars are invariably political in terms of how as an event they are conceived and driven. This is crucial in terms of engineering the politics of legitimacy and support of the sides involved in the conflict. Russia possesses and maintains a physical realm advantage over Ukraine in terms of its operational military capacity and capability. Ukraine is attempting to offset this tangible advantage through the information realm in order to influence the cognitive realm of different global audiences, partly a hearts and minds campaign to garner support and legitimacy and partly to restrict or reduce Russia's operational military choices.

Ukraine has been effectively making use of the mainstream Western media space, including social media, to prime and mobilise publics in the West for Ukraine and against Russia in a storytelling approach, making use of binary opposites that resonate with Western liberal audiences. Stories of Ukrainian heroes fighting against the odds, such as the defenders of Snake Island or the Ukrainian pilot the Ghost of Kyiv, that in the end were fake.[25] The purpose was to boost domestic war morale and to encourage increased and sustained international political, economic, moral and material support for the Ukrainian war effort. There is also a sense of othering Russia as uncivilised and blood thirsty, ignoring the conventions of 'civilised' society by attributing various atrocity stories to it, such as the massacre at Bucha[26] or various claims of an 'inevitable' Ukrainian military victory over Russia (by heroism and/or the gifting of 'wonder weapons' by the West).[27] Ukraine also uses the Internet and social media as a means of trying to stifle or silence dissent from the main Western narrative of supporting Ukraine and opposing Russia through the use of war propaganda and gatekeeping mechanisms, such as the site

[25] S. A. Thompson & D. Alba, *Fact and Mythmaking Blend in Ukraine's Information War*, The New York Times, www.nytimes.com/2022/03/03/technology/ukraine-war-misinfo.html, 8 March 2022 (accessed 29 August 2022).

[26] *CBS News Finds Evidence of Atrocities Near Ukraine's Capital as Russia Is Accused of War Crimes*, CBS News, www.cbsnews.com/news/bucha-massacre-ukraine-russia-atrocities-evidence/, 5 April 2022 (accessed 29 August 2022).

[27] *Updated: Ukrainian Army's Victories*, VOX Ukraine, https://voxukraine.org/en/ukrainian-army-s-victories/, 3 August 2022 (accessed 29 August 2022).

"Myrotvorets" (Peacemaker)[28] that is hosted on a NATO server, which includes the infamous kill list[29] of those potential Western influencers whose views diverge too far and potentially damage the narrative.

Russian communication is seemingly not heavily engaged with Western audiences owing to the stacking of information dominance in favour of the Western narrative by Western governments sanctioning and blocking Russian websites, such as RT and Sputnik. Once beyond the reach of the Western information space, the perceptions and opinions are markedly different in non-Western settings, across the Middle East, Asia, Africa and Latin America, of the war and who is to blame.[30] Russia propagates its own narratives of intended legitimisation of the 'special military operation' (the term used in Russia to denote its military operation in Ukraine) in order to pursue its political and military objectives more freely; the focus on the 'denazification' and demilitarisation of Ukraine feature prominently.[31] Therefore, one may preliminarily conclude that Ukrainian military operations support their information operations, and Russian information operations support their military operations.

Conclusion

In the introduction, the following research question was posed, how do the opposing sides use the social media information domain in order to influence the audience's cognitive domain on a geopolitically contested event? There is an evident highly competitive Web 2.0 environment of conflicting popular geopolitical narratives in the information domain that seek influence and persuasion in the cognitive domain of the audience (the international public with an interest in Ukraine and the conflict that are active seekers of information). To be effective in competitive influence seeking, their message and narrative must resonate and standout; it

[28] The website is found in the following link – https://myrotvorets.center/.

[29] O. Sukharevskaya, *Ukrainian Website Threatens Thousands of People, Including Americans, With Extrajudicial Killings*, Azerbaycan 24, www.azerbaycan24.com/en/ukrainian-website-threatens-thousands-of-people-including-americans-with-extrajudicial-killings/, 26 August 2022 (accessed 29 August 2022).

[30] P. Wintour, *Negative Views of Russia Mainly Limited to Western Liberal Democracies, Poll Shows*, The Guardian, www.theguardian.com/world/2022/may/30/negative-views-of-russia-mainly-limited-to-western-liberal-democracies-poll-shows, 30 May 2022 (accessed 3 June 2022).

[31] PTI, *Special Military Operation Aimed at 'Demilitarisation and Denazification' of Ukraine: Putin's Televised Address*, The Hindu, www.thehindu.com/news/international/special-military-operation-aimed-at-demilitarisation-and-denazification-of-ukraine-putins-televised-address/article65080000.ece, 24 February 2022 (accessed 29 August 2022).

needs to be perceived as being 'legitimate' and this seems to be the motivation for framing the narrative with clashing sets of emotionally laden values and norms as opposed to interests and goals. The competition for influence and persuasion is undertaken to shape perception and opinion; there needs to be a competition for credibility of the messengers in order to win over the publics at the expense of their opponents. This necessitates communication platforms that are competing for influence to not only promote their intangibles (such as reputation, credibility and legitimacy), but also to simultaneously attack the intangible assets of their opponent(s). This is in keeping with the observations made by Bob (2007). It is a question of priming and mobilising the potential audience in favour of one's popular geopolitical narrative whilst simultaneously working at priming and mobilising the audience against opposing geopolitical narrative visions. The ultimate aim is to be able to exploit the information space fully, whilst preventing the opponent from doing the same.

According to Perloff's (2010) understanding of the different effects and desired outcomes of the communication, it was observed that the social media groups and pages were active in attempting mostly to *shape* attitudes and to *change* attitudes of their audience. This is in some part owing to the recent (at the time of enquiry) nature of the event. With more time, the additional function of *reinforcing* attitudes and opinions will likely play a greater role. These sites played a role envisioned by Keck and Sikkink (1998), where their value was found owing to their positioning in an information environment of considerable geopolitical uncertainty and risk, therefore information exchange is a central aspect in the relationships they form. It was observed, of the four grounding principles in social media logic (van Dijck & Poell, 2013), popularity in terms of audience reach and relationship building was the most relevant in the context of social media-based geopolitical advocacy. This fits in with the prediction by Maréchal (2017) of the increasing significance and role of the geopolitics of information.

As noted by Castells (2012), emotion is a key aspect to gaining attention, influencing and mobilising publics. The manner in which this is done within the context of this study indicates an idea of a clash of geopolitical ideas, each vision vying for being seen as the 'best' option for organising society. The pro-Euromaidan platforms stressed the 'spontaneous', grass roots and democratic nature of the movement that overthrew Yanukovych, who was characterized as being someone who was a puppet of Russia and not acting in accordance with the interests of the Ukrainian people. Their (Ukraine's) civilisational choice is a Western model, and not an Eastern (read Russian) one. Anti-Euromaidan social media platforms characterised the events of February 2014 as being a foreign-backed and anti-constitutional coup, and that the 'democratic' forces were in fact riddled with extremist (fascist and neo-Nazi) elements. These are emotional appeals that are designed to appeal to certain values – on the one hand attraction through enthusiasm by the pro-Euromaidan set based upon being just like us (Westerners). On the other hand, the anti-Euromaidan platforms tend to use the emotion of aversion through fear (against the spectre of Nazism). Information in the different segments of the information sphere tended to support the political views and agenda of the site in question.

The nature of the geopolitical information environment on Ukraine seems to follow the changes outlined by Fraser (2009) – from state to individual actors, real world to virtual world mobilisation and power, and old media to new media. This does not mean that traditional media and real-world environments have become totally irrelevant, however their ability to dominate as an effective gatekeeper has been curtailed.

References

Bob, C. (2007), *The Marketing of Rebellion: Insurgents, Media, and International Activism*, New York: Cambridge University Press.

Boréus, K. & Bergström, G. (2017), *Analysing Text and Discourse: Eight Approaches for the Social Sciences*, Thousand Oaks: Sage Publishing.

Boyd-Barrett, O. (2017), *Western Mainstream Media and the Ukraine Crisis: A Study in Conflict Propaganda*, New York: Routledge.

Castells, M. (2012), *Networks of Outrage and Hope: Social Movements in the Internet Age*, Malden: Polity Press.

(2003), *The Internet Galaxy: Reflections on the Internet, Business and Society*, Oxford: Oxford University Press.

Ciuta, F. & Klinke, I. (2010), Lost in Conceptualisation: Reading the "New Cold War" With Critical Geopolitics, *Political Geography*, 29(6), pp. 323–332.

Cunningham, T. (2010), Strategic Communication in the New Media Sphere, *JFQ* Issue 59, 4th Quarter, pp. 110–114.

Dittmer, J. & Gray, N. (2010), Popular Geopolitics 2.0: Towards New Methodologies of the Everyday, *Geography Compass*, 4(11), pp. 1664–1677.

Dittmer, J. & Dodds, K. (2008), Popular Geopolitics Past and Future: Fandom, Identities and Audiences, *Geopolitics*, 13(3), pp. 437–457.

Duhé, S. C. (2007) (ed.), *New Media and Public Relations*, New York: Peter Lang.

Ericson, R. E. & Zeager, L. A. (2015), Ukraine Crisis 2014: A Study of Russian–Western Strategic Interaction. *Peace Economics, Peace Science and Public Policy*, 21(2), pp. 153–190.

Fraser, M. (2009), Geopolitics 2.0, *ARI*, 144.

Hepple, L. W. (1986), The Revival of Geopolitics, *Political Geography Quarterly*, 5(4), pp. 21–36.

Ishchenko, V. (2020), Insufficiently Diverse: The Problem of Non-Violent Leverage and Radicalization of Ukraine's Maidan Uprising, 2013-2014, *Journal of Eurasian Studies*, DOI: 10.1177/1879366520928363.

Jowett, G. S. & O'Donnell, V. (2012), *Propaganda and Persuasion*, 5th Edition, Washington DC: Sage.

Keck, M. E. & Sikkink, K. (1998), *Activists Beyond Borders: Advocacy Networks in International Politics*, Ithaca: Cornell University Press.

Kelly, P. (2006), A Critique of Critical Geopolitics, *Geopolitics*, 11(1), pp. 24–53.

Maréchal, N. (2017), Networked Authoritarianism and the Geopolitics of Information: Understanding Russian Internet Policy, *Media and Communication*, 5(1), pp. 29–41.

Morozov, E. (2011), *The Dark Side of Internet Freedom: The Net Delusion*, New York: Public Affairs.

Ojala, M. & Pantti, M. (2017), Naturalising the New Cold War: The Geopolitics of Framing the Ukrainian Conflict in Four European Newspapers, *Global Media and Communication*, 13(1), pp. 41–56.

Orenstein, M. A. (2015), Geopolitics of a Divided Europe, *East European Politics and Societies and Cultures*, 29(2), pp. 531–540.

Ó Tuathail, G. (1999), Understanding Critical Geopolitics: Geopolitics and Risk Society, *The Journal of Strategic Studies*, 22(2–3), pp. 107–124.

Perloff, R. M. (2010), *The Dynamics of Persuasion: Communication and Attitudes in the 21st Century*, 4th Edition, New York: Routledge.

Richter, H. R. (2012), Web 2.0 and Public Diplomacy in Costigan, S. S. & Perry, J. (eds.), *Cyberspace and Global Affairs*, Farnham: Ashgate.

Staeheli, L. & Nagel, C. R. (2012), Whose Awakening Is it? Youth and the Geopolitics of Civic Engagement in the "Arab Awakening", *European Urban and Regional Studies*, 20(1), pp. 115–119.

Suslov, M. (2016), The "Russian World" Concept in Online Debate During the Ukrainian Crisis in Suslov, M. & Basin, M. (Eds.), *Eurasian 2.0: Russian Geopolitics in the Age of New Media*, Lanham: Rowman and Littlefield, pp. 295–316.

Suslov, M. D. (2014), "Crimea is Ours!" Russian Popular Geopolitics in the New Media Age, *Eurasian Geography and Economics*, 55(6), pp. 588–609.

Van Dijck, J. & Poell, T. (2013), Understanding Social Media Logic, *Media and Communication*, 1(1), pp. 2–14.

10 The Ukrainian and Syrian Conflicts: Civil Wars or Geopolitical Shatterbelts?

Greg Simons

Introduction

At the present point in time, there is a 'recalibration' or 'rebalancing' of power and influence in the international order that can be witnessed through the geopolitical actions of major powers' foreign policy (Klieman, 2015). On the eve of the US-led invasion of Iraq, Cohen noted the newly emerging geopolitical realities facing the United States. "In an increasingly complex geopolitical world, where international relations are influenced but not dominated by forces of globalisation, political power will be more widely dispersed and hierarchy weaker. No single state, region or realm can expect to be dominant, so that a US foreign policy of unilateralism, grounded in the assumption of American global political and economic hegemony, is basically flawed" (Cohen, 2003: 30–31). In 2019, The RAND Corporation (RAND) produced a brief on assessing the impact of different geopolitical costs on Russia as a means of containing and constraining their influence, the proposals included: providing lethal aid to Ukraine; increasing support to the Syrian 'rebels'; promoting liberalisation in Belarus; expanding ties in the South Caucasus; reducing Russian influence in Central Asia; and regime change Transnistria (Dobbins et al., 2019: 4). Local, regional and global instability is on rise; as competition between different-sized powers increases, the incumbent global hegemony weakens.

The intention of this chapter is not to analyse and explore the coding of armed conflicts, but rather how they are publicly represented in open public communication spaces. The communication of words and deeds is a symbolic representation of a constructed reality that is intended to influence audiences' opinion and perception of people, places and events. This is particularly important within the context of armed conflicts, which is heavily influenced by political considerations in addition to the military ones. Consequently, there can be competition in attempts to frame and narrate descriptions and characterisations of armed conflicts, which are based on what variant offers the best operational choices for the definer and what restricts opponents. One result is confusion and definitional uncertainty of the physical event. This gives rise to the research question: how are the public representations of the Ukrainian and Syrian conflicts formulated? Language and representations of people, places and events matter, and they are not random.

This chapter begins with an overview of the role and function of mass media and communication of armed conflicts. Given that the physical and information domains of armed conflict are delineated in the same manner – tactical, operational and strategic levels – it is a relationship and interaction that needs to be understood in terms of motivations for communication and a desired end state of an event or process. The following section concerns defining and understanding the politics of international relations, with a focus on civil war and geopolitics. In the following section, which is divided into two sub-sections, the task is to analyse whether the current and on-going conflicts in Ukraine and Syria can be classified in simple terms as being a civil war or are something else.

Role of Media and Communication in Perception of Conflicts

Information and communication have always formed an essential part of all governments' security and foreign policy promotion. This is clearly expressed in the US National Security Decision Directive (NSDD) (1984: 130), which recognised that international information is an integral and vital component of long-term US national strategy and security policy in influencing the behaviour of other governments. This means that it is used as a key mechanism for shaping fundamental political trends around the globe by affecting foreign audiences in a manner that is favourable to US national interests. International information policy in periods of war and peace is also addressed.

> In view of the importance of psychological factors in maintaining the confidence of allied governments and in deterring military action against US national interests, and in order to be prepared for immediate and effective use of psychological operations (PSYOP) in crisis and wartime, it is vital that the armed forces maintain a strong and active international information capability. (NSDD 130, 1984)

Even though this document was created under the Reagan administration in 1984, it is still valid and demonstrates the importance to all governments of the role of information and communication within the context of crisis and war. It also reveals the offensive (by restricting an opponent's operational choices) and defensive (through preserving unity of an alliance) components of information warfare. Wars and crises are events that possess high political and economic stakes and are a setting where governments seek to use the influence and persuasion potential of mass media due to its potential for public manipulation and control. Such persuasive and manipulative campaigns can be conducted with the rhetorical justification for pursuing the greater national good or interest (Vincent, 2006: 265). This is achieved through effective forms of mass communication, including the use of strategic branding. Such brands can be historically based and are familiar with the target audience. Historical examples should be well known and symbolic, which are accompanied by a short and catchy analogy or metaphor that resonates with the audience.[1] One of the recent branding strategies, that was created within the international brand of the Global War On Terrorism (GWOT), was the term "Axis of Evil", with its associations with the Axis powers of World War Two (Germany, Italy and Japan); here, the value judgement is one of 'evil' within the rational logic of the binary opposite of US 'good'. If successful, the branding creates automatic associations that simultaneously can create legitimacy for one actor and detract from the legitimacy of another. The control or domination of the information environment can be considered as being key for actors to expand their operational choices in the physical environment (and to restrict an opponent) in order to manipulate target audience perception and opinion.

It is necessary to understand how the physical, informational and cognitive components of the human world interrelate and interact with each other. Within the realm of information warfare, there are three domains to be considered: the physical domain, the information domain and the cognitive domain (Alberts et al., 2001: 10).[2] In terms of the search for political and military influence, the domain that they seek to influence is the physical one in order to enable military operations and formulate foreign policy. This is the easiest domain to measure as it is a tangible one. An understanding of the reality and 'ground truth' translates into combat or policy effectiveness and dominance. On the intangible side, information exists and is created in the information domain. It is shared and can be subjected to manipulation, which means that the information in it may not accurately reflect the ground truth. This

[1] J. Holmes, *Strategic Branding in the Beltway Marketplace*, The National Interest, https://nationalinterest.org/feature/strategic-branding-the-beltway-marketplace-24846, 11 March 2018 (accessed 13 March 2018).

[2] For more details on these domains please refer to chapter two in Alberts et al., 2001.

domain concerns the communication of information among and between the various vested actors. The information domain is subject to competition and interference from other actors present, which implies offensive and defensive dimensions to communication activities. The objective is to gain 'Information Superiority' within the information domain over the adversary. The minds of the participants are found in the cognitive domain, which is "where perceptions, awareness, understanding, beliefs, and values reside, and where, as a result of sense making, decisions are made" (Alberts et al., 2001: 13). This is the domain in which physical battles are actually won or lost as it involves such crucial intangibles as leadership, morale, unit cohesion, level of training and experience, situational awareness and public opinion. All content in this domain passes through the filtering process of human perception. The intention is to initially attract the attention of the public, persuade them of the righteousness of one's cause, and to influence their actions and reactions (thoughts and physical actions).

A new conceptual framework for research on public relations, propaganda and promotional culture is in the making. This is Organised Persuasive Communication (OPC), which is "a systematic conceptualisation of different forms of persuasive communication including categories of dialogical, non-deceptive and deceptive OPC as well as persuasion working in relation to socio-political, economic and physical contexts via incentivisation and coercion" (Bakir et al., 2018: 15). However, it would be useful to add a further physical context to the shaping of knowledge and perception through managing the content that appears in the information domain. A distinction needs to be made between "information" and "knowledge", where information is the raw material communicated within the information domain concerning people, processes and events occurring in the physical domain.

Knowledge is when information communicated from the information domain enters the cognitive domain and an individual has made 'sense' of that data according to their world view and the information supplied. This all concerns the ability of an actor in shaping the perception of others. "Perception involves forming a view of something through intuition or interpretation of available knowledge" (Ministry of Defence, 2010: 3–14). Efforts are directed towards the state of gaining Information Superiority, which may be said to have been achieved when one actor is able to create a relative information advantage over their adversary; a mixture of being able to maximise information efficiency whilst simultaneously denying the opponent the ability to do the same. The efficiency component involves and concerns an actor's ability to shape the information space according to the operational requirements and needs (Alberts et al., 2001: 54). This has been impacted by changes in the information environment.

A RAND report (Paul et al., 2018) noted a gradual merging of the wired and wireless worlds, together with a seemingly continuous expansion of the cyber domain. This has presented opportunities and threats for actors in achieving operational and strategic objectives. The spread of information technology has increased the range and scope of information operations and influencing activities but has also permitted new actors (including non-state actors) to enter the market at a relatively low financial cost.

> Even as the information environment becomes more complex and extensive, popular perceptions, attitudes, and behaviours shaped in and through it continue to grow in strategic importance. People, governments, troops, and leaders are connected in ways that were unimaginable two decades ago. The accord of legitimacy to a government or military force's actions is now much more immediate and consequential. No longer can military operations focus strictly on desired physical outcomes. (Paul et al., 2018: ix)

In 2017 the US Department of Defence was spending some US$250 million per year on information operations and informational activities in order to influence efforts at the strategic and operational levels (Paul, 2017: ix), which demonstrates the significant value placed upon shaping the informational and cognitive domains. Thus, the intangible sphere and outcomes are perceived as being at least as, if not more, important as the tangible sphere. All the above considerations fall within the context and application of political warfare, whereby, "political warfare is the marshalling of human support, or opposition, in order to achieve victory in war or in unbloody conflicts as serious as war" (Codevilla, 1989: 77). The threat of political warfare and subversion has long been recognised, and the means to try to counter the effects have been dwelled upon.

Bowen (1997) considered the internal and external nature of a conflict faced by any nation and concluded that it essentially boils down to the definition of identity: the ideologies of political parties; views of constituent parts of the country; views of different social and religious groupings; notions of a greater belonging (Islamic, Christian, European, Asian, African, etc.); and the influences of history. The balance must be struck between a fairly well-elaborated definition of identity and the end result, which needs to be the product of a generally accepted consensus. However, identity must be clearly defined in order to understand what needs to be preserved, and a concept that is understood in both detail and in principle (Bowen, 1997: 3). The logic is to ensure the integrity of a nation through a system of total defence that extends beyond solely military means. The desired outcome of a sustainable and established national identity is a more cohesive, purposeful and resilient nation that acts with a common sense of expectations and purpose through unity. Furthermore, any given country also needs to understand its weaknesses of identity and the types of threat that can take advantage of those weaknesses. The contemporary understanding of a nation's weaknesses, and how to exploit them through political warfare, has been evolving over time.

According to RAND, political warfare has evolved since the period of the Cold War, but there are several attributes that help to define and clarify its contemporary practice:

(1) Non-state actors can conduct political warfare with unprecedented reach.
(2) Political warfare employs all elements of power (DIME – Diplomatic Information Military Economic).
(3) Political warfare relies heavily on proxy forces and means.
(4) The information arena is an increasingly contested battleground, where perceptions of success can be determinative.
(5) Information operations create effects in various ways by amplifying, obfuscating, and, at times, persuading.
(6) Detecting early-stage political warfare requires a heavy investment of intelligence resources.
(7) Political warfare can generate unintended consequences.
(8) Economic leverage is increasingly the preferred tool of the strong.
(9) Political warfare often exploits ethnic or religious bonds or other internal seams.
(10) Political warfare extends, rather than replaces traditional conflict and can achieve effects at lower cost. (Robinson et al., 2018: 219–245)

As noted earlier in human history, by both Sun Tzu and von Clausewitz, war is politics by another means.

Civil War, Geopolitics and the Significance of a Shatterbelt

Wars, crises and other types of armed conflict are found in many different forms and varieties, and with specific circumstances they can easily evolve and transform in their nature and character. In the period 1990–2002, civil conflict was the most common form of armed conflict and accounted for some 90 per cent of civilian and combatant deaths (Lacina, 2006: 276). A definition of a civil war offers a simplistic overview of the process: "A civil war is a fight between factions or regions within an existing state for control over the state. The sides in a civil war often believe they have the 'true' interests of the nation-state at heart" (Flint, 2017: 114). It should be noted that the "true interests" of the people and state can be a projection for the purposes of accumulating political legitimacy, sympathy and support for their cause.

When researching and analysing civil war, the focus can often be related to a country-specific set of factors of a state that is experiencing armed conflict. However, it is argued that individual states do not exist in isolation and are influenced by their exposure and interaction with other states. This takes the form of other countries being involved directly and indirectly in a civil war or the effects of a civil war spilling across international borders (Gleditsch, 2007: 294–295). Sambanis (2004: 8144) argues that there is no consensus on the measurement of civil war. As a consequence, "without ad hoc rules to code its start and end and differentiate it from other violence, it is difficult, if not impossible, to define and measure civil war." Sambanis notes the lack of homogeneity in sparking and ending civil wars.

The likelihood of a civil war is increased by the presence of cleavages in society that can be leveraged by a domestic or a foreign actor. Reynal-Querol's (2002) study of the social and political causes of ethnic civil war resulted in several significant findings. Religious polarisation and animist diversity are critical elements in explaining the incidence of such conflicts. In addition, religious polarisation is a more important factor in creating social cleavages than any linguistic differences that exist in a given society.[3] A third finding was related to the nature of the political system, even though a democracy was more likely to reduce the probability of an ethnic civil war. There needs to be a sense of inclusivity in the political system for the whole of society in order to reduce the sense of alienation or resentment. These factors that predict the likelihood of conflict can also be used to predict the conflict's severity (Lacina, 2006). Furthermore, ethnic civil wars usually begin due to political not economic issues, whereas revolutionary/ideological civil wars are caused by the presence of a low level of democracy and inclusivity in the political system (Reynal-Querol, 2002: 31). The study also supported the idea that conflicts are more likely to be started due to economic opportunities rather than economic grievances. These aspects of the underlying causes coincide with Bowen's (1997) elements of national identity in the

[3] This is due to the possibility that people may speak a different language and have the same religion; hence the world view may be much closer than those with different religions.

creation of a resilient total defence. War and armed conflict are often extreme and deadly forms of competitive politics.

During the Cold War, the superpowers understood and treated civil wars as proxy competitions, where they armed and financed different groups to compete for power (Jones & Stedman, 2017: 33). Some researchers have noted that there has been a decline in interstate armed conflict since the end of the Cold War and a rise in the number of civil wars in the post-Cold War era. There is an assumption that new civil wars differ from their predecessors in that they are criminal, rather than political in nature. However, Kalyvas (2001: 99) challenges the assumption: "the tendency to see fundamental differences between them is based on an uncritical adoption of categories and labels grounded in double mischaracterisation." This was due to incomplete and biased data on recent and ongoing civil wars, ignoring historical research on earlier civil wars, and a lack of clear categories that has resulted in flawed coding. Lacina (2006: 287) notes that an indicator that a civil war will be particularly hard and severe is the availability of foreign assistance to the combatants involved in a conflict. One of the consequences of the end to the Cold War was the lack of competition between the rival superpowers. "Clearly, the disappearance of external sources of legitimation and funding provided by competing superpowers puts a premium on local resources" (Kalyvas, 2001: 117). This had implications for how such wars were funded, and therefore the viability of starting or continuing armed conflicts that included civil wars, as well as the nature of foreign involvement.

Academic research into foreign interventions in civil wars has shown that they are not effective in reducing the level of violence or the duration of those conflicts. This runs contrary to the held beliefs in the policy community. Generally, there is a focus on military and economic interventions; however, the strategic environment is much more complex. Diplomatic intervention can be overlooked in civil wars, but it can be an effective conflict management strategy that has the power to change the course of events. In addition, it can and should be combined with different components in order to make the whole effect greater than the individual parts. There is also the issue of timing, which needs to be right in order to successfully terminate the armed conflict and reach a settlement (Regan & Aydin, 2006: 754). As noted above, the role of foreign powers in civil wars can have a detrimental effect by prolonging their duration and intensity. The diplomatic role of foreign states, for example, which is superficially related to facilitating negotiations, can be misleading. Foreign states when they intervene directly or indirectly in a civil war can pursue an agenda that diverges from the goals of the domestic actors engaged in the conflict. Foreign states can prolong civil wars by blocking resolutions. The motivation to block a resolution is derived from the fact that yet another actor is involved in the settlement process and must approve its terms and conditions. Another point is that foreign actors have less incentive to negotiate than local actors because the perceived costs are lower for them and there is an anticipation of less benefit to be derived from the negotiation process (Cunningham, 2010). In some ways, this reveals the difference between an honest broker and a powerful one in an armed conflict, who may well be driven by geopolitical factors and considerations that are bigger than the localised war.

Geopolitics is a contested term with regard to its definition and perception. When understanding the politics of geopolitics, it is necessary to point to both perception and outcomes. "Geopolitics is not just a way of seeing. It is also the actions and outcomes that simultaneously transform spaces, places and politics" (Flint, 2017: 302). Thus geopolitics can be seen as both a practice and a representation (Flint, 2017: 36). Not long after the end of the Cold War between the United States and the Soviet Union, Brzezinski (1997) noted that the United States needed to gain and retain control of Eurasia in order to assure global dominance. Part of the task was therefore to prevent any one actor or constellation of actors that would be capable of challenging US global hegemony. At the beginning of the new century, Cohen (2003: 33) noted that there is an increase in the number of the world's major and regional powers, and also a strengthening of global and regional organisations. However, he saw this as not being sufficient to eliminate disturbances in the international system, with global terrorism and irredentist wars causing considerable turmoil.

Kelly (1986) noted that many of the twentieth-century wars among major powers originated in shatterbelts, which is a specific although uncommon term in geography and political science. What started as a local turmoil could rapidly evolve into a serious conflict among the powers involved, such as World Wars One and Two, and Vietnam. "Frequently such tension zones are strategically positioned and are relatively close to great power territories or allies" (Kelly, 1986: 162). Cohen (1973: 85) noted that "a shatter belt is caught between the conflicting interests of adjoining Great Powers." In answer to attempting a clear definition of the term 'shatterbelt', Kelly (1986: 164) distilled the varied understandings. He defined shatterbelts as a "two-tiered structure of conflict: the local layer, characterised by political turmoil, social and economic depression and fragmentation, and strategic minerals and passageways; and the international, distinguished by great power competition for footholds among various states of a region." There has been some continuity in terms of the evolution of a shatterbelt and its destabilising effects in the twenty-first century.

The concept of a 'shatterbelt' has been applied to contemporary crises and armed conflicts, including in the Middle East. The Middle East is vulnerable for a number of reasons: the numerous religious and ethnic nationalities living in the region (creating opportunity for division); and the significant energy resources and global transportation routes present (creating a motivation for actor competition to control them) (Zulfqar, 2018: 121–122). Cohen noted that in the wake of the end of the Cold war, a major global and regional geopolitical restructuring took place. He noted that the 1991 Gulf War was a physical manifestation of the Middle Eastern disequilibrium, which since the conclusion of World War Two was a shatterbelt region ensnared in inter-regional tensions and Cold War rivalry between the superpowers (Cohen, 1992). During the period of the Cold War, the Soviet Union exerted a great deal of influence in the Middle East and North African (MENA) region during the era of the bipolar international system. The Soviet collapse witnessed a dramatic reduction in military and economic influence in the region in a unipolar world system (Póti, 2018: 4). The end of the Cold War was seen as a possible opportunity to bring an end to instability in the region through fostering a new regional balance of power and order.

However, order and stability in the region has been severely compromised by specific events and trends, such as the US-led invasion of Iraq in 2003 and the progressive fracturing of the regional order caused by the Arab Spring. The results of these processes and events have been a relative decline of US power and influence in the region, together with an increase in power and influence by China, Iran and Russia (Paraschos, 2017; Zulfqar, 2018). Russian foreign policy interests and goals in MENA have increased markedly from the 1990s, which now play an increasingly important and influential role in the region. Geostrategic interests have been secured through Russia's de-ideologized foreign policy, economic assistance (nuclear power plants and loans) and military equipment sales, and an oppositional stance to US interests (Póti, 2018: 13).

> The modern Middle East and North Africa (MENA) is the epitome of a shatterbelt. One decade ago, the MENA region was relatively stable by today's standards. [...] In stark contrast, the MENA region today is exponentially more volatile: Libya, Syria, Iraq, and Yemen remain mired in violent conflict, generating a degree of insecurity that is unprecedented in the region's modern history and revolutionary in its implications for regional and international order. (Paraschos, 2017: 17)

The US has played a significant role in the MENA region and beyond since September 2001. However, there has been some pointed critique aimed at the US approach to the conflicts in Afghanistan, Iraq and Syria by Anthony Cordesman. This criticism is centred on the fact that there have been three successive presidential administrations fighting wars that have resulted in tactical victories, but without any clear strategy for ending any of the aforementioned wars or bringing about stable peace. "It is dealing with major insurgencies or civil war as if they were limited terrorist movements. [...] It has no grand strategy and is fighting half a war."[4] Therefore, tactical and operational aspects of armed conflicts that the United States is currently entrenched in are harming its strategic interests and capacities.

Major international powers have been actively engaged in efforts to create strategic brands for different events or actors. A recent example of this strategic branding has been employed by the United States against its primary global competitors and opponents. "The United States branded strategic rivals China and Russia 'forces of instability' on Friday, grouping them with Iran and North Korea as countries whose rights abuses amount to a global threat."[5] Attempts at reciprocal negative branding have been undertaken in return, which can be seen with the characterisation of the United States as a self-destructive global power in decline whose excesses need to be checked, in Russia's foreign policy concept (2013). In tandem with the increased use of branding and communication in foreign and security policy is an apparent evolution in how crises and armed conflicts are initiated and managed.

Researchers have noted that the nature of crises and armed conflicts is becoming increasingly blurred, complex and more frequent. At the present point in time, the US possesses the world's greatest tangible military capability. This has enabled the US to pursue its various global security objectives through the projection of military force. In 2016, US Special Operations Forces were deployed to 138 nations world wide.[6] Such use of military force is not fully transparent or accountable, which is an increasing trend in contemporary crises and conflicts. So-called

[4] A. H. Cordesman, *Losing by 'Winning': America's Wars in Afghanistan, Iraq and Syria*, CSIS, www.csis.org/analysis/losing-winning-americas-wars-afghanistan-iraq-and-syria, 13 August 2018 (accessed 23 August 2018).

[5] D. Clark, *US Brands China and Russia 'Forces of Instability'*, AFP in Space War, www.spacewar.com/reports/US_brands_China_and_Russia_forces_of_instability_999.html, 20 April 2018 (accessed 23 April 2018).

[6] N. McCarthy, *US Special Forces Deployed to 70% of the World's Countries in 2016*, Forbes, www.forbes.com/sites/niallmccarthy/2017/02/07/u-s-special-operations-forces-deployed-to-70-of-the-worlds-countries-in-2016-infographic/#7c0ece7f7343, 7 February 2017 (13 February 2017).

grey-zone conflicts, which are neither war nor peace, also provide a means to challenge a stronger opponent via an asymmetric path rather than directly (and likely to lose) or to skirt international norms and conventions. The "defining characteristic is ambiguity – about ultimate objectives, the participants, whether international treaties and norms have been violated, and the role that military forces should play in response."[7] Numerous recent examples exist, such as China's actions in the South China Sea, Russia's actions in Crimea and US actions in Syria. As such, the new realities of the contemporary management of global crises and armed conflicts increase the fog of war, the level of risk and threat through the ambiguity with which they are communicated and prosecuted.

Conflict in Ukraine and Syria: Civil War or Shatterbelt?

Warfare in the closing years of the twentieth century and into the opening of the twenty-first century has often been narrated as a contest and struggle between democracy and non-democracy. There is the theoretical notion of "Democratic Peace", which asserts that "democratic political structures form a pre-condition for stable peace orders in international relations" (Risse-Kappen, 1995: 491). Therefore, it has come to serve as a form of 'conventional wisdom' among Western policymakers. Logic is presented that democracies do not fight each other and bring in an era of peace and stability, and are inherently more peaceful than authoritarian regimes (Risse-Kappen, 1995). The narrative serves two purposes; one is to try and legitimise military action against states that pose no direct threat to the security of the aggressors. A second purpose is to use the narrative in order to create the perception of an inevitability of the triumph of 'democracy' over other forms of governance. Some observers have noted a missionary zeal to aggressively spread liberal democracy to other parts of the world (Woodberry, 2012). In order to achieve this spread, a systematic approach needed to be devised and applied.

Although noted within the context of a case study on Bosnia, neoliberal ideology and practices create a thin layer of accountability, which is enhanced through the use of international organisations, mass media and 'NGOs' that create strategic narratives to influence knowledge

[7] N. Bensahel, *Darker Shades of Gray: Why Gray Zone Conflicts Will Become More Frequent and Complex*, Foreign Policy Research Institute, www.fpri.org/article/2017/02/darker-shades-gray-gray-zone-conflicts-will-become-frequent-complex/, 13 February 2017 (accessed 27 February 2017).

(Kostic, 2017) on events, in order to shape public perception, opinion and consent situated in an environment of information dominance. In this contest and struggle, the role of public perception and public opinion is critical for the above narratives to be effective and to permit an aggressive and expansionist foreign policy done in the name of 'democracy' and 'humanitarianism'. Bliesemann de Guevara and Kostic (2017) argue that the system of neo-liberalism constricts the production of information and knowledge on conflicts and intervention as a means of attempting to control the perception of 'facts' and 'truth' concerning these issues. This means that information producers, such as mass media and journalism form the spearhead of the campaign of influence. "The media, in the modern era, are indisputably an instrument of war. This is because winning modern wars is as much dependent on carrying domestic and international public opinion as it is on defeating the enemy on the battlefield" (Payne, 2005: 81). Mass media and journalism, as with the film industry has the potential power to articulate the affective logic of popular geopolitics to the public; when it does so it can misrepresent the situation in order to sustain an ideology and a policy that may be difficult to otherwise manage. There is the ability to create the narrative and signifiers (Carter & McCormack, 2006) in order to engineer public perception, opinion and consent on potentially controversial foreign policy.

The current situation of military interventions in the name of democracy and human rights has caused some to observe a huge increase in propaganda activities with the ascendency of liberal democracy. In turn, the lack of critical reflection and challenge by mainstream media to this trend have increased their focus on the different forms of propaganda that may manifest in news content (Zollmann, 2017). Mass media and journalism in the West have been an enabler of military interventions, which fits with Payne's characterisation of them as an instrument of war. Waldman (2018) characterises the nature of the post-9/11 US security-driven interventionism as "more persistent, intensive, and vicariously visceral." He describes the current US approach to conflict engagement as vicarious warfare, which is a loose and wide-ranging collection of tools and mechanisms, methods and dispositions that are used in different combinations and contextually applied. This has been observed in some recent conflicts.

Other observers state that the "Libya scenario" cannot be repeated in either Ukraine or Syria as these countries are much more central to the interests of Russia, which may provoke an open confrontation. Therefore, it is 'safer' to use unconventional and hybrid means of indirect warfare to subvert and bring about regime change through Colour

Revolution-type activities (Korybko, 2015). This creates a grey zone not only on the physical field of battle, but also in the informational battlefield too. The actual reasons for armed conflict cannot be openly acknowledged for many years. A rather startling revelation was made by British Prime Minister Theresa May when speaking to Republican policymakers in Philadelphia in 2017, she broke from Tony Blair's 'liberal intervention' approach. "This cannot mean a return to the failed policies of the past. The days of Britain and America intervening in sovereign countries in an attempt to remake the world in our own image are over."[8] This was in reference to the on-going wars in Afghanistan and Iraq, but it is possible to add Libya and other countries too. Therefore, there was some doubt cast upon the public explanation of self-defence or 'humanitarian intervention' as being the actual underlying reasons for the conflict, rather the notion of a 'liberal military intervention.' Thus, the use of deception and misinformation in the information domain is used to favourably shape the cognitive domain of public opinion and perception in order to garner 'legitimacy' for political and military operations in the physical domain.

Ukraine

There is a general lack of consensus in how to define the crisis currently occurring in Ukraine. For example, officially Russia tends to characterise the armed conflict as a civil war that was initiated by the Euromaidan coup, whereas the United States tends to officially emphasise the conflict as being the result of a Russian invasion (Driscoll, 2019). Yakushik (2016) notes the diverse interpretations and characterisations of the conflict that range from a Revolution of Dignity to foreign interventions or separatist movements, to a coup that provoked regionalist movements and a civil war. Political language and the use of symbolism play an important role in the shaping of public perception and opinion of events, but it is also an essential step in the construction of the orthodoxy of knowledge of people, places and events. The dominant narratives and frames of constructed reality, once they become 'sticky', are difficult to challenge once established, hence the competition and conflict in the information domain to subjectively define the 'appropriate' terminology

[8] L. Pasha-Robinson, *Theresa May: US and UK Will No Longer Invade Foreign Countries 'to Remake the World in Their Own Image'*, Independent, www.independent.co.uk/news/uk/home-news/theresa-may-donald-trump-us-uk-no-longer-foreign-intervention-iraq-afghanistan-a7548551.html, 27 January 2017 (accessed 21 September 2018).

that is best suited to the political aims and goals in a manner that seems superficially to be an objective exercise.

Yakushik (2016: 117) observed that "this kind of complex, 'hybrid' political conflict involves several internal and external actors." Sakwa notes that there was a choice forced upon Ukraine to either follow the Western civilisational path and integration (the 'correct' choice) or a more Eastern/Russian civilisational choice (the 'wrong' one), which was applied to a fragile and divided country (2016: 50–80). Smith, in his analysis of the country, notes, "by Huntington's civilizational standard, Ukraine is a severely cleft country, divided internally along historical, geographic and religious lines, with Western Ukraine firmly in the Ukrainian corner and Eastern Ukraine and Crimea firmly in the orbit of Orthodox Russia."[9] The result of this forced choice has been catastrophic for Ukraine and the wider region. In particular, one of the key results is the increased rate of polarisation of the world into politically and economically opposing blocks (Sakwa, 2016: 205–227). Smith ventures so far as to refer to the Ukrainian conflict as a proxy war between the West and Russia, where the goals and outcome of various engagements by each side are unclear and risk a wider and larger conflict.[10] These descriptions by Sakwa and Smith on the origins of the current conflict closely resemble that of a geopolitical shatterbelt. Jalilov and Kelly (2014: 1, 9–11) state that the conflict in Ukraine corresponds to the classical definition of a shatterbelt. There are two evident levels in the conflict: the local one in the form of a civil war; and an international one based on the strategic rivalry for influence in the country between the US-led West and Russia. This second aspect is seen as being symptomatic of a new Cold War.

Unlike previous Colour Revolutions in the former Eastern Bloc, Euromaidan was a violent conflict, and much more so by the time of the regime change (Kudelia, 2018), which seemed to demonstrate a tactical influence inherited from the Arab Spring (Korybko, 2015: 66). The use of the term civil war to describe the on-going armed conflict in Eastern Ukraine is problematic owing to the creation and projection of the brand term "Anti-Terrorist Operation" (ATO) for the Ukrainian military operations against the rebel regions of Lugansk and Donetsk. This is a clear attempt to control the narrative through applying an

[9] W. S. Smith, *Ukraine and the Clash of Civilisations*, The National Interest, https://nationalinterest.org/feature/ukraine-and-clash-civilizations-153636, 12 May 2020 (accessed 3 June 2020).

[10] W. S. Smith, *Ukraine and the Clash of Civilisations*, The National Interest, https://nationalinterest.org/feature/ukraine-and-clash-civilizations-153636, 12 May 2020 (accessed 3 June 2020).

ethical tone to the conflict and to simultaneously affect the brand and reputation of the rebel side negatively, whilst preserving the moral high ground. Thus, it is very problematic from the start to try and frame this conflict as being solely a civil war owing to the competing projections.

Laruelle (2019: 715) remarks that there are two levels to the conflict in Ukraine, with one on the international level involving the aspect of external interference, and one at the domestic level including such aspects as ethnic, regional, linguistic, and economic (one can add religious) tensions. However, the focus is on one or the other level and not each together. Several different factors and aspects are evident within the framework of domestic problems and crises that drive the local level of the conflict in Ukraine. Some researchers (Matsuzato, 2017; Galbreath & Malyarenko, 2020) remarked that in the early stages of the Donbass conflict, researchers tended to focus on the international actors (especially Russia) and tended to neglect the agency of domestic issues and crises, local actors, their intentions and actions; for example, the interactions between the Party of Regions and the Novorussian movement, an alliance of convenience that was built upon regional social discontent, and their ambitions which were later tempered somewhat by Russia in return for support (Matsuzato, 2017: 200–201). Malyarenko and Wolff (2018) acknowledge a number of lines of internal divide and weaknesses that exist in Ukraine that make it prone to being vulnerable to manipulation and the mobilisation of domestic and foreign actors.

Earlier US attempts at regime change in Ukraine have been well documented: the central roles played by US-based foundations and NGOs; the use of political intelligence (such as the use of polls and focus groups to generate data to enable an effective strategy); and attempting to construct a popular public political figure to unite the opposition in order to boost the chances of unseating the incumbent president. The United States is estimated to have spent some US$14 million in the Orange Revolution in 2004 to topple Yanukovych.[11] This was an example of a branded revolution, which had swept aside several governments by reputation and perception management. It was aided using slick, symbolic catchphrases and slogans; communication, recruitment, support and social mobilisation through the Internet and social media (especially Facebook); tactics intended to frustrate and cause the government to become paralysed or overreact in a carefully choreographed public spectacle designed to win hearts and minds, and eventually political power. The façade of the symbolic political space was its projection as

[11] I. Traynor, *US Campaign Behind the Turmoil in Kiev*, The Guardian, www.theguardian.com/world/2004/nov/26/ukraine.usa, 26 November 2004 (accessed 7 September 2018).

a spontaneous local grassroots and democratic movement, but there was an element of foreign support.

> [...] the campaign is an American creation, a sophisticated and brilliantly conceived exercise in western branding and mass marketing that, in four countries in four years, has been used to try to salvage rigged elections and topple unsavoury regimes.[12]

These tactics are selectively communicated and applied according to the façade which projects a hyper-reality of a local grassroots movement and permits the United States and its allies to assume the role of a *powerful broker* in the unfolding process. The information operations form an important aspect of political warfare and subversion through shaping audience perception and opinion in the cognitive domain in order to steer events in the physical domain. It is a matter of attempting to influence the informational flows around the crisis event, and therefore constrain and restrain the target government's operational choices. Social media and online platforms not only permitted the movement against Yanukovych to communicate with each other, but also with the wider global public in a battle for winning public opinion and perception.

In the immediate prelude to Euromaidan the US embassy in Kyiv ran "TechCamps" that were intended to educate NGOs and activists how to use social media in order to achieve optimal impact in their activities. The stated aims of such events are for "supporting Secretary Clinton's Civil Society 2.0, an initiative which builds the digital literacy of civil society organizations around the world."[13] Shortly before Euromaidan, a follow-up full-day interactive TechCamp was held by the US Embassy in Kyiv, and its purpose was explained in a press release.

> The U.S. Embassy in Kyiv in partnership with Microsoft Ukraine hosted TechCamp Kyiv 2.0 on March 1, 2013 at the Microsoft Ukraine Headquarters. TechCamps support the U.S. State Department's Civil Society 2.0 initiative that builds the technological and digital capacity of civil society organizations around the world.
>
> [...] To date, State Department sponsored TechCamps in Ukraine have trained more than 200 civil society organizers from throughout the country and Belarus. The technologies and approaches presented help to build new networks of relationships, enhance skill development, and create new avenues for communication. Adoption of these technologies by civil society organizations

[12] I. Traynor, *US Campaign Behind the Turmoil in Kiev*, The Guardian, www.theguardian.com/world/2004/nov/26/ukraine.usa, 26 November 2004 (accessed 7 September 2018).

[13] *TechCamp Kyiv 2012*, US Embassy Kyiv, www.youtube.com/watch?v=bpIoBUDuL3U, 24 September 2012 (accessed 8 September 2018). The issue was also discussed in the Verkhova Rada, concerning the subversive role and the purpose of the TechCamps – www.youtube.com/watch?v=y9hOl8TuBUM&feature=youtu.be.

will help support the missions of these groups as well as broader social goals of democracy, transparency and good governance in the 21st Century.[14]

Social media formed one of the central mechanisms in the events of Euromaidan that led to the eventual overthrow of President Yanukovych. Protests began and quickly mobilised from 21 November 2013 when Yanukovych announced the decision to suspend preparations for the EU–Ukraine association agreement,[15] crowds gathered in Independence Square (Maidan) and were mobilised by use of the tags #euromaidan and #евромайдан on Facebook and Twitter. Facebook posts of Hromadske TV journalist Mustafa Nayem were widely shared in a short space of time, providing the face of an early and recognisable key influencer to prime and mobilise the crowds.[16] This is a demonstration of the role of key influencers and their intangible power in the information domain in shaping the cognitive domain of their audiences, where they convert the effects of intangible persuasive communication into tangible and physical actions via an induced herd mentality.

The tactics of revolution and subversion of the government of a target country has been undergoing development and a refining process that has been openly observable since the Serbian revolution in 2000 that overthrew Milosevic. This gradual process has created a form of 'blueprint' of political warfare and subversion that can be applied to other target countries. As such there is a transfer of knowledge and people in and between the various iconic events within the revolutionary chains (in and among the various Colour Revolutions and Arab Spring country cases). Social media provides a means to share and spread knowledge and tactics, and to coordinate political warfare. During the Arab Spring in Egypt, anti-government movements sought to learn from the

[14] *U.S. Embassy Hosted TechCamp Kyiv 2.0 to Build Technological Capacity of Civil Society*, US Embassy Ukraine, http://archive.is/itUBA#selection-1327.0-1327.318, 1 March 2013 (accessed 8 September 2018).

[15] *Ukrainian Parliament Fails to Pass Law on Tymoshenko Release*, Yahoo News, www.yahoo.com/news/ukrainian-parliament-fails-pass-law-tymoshenko-release-095738223.html, 21 November 2013 (accessed 8 September 2018).

[16] K. Kapliuk, *Role of Social Media in Euromaidan Social Movement Essential*, Kyiv Post, www.kyivpost.com/article/content/euromaidan/role-of-social-media-in-euromaidan-movement-essential-332749.html, 1 December 2013 (accessed 8 September 2018); P. Barberá and M. Metzger, *How Ukrainian Protestors Are Using Twitter and Facebook*, Washington Post, www.washingtonpost.com/news/monkey-cage/wp/2013/12/04/strategic-use-of-facebook-and-twitter-in-ukrainian-protests/?utm_term=.ecfff47c2941, 4 December 2013 (accessed 8 September 2018).

Ukrainian experience in Euromaidan and apply those lessons there.[17] The foreign influence in driving and maintaining subversive information warfare proved to be a factor in pushing other foreign stakeholders to intervene.

Bunce and Hozic (2016) admit that there are the realist factors of international relations at play in prompting Russia to intervene in Ukraine (which they refer to as an "on-going invasion"), but of possibly greater significance and importance for the Russian authorities is the domestic situation. Russia's engagement, according to this logic, is about diffusion-proofing Russia from experiencing a foreign-supported revolution that could overthrow the government. The notion and threat of direct Russian military intervention in Ukraine also featured in Western mainstream media.[18] Ben Aris was critical of this use of terminology, where he defines an invasion as "a military operation designed to capture territory" and argues that this is not Russia's goal, but rather it is to force a negotiated peace through its "incursion".[19] An independent investigative journalism outlet not only dismissed the idea of "a mysterious invasion", but that "official Washington draws the Ukraine crisis in black and white colours with Russian President Putin the bad guy and the US-backed leaders in Kiev the good guys. But the reality is much more nuanced, with the American people consistently misled on key facts."[20] Mearsheimer laid the blame for the results of the situation in Ukraine largely with the United States and its European allies for encroaching on Russia's core strategic interests, which resulted in Russian "pushback".[21] He added that this was proof that realpolitik remains relevant in the twenty-first century.

Some have simplistically labelled the situation as a "return of geopolitics" of a black and white world of 'aggressive' Russian expansion versus

[17] N. Ketchley, *How Social Media Spreads Protest Tactics from Ukraine to Egypt*, Washington Post, www.washingtonpost.com/news/monkey-cage/wp/2014/02/14/how-social-media-spreads-protest-tactics-from-ukraine-to-egypt/?noredirect=on&utm_term=.e2c4aa77e7f7, 14 February 2014 (accessed 8 September 2018).

[18] A. E. Kramer, *Russian Intervention in Ukraine Is Likely, NATO Says*, The New York Times, www.nytimes.com/2014/08/12/world/europe/russian-intervention-in-ukraine-is-likely-nato-says.html, 11 August 2014 (accessed 15 August 2014).

[19] B. Aris, *Moscow Blog: Russia Invades Ukraine (Not)*, Business New Europe, www.bne.eu/content/story/moscow-blog-russia-invades-ukraine-not, 28 August 2014 (accessed 29 August 2014).

[20] R. Parry, *Who's Telling the 'Big Lie' on Ukraine?* Consortium News, www.consortiumnews.com/2014/09/02/whos-telling-the-big-lie-on-ukraine/, 2 September 2014 (accessed 8 September 2014).

[21] J. Mearsheimer, *Why the Ukraine Crisis Is the West's Fault*, Foreign Affairs, www.foreignaffairs.com/articles/141769/john-j-mearsheimer/why-the-ukraine-crisis-is-the-wests-fault, September/October 2014 (accessed 20 August 2014).

Western liberalism's 'liberation' that demands a "more muscular liberalism" to engage in geopolitical conflict with Russia. This is being exemplified by the conflict in Ukraine (Auer, 2015). Celso (2018: 99–101) also formulated Russia's actions as a type of geopolitical aggression against the West and NATO, within the frame of being a form of superpower hybrid warfare. It has been noted that there is a tendency to project and promote a subjective construction of reality that is intended to promote the righteousness of the West's actions while demonising Russia, which does little to actually understand the drivers and sources of motivation for Russia's policy and actions in Ukraine (Smith, 2017). Malyarenko and Wolff (2018) contend that Russia's goal concerns the strategy of competitive influence seeking where the goal is to maintain a consistent level of instability to impede a Ukrainian government's capability and capacity owing to its anti-Russian and pro-Western inclinations.

Geopolitics and various national interests can be masked and obscured by rhetorically elegant and worthy goals, which can cause such widely diverse interpretations of the physical domain's events and processes. Mullerson (2014) notes that the invocations of international law and morality are often used as a smokescreen for geopolitical competition. Furthermore, unlike in the Cold War where the geopolitical confrontation was based upon the notion of clashing political ideologies, the current struggle of which Ukraine is part, concerns unipolar versus multipolar visions of the world order and the norms and values that accompany the winner.

This chapter was written well before the 2022 Russia–Ukraine War began; this brief commentary provides an overview of events from February 2022 in light of the war and how this fits in the context of this chapter. After an intense and increasing period of tension and hostility from November 2021 until February 2022, the Russian overt military attack on Ukraine commenced. This was simultaneously both expected and unexpected, given the war of words and nerves that was waged before the attack, together with the warnings and opportunities to prevent the outbreak of war (Simons, 2022a). Initially the war was presented in a narrative of binary construction of liberal values and norms. Namely, it was Ukraine's sovereign choice to join NATO, even if it was perceived as impinging on Russia's sense of national security. This is set within an analytical context of a string of historical broken promises and warnings from 1990 by the West, and from Russia on the hazards and risks of expanding NATO eastward and disregarding Russian security interests, which were ignored, selectively interpreted or misrepresented. A similar blueprint for conflict is being deployed in a scenario in the Pacific region between China and Taiwan (Simons, 2022b), the former being another competitor and challenger of US global hegemony. Open and public rhetoric by high-ranking US officials, such as Victoria Nuland, Secretary of State Anthony Blinken and Secretary of Defence Lloyd Austin, hint at less benevolent reasons for supporting Ukraine. Austin has spoken publicly on using the war in Ukraine to weaken Russia.[22] The problem for Ukraine and its political leadership is that it has become a dependent client state of the US-led West during a period of instability created by global geopolitical transformations taking place; consequently it cannot afford to keep fighting a war of attrition intended (by the US) to weaken and contain Russia at the expense of Ukrainian blood, but it also cannot stop the fighting that may lead to political instability and regime change in Ukraine at the hands of vested domestic and foreign actors.

Syria

The war in Syria has attracted a lot of media attention, which often paints the conflict as a Manichean struggle between 'good' and 'evil', 'right' and 'wrong' and this hides the more complex nature of it. "Western news media outlets have paid considerable attention to the civil war in Syria, but much of the coverage is simplistic and melodramatic" (Carpenter, 2013: 1). Through a reliance on the mainstream media outlets' characterisation of the Syrian conflict, Flint (2017: 114) concludes that in Syria "demonstrations and protests became civil war as rebels and government forces fought." He also links the early stages of the conflict in Syria in March 2011 to the Arab Spring that is characterised as being a call for political change. However, this ignores the role of the United Kingdom as being one of the supporting actors of the United States in the actual and attempted regime changes undertaken under the cover of the Arab Spring, and the support provided by mainstream mass media, such as the BBC, to this form of foreign policy. Ettinger (2017) notes that in the aftermath of the 11 September 2001 attacks on the United States, their interests in the Greater Middle East have come to include nation-building in Iraq and Afghanistan, counterterrorism and counterinsurgency, nuclear non-proliferation with Iran and Russia's overt engagement in the Syrian war. The result has been that the Middle East has once more become the stage for great power politics.

Some authors noted an evolution in the nature of the Arab Spring conflicts, from revolution to civil war. Bhardwaj (2012: 90) observed that

[22] M. Ryan & A. Timsit, *US Wants Russian Military 'Weakened' from Ukraine Invasion*, Austin Says, The Washington Post, www.washingtonpost.com/world/2022/04/25/russia-weakened-lloyd-austin-ukraine-visit/, 25 April 2022 (accessed 29 August 2022).

"international and regional actions have been determined as lynchpins in driving conflict to civil war." In 2012, Syria was assessed by Bhardwaj as being in a state of conflict that fell just short of being classified as a civil war, but was on its way to approaching this state of being. Early on in the on-going conflict, Malantowicz (2013: 60) noted that the Syrian civil war had deeply embedded ideological and political aspects. However, the war has grown increasingly more complex and destructive, with the gradual entry of other actors and other interests. Galariotis and Ifantis (2017: 1) characterise the civil war in Syria as having international implications "due to the on-going multisided character of the sectarian conflict and the involvement of all major global and regional powers." In 2014 Hughes (2014: 524) predicted that "it is likely that the Syrian civil war will become a proxy conflict, due to its intersection with regional rivalries and power political disputes." This implies a conflict that has grown, evolved and transformed in terms of its complexity and its impact on a local, regional and global basis, and in the means and mechanisms employed by different actors.

An article appearing in *Foreign Policy* noted a more nuanced state of conflict existing in Syria. Spyer notes that "the country's [Syria] civil war is over – and an entirely new one has started with the United States at the centre of it." He added that "since mid-2014, there have been two parallel wars taking place on Syrian soil."[23] This includes one dimension involving domestic actors, and another of regional and international powers. There are several seemingly contradictory actions by a vast and complex array of varied strength alliances of convenience and necessity at play in Syria, which means that the domestic level of the conflict is entirely intertwined with the regional and international dimensions.

> The new contests in Syria derive not from internal Syrian dynamics, but from the rival interests of outside powers pursued over the ruins of Syria: Turks against Kurds, Israel against Iran and its proxies, the United States against Iran, and now, potentially, Ankara against Washington. These external forces are all determined to gain advantage over one another in Syria. And so, even as Syria's two longstanding conflicts wind down, war and strife are not departing the area.[24]

There are numerous overt and obvious references to the geopolitical understandings and logic of the armed conflict in Syria through political statements. What started as a 'simple' civil war had become, by 2019,

[23] J. Spyer, *Welcome to Syria 2.0*, Argument, Foreign Policy, https://foreignpolicy.com/2018/01/25/welcome-to-syria-2-0/, 25 January 2018 (accessed 29 July 2018).

[24] J. Spyer, *Welcome to Syria 2.0*, Argument, Foreign Policy, https://foreignpolicy.com/2018/01/25/welcome-to-syria-2–0/, 25 January 2018 (accessed 29 July 2018).

three civil wars in one according to Spyer,[25] where the fight to depose Assad was essentially over and the fight to carve out and hold territory began. Three primary actors are involved that have both domestic and foreign connections: 1) the Syrian government; 2) the US-aligned SDF and 3) the Turkish-Sunni Islamist zone (smallest of the three in territory). This complexity has not decreased.

Initially the United States under the Obama presidency communicated strongly for regime change as a precondition for any form of 'peaceful' settlement of the conflict (Blanchard, Humud & Nikitin, 2014).[26] The regime change policy was continued by other allies, including Turkey, even after the balance of power on the battlefield swung in favour of the Syrian government.[27] It is difficult for a US administration to justify itself to and convince public opinion of the need to intervene directly and overtly in Syria using military enforcement to effect the desired regime change. Therefore, more indirect and covert paths are sought, and at times these are packaged within a geopolitical message. Rhetorically, the United States and its allies try to justify the illegal presence of military forces in Syria as being necessary to fight ISIS. There is growing evidence to suggest that the US government has been using terrorist proxies as a means of forcing regime change and seeming to legitimise direct military intervention in Syria.[28]

However, when President Trump indicated that he would like to see the rapid withdrawal of US military forces from Syria, a more realistic and honest answer was provoked. "'He's not going to pull out until he has a solution that won't cede the field to Russia and Iran,' said Jim Hanson, the president of the Security Studies Group, a conservative think tank with close ties to the White House."[29] These exact thoughts

[25] J. Spyer, *Syria's Civil War Is Now 3 Civil Wars*, Foreign Policy, https//foreignpolicy.com/2019/03/18/syrias-civil-war-is-now-3-civil-wars/, 18 March 2019 (accessed 24 March 2019).

[26] For example, refer to C. Kanthan, *Syria: 40 Links That Prove Regime Change Plans*, World Affairs, https://worldaffairs.blog/2017/04/18/syria-20-links-that-prove-regime-change-plans/, 18 April 2017 (accessed 16 March 2018); Memo on Syria, US Department of State Case No. F-2014-20439 Doc. No. C05794498, dated 30 November 2015.

[27] A. Barnard, *Assad Must Go, Erdogan Says, as Syria War Winds Down*, The New York Times, www.nytimes.com/2017/12/27/world/middleeast/syria-evacuations.html, 27 December 2017 (accessed 29 December 2017).

[28] For example, the audiotapes of then Secretary of State John Kerry communicating with such groups in this video: www.youtube.com/watch?v=t3KfmjdviHM&fbclid=IwAR3_mnNocQizU1RiXHoMK6SY6asa-p1UnMChdcwTpfzWp_TP2KM5AOUZgCM or in this second video https://www.youtube.com/watch?v=e4phB-_pXDM.

[29] D. S. Cloud, B. Bennett & T. Wilkinson, *Trump Orders Pentagon to Prepare for Withdrawing U.S. Troops from Syria, but Some Advisors Push Back*, LA Times, www

and justifications are found in a report that claims it is unlikely that the United States could persuade Russia to distance itself from Iran and to create a wedge between the countries. In addition, Iran's operations in the area are likely to significantly alter the region's balance of power through their increasing capacity and success in waging "hybrid" warfare (Bucala, 2017). Thus, the underlying reason is not about defeating ISIS, for which no mandate exists that allows the legal deployment of military forces there, but as a means to counter possible Russian and Iranian influence in the country. The United States and Iran have been engaged in a series of tit-for-tat accusations concerning the legality of the presence of Iran's forces in Iraq and Syria, where the US goal is to counter Iranian influence in Syria (Edelman & Wald, 2017). US Secretary of State Mike Pompeo argued that it was illegal and subversive and Iran countered by saying they were invited by the government of the country and it was the United States that was unwelcome.[30] Although, interestingly, a 2017 report from the Institute for the Study of War (Cafarella, Kagan & Kagan, 2017) argued that the US strategy was relying on proxies and air strikes to defeat Islamic State, but called on a temporary alliance of convenience on US terms with Russia and Iran to defeat the terror group. So, it is not about a humanitarian concern nor counterterrorism, but rather an attempt at the denial of geopolitical influence to the long-term allies of Syria.

The United Kingdom was also a vocal supporter and proponent of regime change from the onset of the so-called Arab Spring in Syria, calling for the abdication of President Assad and the coming to political power of the Syrian 'opposition'. After the manipulation of Iraq (2003) and Libya (2011), the narrative was that the West was not actively seeking war (and regime change), but rather war was coming to the West ("unavoidable") (RUSI, 2012). A reason given later, to motivate a more forceful policy in Syria, was to prevent the spread of Russian and Iranian influence (Chalmers, 2015: 6). There was a desire to be involved in direct military intervention against the Syrian government, which was made difficult owing to the strong public opinion against yet another 'humanitarian' war in the wake of Iraq in 2003 and more recently in Libya in 2011 (House of Commons Foreign Affairs Committee, 2016). When a parliamentary vote in August 2013 on direct military

.latimes.com/nation/la-fg-pol-syria-trump-20180404-story.html#nws=mcnewsletter, 4 April 2018 (accessed 5 April 2018).

[30] T. O'Connor, *Iran Tells US: We Were Invited to Iraq and Syria, What About You?*, News Week, www.newsweek.com/iran-tells-us-we-were-invited-iraq-syria-what-about-you-945166, 25 May 2018 (accessed 29 May 2018).

intervention, which was in the wake of a chemical weapons event, failed to pass (Chalmers, 2015: 3), another approach was needed. This came in 2015 with the terror attacks in Paris and the excuse narrative calling on the duty to respond militarily to ISIS in Syria; this was so that the United Kingdom was able to demonstrate its reliability as an ally and partner of France (Chalmers, 2015). This is a good example of the use of a false logic, because regime change in Syria had little direct connection to the terror attacks in Paris, and successful regime change was likely to increase the risk and likelihood of future terror attacks across the West. As the prospects for a quick regime change vanished and the ability to employ a Libya scenario of large-scale and direct Western military intervention was risky, other options began to be considered.

There are a multitude of parties and interests supporting the Syrian government against the attempted regime change challenge. Although not yet directly involved, China has been steadily developing its scientific and military relations with Syria. One of the aspects that ties China to the conflict is the participation of Chinese Uyghurs in the fighting to topple Assad. Beijing perceives the conflict as a war against terrorists. In an interview in August 2018, the Chinese Ambassador to Syria, Qi Qianjin, told Syrian media that "Beijing is prepared to aid the government's push to retake territory throughout the country."[31] Iran has been actively involved in the Syrian conflict, and has had a long-standing relationship with Syria, which was the only Arab country that supported Iran during the 1980–88 Iran–Iraq War (Ansari & Tabrizi, 2016: 3). There are three objectives to Iran's involvement in the Syrian conflict: to defeat ISIS and Jabhat Al-Nursa; restoration of the *status quo ante*; and the preservation of Syria's state institutions (Ansari & Tabrizi, 2016: 6–9). Iran also has a geopolitical interest to protect and to keep the lines of communication open to its ally Hezbollah and religious minorities in the Arab world, as well as preventing the gradual build-up of a radical Sunni threat emerging from the West (Chalmers, 2015: 7). Russia is one of the international actors that has managed to make a significant impact on the course of the conflict.

Russia (and the Soviet Union) has held a long-term diplomatic and defence relationship with Syria (such as the lease of the Soviet-era naval base at Tartus). Russia has been involved indirectly, and then later directly, from an early stage in the current Syrian conflict, supplying weapons, advising and training of military personnel, and conducting

[31] Asia Times Staff, *China Says Willing to Team With Syria's Assad in Push to Retake Territory*, Asia Times, www.atimes.com/article/china-says-willing-to-team-with-syrias-assad-in-push-to-retake-territory/, 3 August 2018 (accessed 5 August 2018).

diplomacy. The spectre of foreign-backed regime change operations in Syria was treated initially with suspicion (Nikitina, 2014; Dias & Freire, 2019), not least owing to the connections and interactions with the New Cold War concept (Perra, 2016; Aksenyonok, 2019). Moreover, 2013 saw a diplomatic coup when Russia and the United States brokered a deal that saw Syria divest itself of chemical weapons stocks and therefore avoid direct US military action. On 30 September 2015 Russia became directly and overtly involved militarily and saw off what seemed to be an inevitable collapse of the Syrian armed forces (Lain & Sutyagin, 2016: 17). Russia's geostrategic calculus after the regime change in Libya in 2011 meant that they were forced to act in Syria in order to preserve their influence and credibility as an international actor (Póti, 2018). In addition to the geopolitical and geostrategic interests in the MENA region, a collapse of the Syrian government would represent a significant domestic security risk to Russia with a possible resurgence of terrorism (Kozhanov, 2016; Zulfqar, 2018). This demonstrates the nature of the complicated and intertwined conflicts and tensions around the globe currently.

Several researchers have attempted to characterise the events in Syria as being 'grassroots' and 'spontaneous' as well as being part of a bigger picture of a popular desire and demand for democracy, thereby emphasizing the internal and civil dimensions of the conflict. However, one can still note the obvious geopolitical dimensions of external actors who are either supporting or subverting the Syrian government. An increasing number of countries, in the Global South especially, see the mechanism of Responsibility to Protect (R2P) as a means and an excuse for the Global North to maintain its global hegemony (Morris, 2013; Thakur, 2013; Kassim, 2014: 15–22). When Rex Tillerson was Secretary of State, he promised increased diplomatic action and placed the focus on a 'political solution' to what he described as a civil war. However, the Pentagon admitted to having some 2,000 US troops in Syria (in contravention of international law). He also stated that "a total withdrawal of American personnel at this time would help Assad [and a continued US presence will] help pave the way for legitimate local civil authorities to exercise responsible governance of their own liberated areas."[32] Therefore, the United States has placed itself as a powerful broker in a regime change policy that is based upon subjective notions of what is righteous, while other such value- and norm-based arguments are in

[32] T. Wilkinson, *Tillerson Says U.S. Military Should Stay in Syria and Vows Diplomatic Push to End Civil War*, Los Angeles Times, www.latimes.com/nation/la-fg-tillerson-syria-20180117-story.html, 17 January 2018 (accessed 18 January 2018).

reality breaking international law in something much more complex than just a civil war, although it is framed as being one.

The Syrian conflict is highly complex and volatile with a mix of numerous competing and contrasting actors, interests and goals. For example, Israel secretly armed and funded twelve Syrian rebel groups via the Golan Heights border. When the Syrian Arab Army took back the territory, these forces expected Israel to come to their aid with direct military support and were disappointed that they did not.[33] United Nations expert Richard Gowan described the complexity of the Syrian conflict as "the effect of unverified social media posts and slick state propaganda on a civil war wrapped in a proxy war inside a great-power war" that has resulted in a deluge of information of dubious quality.[34] Gregory Shupak from FAIR also characterised the Syrian War as a proxy fight at two different levels – among regional actors and global actors.[35] Just to what level the conflict remains a proxy war is becoming less clear, especially as the United States is said to control 28 per cent of Syrian territory.[36] As early as July 2012, RUSI (2012: 1) referred to the Syrian conflict as "an arc of proxy confrontation" across the MENA region. One can also argue that the conflict is no longer a proxy war with the presence of US, UK, Russian and Iranian troops all populating the same physical battlespace.

The year 2015 proved to be a key point in the Syrian conflict when Russia became militarily involved in the war directly and overtly. In the summer of 2015, the Syrian government was clearly in a strategic defensive position and at the point of collapse. Just slightly before the Russian intervention, predictions were made that began to speak of a shatterbelt at a tactical and operational level. "Any changes to the map of Syria's conflict in the rest of 2015 will almost certainly occur in its 'shatterbelt': those areas caught between the regime, armed opposition, and self-proclaimed Islamic State."[37] This labelling of the actors simplifies a

[33] *Israel Secretly Armed and Funded 12 Syrian Rebel Groups, Report Says*, Haaretz, www.haaretz.com/middle-east-news/syria/in-syria-israel-secretly-armed-and-funded-12-rebel-groups-1.6462729, 8 September 2018 (accessed 17 September 2018).

[34] U. Friedman, *The 'CNN-Effect' Dies in Syria*, The Atlantic, www.theatlantic.com/international/archive/2018/03/cnn-effect-syria/554387/, 1 March 2018 (accessed 7 March 2018).

[35] G. Shupak, *Media Erase US Role in Syria's Misery, Call for US to Inflict More Misery*, FAIR, https://fair.org/home/media-erase-us-role-in-syrias-misery-call-for-us-to-inflict-more-misery/, 7 March 2018 (accessed 10 March 2018).

[36] J. Spyer, *Welcome to Syria 2.0*, Foreign Policy, https://foreignpolicy.com/2018/01/25/welcome-to-syria-2-0/, 25 January 2018 (accessed 3 July 2018).

[37] Y. Sayigh, *Redrawing the Lines in Syria's Shatterbelt*, Carnegie Middle East Centre, http://carnegie-mec.org/2015/06/25/redrawing-lines-in-syria-s-shatterbelt-pub-60484, 25 June 2015 (accessed 29 May 2018).

much more complex reality in the physical domain. Steele advocated for Western governments to end the war in Syria and to push for peace, blaming them as being at least partially responsible for prolonging the conflict. He summarised the situation by stating: "the Syrian conflict was never a simple binary struggle between supporters and opponents of Assad. Millions of Syrians had little or no faith in either side but deplored the militarisation of what had started as a non-violent uprising and became a proxy war in which outside states used Syria as a battleground for their own interests."[38] This aligns with other more critical analyses of the West and its wars of regime change and convenience, and the effects of these conflicts. "We are facing a huge, potentially divisive, turning point, the most significant moment in global affairs since the end of the Cold War. The causes have been well rehearsed. Western powers, led by the US, weary of intervening after their experiences in Iraq and Afghanistan – and Libya – are reluctant to commit themselves, even though their interventions in the past provoked the crisis [Syria] in the first place."[39] Other characterisations of the Syrian conflict include referring to it as superpower hybrid warfare, with a focus on the United States versus Russia in particular (Celso, 2018). In reality, it is a very broad array of armed groups supported by an equally wide array of foreign powers that possess different interests and goals, despite such simplified value- or norm-symbolic slogans as "fighting ISIS", "containing Iran or Russia" or "supporting democracy."

Conclusion

In the introduction of this chapter, the following research question was posed, *how are the public representations of the Ukrainian and Syrian conflicts formulated*? Before answering this question, it is necessary to contextualise these conflicts in a wider frame. Even though the armed conflicts in Ukraine and Syria are in different geographical regions (Europe and the Middle East), they do have several aspects in common. Both conflicts started relatively close to each other, Syria in 2011 and Ukraine in 2013, neither have ended at this stage and both seem to have no immediate prospects of ending. Syria and Ukraine are both highly complicated and multi-layered conflicts, which have drawn in actors from the local,

[38] J. Steele, *If Ending Syria's War Means Accepting Assad and Russia Have Won, So Be It*, The Guardian, www.theguardian.com/commentisfree/2018/sep/21/syria-war-russia-assad-western-proxies, 27 September 2018 (accessed 22 September 2018).

[39] R. Norton-Taylor, *Syria: A Geopolitical Earthquake*, The Guardian, www.theguardian.com/news/defence-and-security-blog/2015/oct/06/syria-a-geopolitical-earthquake, 6 October 2015 (accessed 29 May 2018).

regional and global levels. These conflicts have both been framed within revolutionary brands; Syria within the wider context of the Arab Spring and Ukraine within the country-specific brand names of Euromaidan or the Revolution of Dignity. These brands are embedded with political and social symbolism of the 'revolutionaries' cause and a sense of expectation management of the outcome or fate of the country and its people.

The public representations and descriptions of the Ukrainian conflict are varied and contested. Specific terms are used by specific actors, which indicate the instrumentalisation of language for political and/or geopolitical aims and objectives. From a Ukrainian and an official US/EU narrative position, the conflict started as a popular revolution by the grassroots for a Western-oriented future (as opposed to Yanukovych's 'Russian' future). The situation deteriorated with the Russian annexation of Crimea and the beginning of the "Anti-Terrorist Operation" (ATO) in Donbas. It has since evolved into the current iteration of a Russian invasion and the Russian–Ukrainian War. From a Russian narrative perspective, the current situation began with a US-promoted and supported civil insurrection that was followed with regime change by a coup using far right forces against the legitimate Ukrainian government of Yanukovych. This was then followed by a civil war, based upon various internal cleavages found in Ukrainian society that were exacerbated by Western interference. The current phase is being understood as an attempt by the United States to take Ukraine completely into the political and security structures as a means to further isolate and encircle Russia in a New Cold War. Therefore, the sum of the narrative is that the United States instigated a successful regime change operation and is now defending and attempting to stabilise their client government by increasing its capability and capacity to effectively govern its national territory against Russian attempts to destabilise the situation through proxy forces and other means.

Some inverse similarities exist in the characterisation of the Syrian conflict and its genesis. According to a more standard Western mainstream understanding of the beginning of the Syrian conflict, it started with a 'grassroots' movement of people seeking a better and freer life from a 'tyrant'. This is based on opposing visions and versions of norm- and value-based ideological choice of freedom versus oppression, whereas Ukraine also had the additional layer of a civilisational choice (Western versus Russia). Both are rhetorically loaded constructions intended to create a constructed logic of a 'right' choice and a 'wrong' choice. Syria has been largely described as being a civil war, although there are some mentions of a proxy war. Direct military intervention in

Syria against the Syrian government and military is problematic after the deceptions used to become involved in Iraq in 2003 and Libya in 2011; the 'fighting ISIS' logic includes the role of air strikes, and the illegal occupation of parts of Syria (including the theft of oil). However, there are periodic glimpses to the geopolitical dimensions that infer the existence of a shatterbelt, such as the logic of Western military operations to contain the influence of Iran or Russia or even the Syrian government. Unlike the situation in Ukraine, Syria is an example of an unsuccessful attempt (so far) at regime change by the US-led West. Russia and the allies of the Syrian government have a different narrative perspective, which includes the role of US-led subversion of the Syrian government by indirect means via information technologies that are demonstrated by the Colour Revolution-like approach to information and political warfare using a 'Libya scenario'. Those opposing the Syrian government are characterised as being terrorist proxies (of the United States or Saudi Arabia for example) that are paid, armed and supported by foreign powers, rather than holding a genuine belief in any ethical and moral justifications. Therefore, the overall and combined understanding involves the 'reality' of US-led regime change operation that involves proxy forces within the context of a foreign-fuelled civil war, where the Syrian government and its allies are fighting for regime stability and territorial integrity, and additionally are attempting to reduce the further spread of insecurity beyond the border, as was the case when the Libyan government was deposed with NATO help.

Superficially, terms and definitions are intended to provide clarity of understanding complex and complicated phenomena that occur in the human world in an objective manner. That is, they give life to a clear understanding and comprehension to the conflict event by creating a cognitive lens or tool that provides the necessary translation. Often though, concepts are a subjective political means of projecting a desired constructed reality to an audience to influence their perception and opinion of the event. This is related to the goal of creating the orthodoxy of knowledge (the result of the spiral of silence, which excludes competing narrative and conceptual accounts), which when achieved is tantamount to information superiority and giving the actors that possess it a great deal of power and ability in attaining perception management. This form of communication is about attempting to convey the 'legitimacy' of one's own political or geopolitical cause versus the 'illegitimacy' of an opponent's via the careful and deliberate use of terminology and language in the information domain in order to influence the audience's cognitive domain. Both conflicts, rather than viewing them as either civil war or shatterbelt, possess the characteristics of these terms. In fact, the weaknesses of the state and society that enabled a shatterbelt to form and consolidate also created an 'ideal' environment for initiating and maintaining a civil war that involves foreign actors as powerful brokers for the opposing parties.

References

Aksenyonok, A. (2019), *The Syrian Crisis: A Thorny Path from War to Peace*, Valdai Papers #104, Moscow: Valdai Club.

Alberts, D. S., Garstka, J. J., Hayes, R. E. & Signori, D. A. (2001), *Understanding Information Age Warfare*, Washington DC: CCRP Publication Series.

Ansari, A. & Tabrizi, A. B. (2016), I. The View from Tehran in Tabrizi, A. B. & Pantucci, R. (Eds.), *Understanding Iran's Role in the Syrian Conflict*, Occasional Paper, London: Royal United Services Institute, pp. 3–9.

Auer, S. (2015), Carl Schmitt in the Kremlin: The Ukraine Crisis and the Return of Geopolitics, *International Affairs*, 91(5), pp. 953–968.

Bakir, V., Herring, E., Miller, D. & Robinson, P. (2018), Organised Persuasive Communication: A New Conceptual Framework for Research on Public Relations, Propaganda and Promotional Culture, *Critical Sociology*, 45(3), 2019, pp. 311–328.

Bhardwaj, M. (2012), Development of Conflict in Arab Spring Libya and Syria: From Revolution to Civil War, *The Washington University International Review*, 1(1), pp. 76–96.

Blanchard, C. M., Humud, C. E. & Nikitin, M.-B. D. (17 September 2014), *Armed Conflict in Syria: Overview and U.S. Response*, Congressional Research Service, RL33487.

Bliesemann de Guevara, B. & Kostic, R. (2017), Knowledge Production In/About Conflict and Intervention: Finding 'Facts', Telling 'Truth', *Journal of Intervention and Statebuilding*, 11(1), pp. 1–20.

Bowen, K. (1997), Total Defence: What Is Being Defended and What Are the Threats?, *OR Insight*, 10(1), pp. 2–7.

Brzezinski, Z. (1997), *The Grand Chessboard: American Primacy and Its Geostrategic Imperatives*, New York: Basic Books.

Bucala, P. (February 2017), *Iran's New Way of War in Syria*, A Report by the Critical Threats Project of the American Enterprise Institute and the Institute for the Study of War.

Bunce, V. & Hozic, A. (2016), Diffusion-Proofing and the Russian Invasion of Ukraine, *Demokratizatsiya: The Journal of Post-Soviet Democratisation*, 24(4), pp. 435–455.

Cafarella, J., Kagan, K. & Kagan, F. W. (2017), *U.S. Grand Strategy: Destroying ISIS and Al Qaeda, Report Four: America's Way Ahead in Syria*, Washington DC: Institute for the Study of War.

Carpenter, T. G. (2013), Tangled Web: The Syrian Civil War and Its Implications, *Mediterranean Quarterly*, 24(1), Winter, pp. 1–11.

Carter, S. & McCormack, D. P. (2006), Film, Geopolitics and the Affective Logics of Intervention, *Political Geography,* 25(2), pp. 228–245.

Celso, A. N. (2018), Superpower Hybrid Warfare in Syria, *MCU Journal,* 9(2), Fall, pp. 92–116.

Chalmers, M. (November 2015), *UK Air Strikes in Syria: Time for a Decision?,* Briefing Paper, London: Royal United Services Institute.

Codevilla, A. M. (1989), Political Warfare in Lord, C. & Barnett, F. R. (Eds.), *Political Warfare and Psychological Operations: Rethinking the US Approach,* Washington DC: National Defence University Press Publications.

Cohen, S. B. (2003), Geopolitical Realities and the United States Foreign Policy, *Political Geography,* 22(1), pp. 1–33.

(1992), Middle East Geopolitical Transformation: The Disappearance of a Shatter Belt, *Journal of Geography,* 91(1), pp. 2–10.

(1973), *Geography and Politics in a World Divided,* 2nd Edition, New York: Oxford University Press.

Concept of the Foreign Policy of the Russian Federation, Unofficial Translation (18 February 2013), The Ministry of Foreign Affairs of the Russian Federation (official site), www.mid.ru/en/foreign_policy/official_documents/-/asset_publisher/CptICkB6BZ29/content/id/122186.

Cunnigham, D. E. (2010), Blocking Resolution: How External States Can Prolong Civil Wars, *Journal of Peace Research,* 47(2), pp. 115–127.

Dias, V. A. & Freire, M. R. (2019), Russia and the Arab Spring: A Counter-Revolutionary Power in the MENA Region in Cakmak, C. & Özcelik, A. O. (Eds.), *The World Community and the Arab Spring,* London: Palgrave Macmillan, pp. 161–183.

Dobbins, J., Cohen, R. S., Chandler, N., Frederick, B., Geist, E., DeLuca, P., Morgan, F. E., Shatz, H. J. & Williams, B. (2019), *Overextending and Unbalancing Russia: Assessing the Impact of Cost-Imposing Options,* Santa Monica: RAND Corporation. www.rand.org/pubs/research_briefs/RB10014.html.

Driscoll, J. (February 2019), *Ukraine's Civil War: Would Accepting This Terminology Help Resolve the Conflict?* PONARS Eurasia Policy Memo No. 572.

Edelman, E. & Wald, C. (2017), *Countering Iranian Expansion in Syria,* Washington DC: JINSA Gemunder Centre Iran Task Force.

Ettinger, A. (2017), The Patterns and Implications of American Military Interventions in the Post-September 11 Era, *Canadian Foreign Policy Journal,* 23(3), pp. 205–217.

Flint, C. (2017), *Introduction to Geopolitics,* 3rd Edition, London: Routledge.

Galariotis, I. & Ifantis, K. (Eds.) (2017), *The Syrian Imbroglio: International and Regional Strategies,* Florence: European University Institute.

Galbreath, D. J. & Malyarenko, T. (2020), Evolving Dynamics of Societal Security and the Potential for Conflict in Eastern Ukraine, *Europe-Asia Studies,* DOI 10.1080/09668136.2019.1705964.

Gleditsch, K. S. (2007), Transnational Dimensions of Civil War, *Journal of Peace Research,* 44(3), pp. 293–309.

House of Commons Foreign Affairs Committee (14 September 2016), *Libya: Examination of Intervention and Collapse of the UK's Future Policy Options*, Third Report of Session 2016–17, HC 119.

Hughes, G. A. (2014), Syria and the Perils of Proxy Warfare, *Small Wars & Insurgencies*, 25(3), pp. 522–538.

Jalilov, M. & Kelly, P. (2014), *The Ukrainian Shatterbelt: A New Cold War?* I Report dell' IsAG.

Jones, B. D. & Stedman, S. J. (2017), Civil Wars & the Post-Cold War International Order, *Daedalus, the Journal of the American Academy of Arts & Sciences*, 146(4), Fall, pp. 33–44.

Kalyvas, S. N. (2001), 'New' and 'Old' Civil Wars: A Valid Distinction?, *World Politics*, 54(1), pp. 99–118.

Kassim, Y. R. (2014), *The Geopolitics of Intervention: Asia and the Responsibility to Protect*, New York: Springer.

Kelly, P. L. (1986), Escalation of Regional Conflict: Testing the Shatterbelt Conflict, *Political Geography Quarterly*, 5(2), pp. 161–180.

Klieman, A. (2015), Pushing Back: The Balance and Balancing of Power in Klieman, A. (Ed.), *Great Powers and Geopolitics: International Affairs in a Rebalancing World*, New York: Springer, pp. 11–29.

Korybko, A. (2015), *Hybrid Wars: The Indirect Adaptive Approach to Regime Change*, Moscow: Russian University of Friendship Peoples.

Kostic, R. (2017), Shadow Peacebuilders and Diplomatic Counterinsurgencies: Informal Networks, Knowledge Production and the Art of Policy Shaping, *Journal of Intervention and Statebuilding*, 11(1), pp. 120–139.

Kozhanov, N. (2016), *Russia and the Syrian Conflict: Moscow's Domestic, Regional and Strategic Interests*, Hamburg: Gerlach Press.

Kudelia, S. (2018), When Numbers Are not Enough: The Strategic Use of Violence in Ukraine's 2014 Revolution, *Comparative Politics*, 50(4), pp. 501–521.

Lacina, B. (2006), Explaining the Severity of Civil Wars, *Journal of Conflict Resolution*, 50(2), pp. 276–289.

Lain, S. & Sutyagin, I. (2016), 'III. The View from Moscow' in Tabrizi, A. B. & Pantucci, R. (Eds.), *Understanding Iran's Role in the Syrian Conflict*, Occasional Paper, London: Royal United Services Institute, pp. 17–24.

Laruelle, M. (2019), Introduction to the Special Issue: The Donbass Conflict, *Nationalities Papers*, 47(5), pp. 715–718.

Malantowicz, A. (2013), Civil War in Syria and the New Wars Debate, *Amsterdam Law Forum*, 5(3), Summer, pp. 52–60.

Malyarenko, T. & Wolff, S. (2018), The Logic of Competitive Influence-Seeking: Russia, Ukraine, and the Conflict in Donbas, *Post-Soviet Affairs*, DOI 10.1080/1060586X.2018.1425083.

Matsuzato, K. (2017), The Donbass War: Outbreak and Deadlock, *Demokratizatsiya: The Journal of Post-Soviet Democratization*, 25(2), pp. 175–200.

Ministry of Defence (December 2010), *Understanding*, No. 4 (JPD04), Shrivenham: Joint Doctrine Publication

Morris, J. (2013), Libya and Syria: R2P and the Spectre of the Swinging Pendulum, *International Affairs*, 89(5), pp. 1265–1283.

Mullerson, R. (2014), Ukraine: Victim of Geopolitics, *Chinese Journal of International Law*, 13(1), pp. 133–145.

Nikitina, Y. (2014), The 'Colour Revolutions' and 'Arab Spring' in Official Russian Discourse, *Connections*, 14(1), Winter, pp. 87–104.

Paraschos, P. E. (March/April 2017), *Geopolitical Risk in the Middle East and North Africa: Shatter Belts & Great Power Rivalry*, Japan Spotlight, pp. 17–20 (accessible from www.jef.or.jp/journal/pdf/212th_Cover_03.pdf).

Paul, C., Clarke, C. P., Schwille, M., Hlávka, J. P., Brown, M. A., Davenport, S. S., Porche III, I. R. and Harding, J. (2018), *Lessons from Others for Future U.S. Army Operations in and through the Information Environment*, Santa Monica: RAND Corporation.

Paul, C. (2017), *Assessing and Evaluating Department of Defence Efforts to Inform, Influence and Persuade* (Worked Example), Santa Monica: RAND Corporation.

Payne, K. (2005), The Media As an Instrument of War, *Parameters*, Spring, pp. 81–93.

Perra, A. (2016), From the Arab Spring to the Damascus Winter: The United States, Russia, and the New Cold War, *Contemporary Review of the Middle East*, 3(4), pp. 363–386.

Póti, L. (July 2018), *Russian Policies towards the MENA Region*, Middle East and North Africa Regional Architecture, Working Papers, No 9

Regan, P. M. & Aydin, A. (2006), Diplomacy and Other Forms of Intervention in Civil Wars, *Journal of Conflict Resolution*, 50(5), pp. 736–756.

Reynal-Querol, M. (2002), Ethnicity, Political Systems, and Civil Wars, *Journal of Conflict Resolution*, 46(1), pp. 29–54.

Risse-Kappen, T. (1995), Democratic Peace – Warlike Democracies? A Social Constructivist Interpretation of the Liberal Argument, *European Journal of International Relations*, 1(4), pp. 491–517.

Robinson, L., Helmus, T. C., Cohen, R. S., Nader, A., Radin, A., Magnuson, M. & Migacheva, K. (2018), *Modern Political Warfare: Current Practices and Possible Responses*, Santa Monica: RAND.

RUSI (25 July 2012), *A Collison Course for Intervention*, Syria Crisis Briefing, London: Royal United Services Institute

Sakwa, R. (2016), *Frontline Ukraine: Crisis in the Borderlands*, London: I. B. Tauris.

Sambanis, N. (2004), What Is Civil War? Conceptual and Empirical Complexities of an Operational Definition, *Journal of Conflict Resolution*, 48 (6), pp. 814–858.

Simons, G. (2022a), Ukraine and the 'Imminent' Russian Invasion: New Mass Media as a Means of Propaganda and Manipulation, *Bulletin of Moscow Region State University*, (2). (In Russian.) https://doi.org/10.18384/2224-0209-2022-2-382.

(2022b), 'Inevitable' and 'Imminent' Invasions: The Logic Behind Western Media War Stories, *Journal of International Analytics*, 13(2), pp. 43–58. (In Russian.) https://doi.org/10.46272/2587-8476-2022-13-2-43-58.

Smith, N. R. (5 April 2017), What the West Can Learn from Rationalising Russia's Action in Ukraine, *Foreign Policy Research Institute*, Summer, pp. 354–368.

Thakur, R. (2013), R2P After Libya and Syria: Engaging Emerging Powers, *The Washington Quarterly*, 36(2), pp. 61–76.

US National Security Decision Directive (NSDD) (1984), Number 130, UN International Information Policy (March 6, 1984).

Vincent, R. C. (2006), Global Communication, and Propaganda in Kamalipour, Y. R. (Ed.), *Transnational Media and Global Communication*, 2nd Edition, Belmont: Wadsworth Publishing Co, Chapter 11, pp. 232–270.

Waldman, T. (2018), Vicarious Warfare: The Counterproductive Consequences of Modern American Military Practice, *Contemporary Security Policy*, 39(2), pp. 181–205.

Woodberry, R. D. (2012), The Missionary Roots of Liberal Democracy, *American Political Science Review*, 106(2), pp. 244–274.

Yakushik, V. (2016), Competing Interpretations of the Socio-Political Crisis in Ukraine in 2013–2016, *Argumentum*, 8(3), pp. 105–121.

Zollmann, F. (2017), Bringing Propaganda Back into News Media Studies, *Critical Sociology*, DOI: 10.1177/0896920517731134.

Zulfqar, S. (2018), Competing Interests of Major Powers in the Middle East: The Case Study of Syria and Its Implications for Regional Stability, *Perceptions*, XXIII(1), pp. 212–148.

11 The Impact of the COVID-19 Pandemic on Humankind, Society, and Politics

Iulian Chifu

Introduction

The Coronavirus pandemic has already brought major changes to human behavior, society, politics, leadership, democracy, and the perspective of international relations, since the first wave in spring 2020.[1] At the level of the individual and human society, the changes are relevant to the psycho-sociological dimension, the state of emergency and isolation at home creating various behaviors and ways of reaction. The repetition of this state brings permanent changes in society, in the multiplied individual preferences and behaviors, calibrated on the themes and directions of adaptation to the pandemic. In addition, free time and the proximity of the computer and the Internet allowed the development of an 'info-demic', information warfare and propaganda, the uncontrolled spread of conspiracy theories, based on lack of facts and lack of societal cohesion, and the absence of direct dialogue between members of society. Also, the inconsistency of the authorities, who were forced to communicate in times of crisis without exact data and clear information on COVID-19, amplified the conditions in which this infodemic managed to spread (Chifu & Şaranuţa, 2020).

At the level of society and politics, the Coronavirus pandemic has revealed the strengths of democratic governments in the face of technological and social media developments; these are developments that have had an unexpected impact in promoting populism and extremist approaches, popular, overconfident and shocking claims, extraordinary speculation, even if unproven and against expert advice, or adopted a

[1] Caveat: This first part of the assessment is made in the middle of the initial wave of the crisis, when in Russia, Great Britain, Turkey, Iran, and the USA numbers of cases were still increasing very fast. At that time, the likely outcome of the second wave, in the autumn/winter, the Coronavirus mutations, the prospects of safe treatment, or finding an effective vaccine were unknown. The likelihood of perpetuating social/physical distancing measures have not, however, radically changed the current assessment, so that is why we return with revisions as knowledge about COVID-19 advances.

balanced approach. Against this background, the current pandemic has highlighted the need for thoughtful and professional people to return to the forefront and the need for political elites – previously constrained, blasphemed, and largely unacceptable in Western states – to reconnect with the true professional elites in society. The recourse to expertise and the need for professionalism drove, at the political level, populism and extremism of all kinds in favor of established classical ideological approaches and rationality versus excessive emotionality, provoked and amplified through social media.

Although we do not know how long these changes will last and how permanent they will become (Chifu, 2020a; DiGeorgia, 2020), at the level of international relations there are great changes that show a shift in trade and economic and power relations with China, and its decoupling from global trade, with or without President Trump's re-election. In this context of overlapping crises, caused by the Coronavirus pandemic, authoritarian states and mainly those with populist leaders seemed to be the big losers in such a period where actions are the only things that matter, with chasing approval and political advantage are punished by the public. China is the main target, through the guilt attributed to it due to the perceived origin of the virus, together with Donald Trump's campaign strategy. Even Russia, although it now has the opportunity to realign, is being pushed off the map to marginal areas of low crisis management, collateral error, and the impact of the crisis, while President Putin's uncertain future makes it less present on the negotiating table of relevant aspects of the world of tomorrow.

Is the Pandemic Threatening Humanity? The COVID-19 Psychosis in Various Forms

The major impacts of the Coronavirus pandemic have been studied, giving rise to multiple theories. The evolution of the pandemic crisis has affected the individual, society, politics, and international relations alike. Fundamental emotions have been identified and exploited by the infodemic: information warfare; the speculation of opportunities generated by the pandemic; how to limit the effects of isolation at home; social distancing; and wearing a mask – all leading to a greater inclination to spend time in the virtual space, on the Internet. But the Coronavirus itself, and the isolation methods adopted to limit the spread of the COVID-19 virus, have generated significant changes in human behavior that, if such measures return, are likely to last longer, giving rise to new ways of living after the end of the pandemic. Without claiming professionally to be a psycho-sociologist, I reveal below what a research agenda would look like and how the effects of the pandemic on the individual and human society should be investigated.

The principal method of preventing the spread of the virus was isolation. Isolation as an individual from a family, in a broad sense, sometimes from a small family, occurred when the person in question was caught up in developments away from home, at school, or at work, and from where they could not return. Then there was the isolation from friends, colleagues, society, and comfort zone. The dramatic reduction of physical and direct social interactions – replaced by electronically mediated ones – is not necessarily a phenomenon that proportionately maintains the same levels of social comfort. For the most isolated and austere, introverted and antisocial, it can be a normal situation. For the sociable, who draw their sustenance from social interactions, this change was major and impactful.

The main impact was on the lonely. Alone in the house, in everyday life, but not alone in the world. The loneliness that came from isolation and the lack of banal contact, of ordinary communication with those known and close, but with whom they did not live, led to the first category of psycho-sociological effects. The accentuation or revelation, for some, of loneliness, of a life in which, in the end, they are alone, had important psychological effects. This was a discovery for some about their own lives.

Just as we saw dramas within formal families who found themselves locked together to reveal and discover the essence of their relationship. Normal coexistence is hurried with relatively short time spent together. Lockdown has led to time spent together, in the family, more time for socialization and rediscovering each other. Alienated or incompatible spouses, ignored and unwanted, tolerated partners with whom they found themselves under the same roof for days, weeks, months, without any other distractions in alternative spaces or company, or the ability to share problems or become absorbed in work that allowed them to forget the less-than-happy relationships at home.

We also had positive effects resulting from the isolation and fear of the pandemic. Entire families managed to rediscover themselves, to come alive, to reconnect and reduce the stresses and strains of everyday life that normally only happened during holidays and late nights. Some have rediscovered their spouse and revived or relaunched even deeper relationships. Others had time to get to know their children even better, to rediscover them with all their anxieties, originality, intelligence, and happiness, seeing their passions and needs. Conversely, children benefited from the presence of their missing parents, reconnected in deeper ways, and re-established family connections that were more than simply cohabitation and family or economic symbiosis. Many enjoyed the quality time spent with the family. And this has had positive effects on limiting the effects of fear, panic, unrealistic expectations, or the stress of isolation.

Perhaps the most important achievement of the pandemic is a significant decrease in stress and pressure, slowing down the continuous and endless daily routine of work, and in everyday existential concerns. Complete relaxation of routine was imposed, creating an unexpected time for reflection and for each of us. This period of introspection was extremely helpful to everyone.

The future impact is difficult to predict, but there are known aspects and elements. First, it is hard to tell if it was just an episode, if the impact is partial, temporary, with possible effects over time, if it was a unique two-month experience, or if it will return. If repeated and periodic, unexpected isolation will lead to permanent changes. In any case, it has introduced flexible working hours, paved the way for free and adjustable forms of work programs for families, children, and domestic or administrative activities that could be the foundation for future employment contracts.

Of course, the impact is not uniform, nor even simple to generalize in its main lines, which dictate social trends. The impact depends on the personality of the individual going through such an experience, on life habits, on their different emotions and feelings, also individual in their form of manifestation and perception, on the behavior of each person in a situation of prolonged crisis or isolation, of the unique identity of the individual, of each of us, with a distinct character and different behavior when facing a unique, unpredictable, crisis. Everything relates to distinct typologies of people, and how these characteristics affect us; they are not permanent components, because the human always learns and adapts – with ease or difficulty – and everyone's personality adjusts to the events of life they are facing.

The crisis accentuates and accelerates pre-existing behaviors and reveals hidden or indistinguishable elements in particular individuals. Diseases, disorders, and inclinations that are revealed and accentuated by the pandemic crisis are more visible and easier to diagnose. Coronavirus has brought manifestations of more obvious behavior that can no longer be hidden, made up, or controlled in its sensitive and inappropriate areas, and that constantly flare up.

Of course, beyond calming work stress, agitation, the continuous movement, speed and pressure of daily life, through periods of isolation other effects of the crisis itself come into play. People experience different reactions, and distinct behaviors, and their capacity and resilience during a crisis varies from person to person. Besides, the crisis forces and exposes the ability to manage individual and collective emergencies by individuals, societies, and decision-makers, while exposing the capacity and fairness of political choices that democratic societies make under normal conditions for the representatives that lead them.

The pandemic has generated several actions and social measures beyond isolation at home, where it exists. The isolation produced various effects according to the quality and size of buildings, and the number of family members forced to live there together for two months almost non-stop. It is about wearing a mask and social distancing, those physical aspects that can have an effect. In this respect, we can observe several psychological effects of wearing a mask, which we see as the motivation for certain marches and demonstrations that have challenged measures to combat the spread of Coronavirus.

We are talking primarily about classical elements which are completely related to the individuality of the human being and their recognition in society, fundamental needs that we find were already a major concern – as shown in the book on identity by Fukuyama (2019). Identity, dignity and recognition become fundamental human needs that are required to be transposed into rights quantified in the laws and norms of behavior inside the society.

Wearing a mask creates problems with identity, recognition, and personal image. The presence of characteristic facial expressions in communication and being able to read the reactions on the face of the person with whom you are communicating are important for understanding, non-verbal communication signals, and especially for relating to the other in society. Wearing a mask in a uniform fashion has blurred individual characteristics, image, and direct physical recognition. It is also true that a real mask industry has emerged in which creativity has tried to outweigh the need for identity, originality, and difference, but it does not fully compensate for these differences that have disappeared through the wearing of masks in enclosed public spaces. In the short term, the problems are minor, but if this is not just a temporary measure, it will lead to profound and permanent change.

We also have several effects that come from social distancing. The elements of closeness, touch, or embrace are missing; an increasing number of senses and elements of non-verbal communication disappear from the space of knowledge or recognition, of individual identity. Handshakes, hugging, touch communication of any kind disappears, even though these are purely human and widely understood characteristics of sociable human typologies and capable of displaying empathy, which is an essential component as a binder of society, for emotional and identity bonding and societal cohesion.

At a general level, the impact of socially restrictive measures is easy to understand, but the banning of travel of any kind also gives rise to deep trauma, behavioral changes, and accelerations or accentuations of character traits that were not previously apparent. It creates divisions and alienation, not only within states, but more so with national and international restrictions. This aspect is also relevant at a commercial and economic level – the isolation of partners, friends who can provide support, customers, and far-away suppliers. Therefore, there is a need to recalibrate and identify alternative contacts (indirectly, e.g., via the Internet); however, there are potential costs in terms of losing direct contact with business partners with whom relationships have been based on personal credentials, symbolism of face-to-face discussions at the partner's headquarters, direct knowledge, and guarantees of commitment provided by a handshake.

The Superficiality of Social Media

The pandemic and the solutions identified to reduce the impact of the virus's multiplication have led to a profound undermining of social relations. The transfer of meetings and contacts from the physical space to the virtual one comes with countless shortcomings and negative side effects. First, the superficiality of relationships in the virtual space, the ease of their elimination or termination – for example, unfriending or blocking on social media – or especially the playful elements in the virtual space that multiply and justify their presence, can very easily become problematic.

Social relationships are about the social and emotional investment of the actors/individuals that generate and maintain such a relationship. However, they are also about those real and apparent exhibited characteristics and about false identities. In addition, they concern playing games with the other person, fooling them, but also about the more serious aspects: altering reality through the prism of the desired image; fraud and crime-related activity to identity theft of or false identity; anonymization in the case of social or economic and commercial relations; as well as about deception or other even more complicated effects that undermine the transactional trust of the actors at the economic level.

Superficial social media, the penchant for play and playfulness, taking life and relationships lightly, can have major effects on relevant categories of the population. Moreover, studies related to the impact of technology, specifically social media and access to virtual space and the doubling of personality, on the individual and social relationships, exist and are well-founded. From here it is only a step towards understanding the relevance of such an impact when entire categories of social relations move into social media and presumptions have been made about pre-existing social partners, who have previously been given a degree of trust, but without physical knowledge about them – due to a lack of personal encounters and familiarity with the individual concerned.

Withdrawal from the physical social space gives rise to an increased inclination towards introverted characteristics. Lack of socialization has led to introspective behavior, less visible and expansive and less warm that in normal times; less human too. However, this perspective creates a special kind of stress for extroverts, for sociable people, tricksters and con artists. In addition, isolation comes with the feeling of seclusion. The feeling of imprisonment is also extremely deeply, carefully and widely studied. Such feelings have a major impact. The isolation and limitation of freedoms give rise, for more sensitive individuals with past traumas or complicated histories, to the feeling of imprisonment, relevant beyond the dramatizations of conspiracy theories that are sustained on social media. The questions remain as to how permanent the change is given the limitations of movement and how concerned we should be about repeating these episodes at certain intervals.

We cannot ignore here the secondary crises of the coronavirus pandemic, whereby major concerns other than the medical emergency generated by the coronavirus have been created. It is a different type of stress; no longer that of haste, agitation, and a constant rush to work that is part of modern-day existence, but of other life concerns – health in the broadest sense (not necessarily related to Coronavirus), issues related to the economy and its impact on the family, job security and future income, the recalibration of life and fundamental values affected.

Best Case Scenario: The Michelangelo Effect

The "Michelangelo effect" translates into individual development in the direction of building the type of person we would like to be. This type of effect or syndrome is manifested in the fact that isolation is possible if we are with the partner with whom we have a close romantic relationship during the pandemic. Natural and accepted changes range from the desire to please or meet the partner's expectations or even from the desire to be better oneself, closer to an ideal presented to them as a mechanism for growth, individual development, and positive self-improvement.

This is the best-case scenario of the effects and impact of solutions found internationally to limit the spread of Coronavirus. Staying with a desired and loved one for a long time gives rise to such a phenomenon that is difficult to achieve in other conditions. It compensates for many of the shortcomings of isolation. But not every individual necessarily reaches the position of being with their loved one, which impels a certain desired or hoped-for transformation.

However, quarantine and pandemic isolation can lead to the destruction of routines in individual and social life. For the individual, the routines created about daily life support, relational clarity and one's position in society, in the social hierarchy and aspirations towards the peak elite – where appropriate – the order of life, the known rhythm. This formula is deeply troubled by crisis and isolation. Instead, there are positive elements of individual clarifications of the meaning of life, of what we want or do not want to do with our lives, there are elements of clarification of future plans, options, and perspective strategies at the individual, family, and professional levels.

Alternatively, a period of isolation of this type and for so long, possibly repeated, leads to individual, couple, or family debates about the meaning of life, happiness, and relevant feelings to each other. And a return to everyday activity also brings new relevant changes with other options beginning to emerge. In particular, the long periods of isolation with a loved one or the family accentuate the inclination to make time to live, the rediscovery of the fact that there are things outside of work, occupation, and everyday madness. This affects the efficiency and amount of work done for corporations or the extreme wear and tear and urgent choices that recalibrate life to more flexible areas and that allow the rebalancing of professional life with family and individual priorities. Not infrequently, leisure time is usually sacrificed at the altar of occupation and other pressing problems.

We will not discuss here the psychological effects and stress of one's fear of catching the disease, or that of a loved one catching it or, God forbid, the loss of a loved one. There is a rich literature in this area, and solutions range from religion to psychology and social reintegration beginning with the family and external support groups.

Job loss, declining incomes, and economic degradation that are the result of the pandemic and secondary economic crisis introduce a new type of stress and create complicated, large-scale personality problems. This is particularly so for those first affected at the very beginning of the crisis, who lived through the stress without any help during the whole period of isolation, and to those over 50 years old who lost their livelihoods, with lower chances of being re-employed. Less for those who are relevant and well situated in society. The severity of the impact generally increases for those at the middle and lower levels of society.

However, tolerating pressure and limiting the impact of this type of stress fare better in the EU than in the US and the UK, where support formulas (including health) exclude the social model and the welfare state, that is, broad societal aid versus liberalism and stricter conditioning in terms of resources. At the EU level, support schemes follow the European social model far beyond the harsh and pure liberalism of competition that leaves losers behind. In the US, the number of evicted and homeless people, who end up on the streets, is increasing.

The new stress of returning to the lifestyle before the pandemic has been mentioned. After two months of slowing down, limiting stress and any need for quick action, living with family and loved ones for much longer, the move to normal pre-crisis activity creates the need to manage change, to foster the inertia of a slower rhythm and staying closer to loved ones. Thus, it creates a new category of stress that is worth investigating. Humans are very adaptable, but there are categories and percentages of the population that have a lower propensity for adaptation and feel the impact and stress of returning to work and previous responsibilities. Sometimes there are individuals and cases that can never recover.

Some resist better, others have a harder time during a crisis, due to the stress, pressure, panic, and the fear of its impact on oneself or those closest or, conversely, the silence and introspection that occur in the case of isolation. Not all people are well and easily able to accept the reality of their own life and individuality; they accept themselves and can reflect without fear on themselves and their own lives. Some have a major problem accepting themselves, preferring stress and distraction with a lot of work, sleep without dreams and with minimal periods of reflection on the meaning of their own life. Adaptation is a characteristic of the human being and is extremely beneficial if a new period of isolation does not return too quickly, or last too long, or if it is repeated too often. Moreover, the cyclical and long-term resumption of such isolation can lead to the perpetuation of the changes we identified during the pandemic.

The Time of Earnest People: The Return of Professional Elites to the Rule of the Democratic State

The Coronavirus crisis, like any crisis, reveals well-hidden or less visible aspects of society and accelerates trends that pre-existed but developed in a time that made them imperceptible, under the radar of public interest and attention. That is why we discover today that, among the multiple concurrent and successive crises of the medical emergency caused by the pandemic, there are, in addition to those of the health system, information war, society, human behavior and economic crisis, some that directly address the essence of democratic societies: the crises of liberal democracies, leadership, and of the political class.

All these processes were pre-existing. Not infrequently, the debate of recent years has lamented the recent developments generated mainly by new technologies that have brought to the fore populist personalities and extremist discourses of any kind because they are adored, enticed and promoted by social media, even the traditional media, as they draw audiences. The words, or the illusion of strength, were more popular than the coldly calculating or pragmatic analysis, and the less spectacular professionalism.

The crisis of liberal democracy is represented – at the European and also transatlantic level – by the success, through democratic rules in conditions of the pre-eminence of social media and the post-truth world, and the promotion of a certain type of leader: charismatic; special; who attract audiences; but who do not solve concrete problems. They bring votes – which is no small thing – but in times of crisis, they are part of the problem rather than the solution. Even during the current pandemic, we felt the need to return to meritocracy and professional persons.

The crisis of democracy overlaps with that of political parties, of politics in general, which involve to a lesser extent the battle for professionals, for prominent minds, and more for people loyal to the party, compliant "mouthpieces" or people pushed forward by party financiers. The result was, not infrequently, the process of isolating parties from reality, moving away from the traditional partnerships within society – the left with trade unions, liberals with businessmen and civil society, and Christian Democracy with religious groups – from which they recruited professionals, resources, and non-political support.

On the contrary, we have witnessed a growth in many states of a partitocracy, in which the three, five, seven, or nine per cent of the citizens of a country, the party members, become the only ones eligible to lead the executive structures of the state. When professionals and the natural elite run away from party membership and are excluded from the leadership of public affairs and democratic states through such a mechanism, parties end up full of people without a profession, without experience, with no past qualifications other than being a party member, regardless of whether they are graduates of higher education earlier or later in life.

Obviously, this situation could only lead to removal from the leadership of the democratic states of most of the professional and natural elites in each state. And, furthermore, to a leadership crisis: how many times does one hear repeated complaints that we no longer have leaders like those from history, or where is there a Margaret Thatcher, a Francois Mitterrand, or a Winston Churchill (although for Romania his personality carries negative elements and memory, from the famous Yalta napkin). Even a Helmuth Kohl, Jacques Chirac, Jacques Delors, or Jose Maria Aznar (let's put aside the blunder with the attack at Atocha station) are no longer found today as part of the leadership.

The crisis has revealed the need to return to the forefront of serious people, with specializations, with professional careers, that is, of the meritocracy. The return to management positions of those who know the fields in which they serve and do not just learn while in office or adapt along the way. Amongst professionals, the parties do not have a good track record of picking the best; it is usually said that when you do not have a career or profession, you enter politics to get ahead. The observation is not generally valid, as there are many professionals in political life, but few and far between from the decision-making era of the past.

Relevant, massive changes are also seen at the level of national institutions and their roles. But the debate between legitimacy, credibility, and professionalism is seen more and more often in times of crisis when public pressure is towards clarifications, guarantees, and figures that inspire trust. By no means in political figures. The ability of political parties and governments to bring to the forefront, directly or in partnership, credible professionals and established elites of society, to assume and fulfil the roles of decision-makers in times of peace, not only in war and crisis, will become a point of reference and desireability or rejection for the future choices of the democratic public.

The visible failures of populism in the current crisis have been intentionally left to the end, regardless of where they come from, be it liberal democracies, autocracies, illiberalism, or isolationist nationalism. The closed borders did not protect anyone; it was not the foreigners who invaded the country but rather their co-nationals. The solution of getting someone to choose whether to stay where they are or come back to their country, in a time of crisis, is related to civic spirit and conviction, reason and responsibility – and also objective risks of living in the host country – and less about enforcement, blockages, or constraints on their citizens.

But perhaps the biggest problem remained the invalidation of triumphalist, snoring, bombastic and absolute approaches, and extremist in form and empty in substance. From President Trump's first approach ("there are about 15 cases and they've disappeared"), to those of Vladimir Putin ("we have solved the crisis," a statement made before the actual outbreak of the pandemic), from Lukashenko's vodka bucket and sauna and the healing plants of Berdymhamedov (who banned the word Coronavirus in Turkmenistan), to Kim Jong Un's North Korea (which closed its borders and declared no infection cases), which all proved to be failures due to lack of solutions. The virus knows no nationality or borders, only people, of all categories and colors, which it infects. Therefore, dealing with that requires thinking, skill, experience, expertise, knowledge, and science – not empty statements!

All this emphasizes that the time has come for principled people to emerge. Decent, without PR ambitions, with a clear profession as a base, recognized in their fields, who have done something for their nations and for the fields in which they work, who have already proven their skills. They must return to the forefront, a return to the established professional elites; most have retired to ivory towers, driven away by the reprehensible reputation of political life, or, directly, by the lobby of campaign contributors and loyal mouthpieces who populate parties and public institutions.

The political class and the professional elites must meet somewhere in the middle. Elites, professionals, principled people must support the political structure of democratic states, accepting decision-making, leadership, and important roles, just as the political class that ensures legitimacy and gathers votes must open its doors to them, bring in these professionals. In the era of social media, this meritocracy will balance the slide of democracy towards populism and its extremes, partocracy, and especially exceptionalism or unipersonal approaches of the *sic volo* type – because that's what I want!

The World after COVID-19: The Impact of the Pandemic on International Relations and Global Security

The world began to change rapidly and a number of trends at the beginning of the crisis were observable. The crisis caused by the Coronavirus pandemic is already having major effects. We will leave aside, for the moment, the effects on poor or excessively oil-dependent countries, with Venezuela's case being the most eloquent on the major impact of the pandemic on global consumption, energy products, and oil-dependent or other commodity-dependent states. We will focus instead on the power shifts in the world, on the areas where the major global actors, or actors with relevant potential and specific strategic weight, matter (Roger, 2019).

The Coronavirus crisis has several special features. The main problem is that it tests the leadership and system of a state, not just its health system, in many ways, especially since the pandemic comes with multiple overlapping crises, being, in fact, a crisis of multiple crises. In addition to the current medical predicament, the following are noteworthy:

(1) first is about the information warfare, propaganda, and exploitation of the crisis in the battles for prestige and image;
(2) second is the crisis of confidence of the society, in itself, its leaders, decision-makers, and, at the same time, about panic, credibility, and crisis decision-making;
(3) third is about leadership and the quality of the political class;
(4) fourth concerns the quality of liberal democracy in the social media era and populism, and extreme ideologies being promoted versus professionalism, meritocracy and the return to the leadership of qualified elites in a society;
(5) there is the crisis and the resettlement of the medical/health system, with all its current strengths and imbalances, with the difficult test that it is failing today;
(6) finally, it is the global economic crisis caused by the Coronavirus, including supply chains, and the trade crisis, as well as the foreseeable decoupling of China.

The economic crisis seems the most important because it determines the readjustment of the world of tomorrow; with its existential threats, but also with opportunities to take control of global governance and the management of tomorrow's globalization. This is not a first, the world already had to deal with these impending crises, with or without the Wuhan Coronavirus explosion and its spread around the world. It is just that COVID-19 accelerated their evolution and brought them to the forefront. Technological developments had long since moved things on to a distinct spectrum, but no one has dealt with the trending effects of these changes on humankind, society, politics, and international relations.

The solutions are few, and are often the same, so we do not have to reinvent the wheel, especially since the world has neither changed so dramatically overnight, nor have our minds or those of the advisers sitting alongside the decision-makers evolved. There will be no dramatic changes over a few months.

(1) The G0 world – of no one, No One's World – the anarchic world without leadership, after Trump USA's withdrawal from the world scene. If not in splendid isolation, in a more self-preoccupied posture; "America first, Great again!", rather than ensuring global leadership.
(2) Returning to the world with US leadership – even if more nuanced, changed, and with more limited tasks, with or without the current president, with or without the current US leadership.
(3) The G2 world, the globalized world along the main lines of the big players, the US and China. If they get along. If rivalries and the

prospect of confrontation fade. If they collaborate. If China accepts the rules and follows them. If they don't end up fighting. Or at least not breaking the global trade, polarizing it into two sides, as they started to do. Slim chance! It would take at least half a miracle for this, or major global pressure.

(4) The US–China–EU tripod – is rather a European ambition, a visionary assumption of Emmanuel Macron on the global role of the EU between the two seas, balancing them, but without the resources and agreement of European economic engines. Again, it is difficult to predict the likelihood of this scenario in the short and medium term.

(5) P5 – the group of permanent members of the Security Council, as leaders in the debate on the future of the world and the management of globalization. Again, with very big differences in quality, lift, manners, fundamental values, and specific weight between the actors and with major differences between them. But with an initiative already on the table: France/EU–US–Russia. We will see the leaders in action, and their results, if this initiative will work.

(6) G7 (G8) – is a natural framework for discussing the major issues of the world, the G8 variant being the desire and aspiration of Russia, which is not in the category of the most industrialized states in the world but yearns a global role.

(7) G20 – a broader framework, which blurs the ambitions and trends of the Great Power policy that all the other projects so far reveal. It was proposed as a sketch in a letter published as an editorial in The Washington Times by Mevlut Ceavusoglu, Turkey's foreign minister. This is also a *pro domo* plea, but it has its substance and relevance. (France24, 2020)

Our analysis has led to several scenarios, none of which are likely to be effective and salutary in the current context. All of them bring to the fore rather the continuation/acceleration of globalization as an objective process, and the effects of the Coronavirus crisis on the populist, nationalist, and isolationist options, which we will analyze below.

Worst case scenario remains the politics of power; that is, the inclination to use force, war, aggressive influence to achieve political goals, respectively the policy of Great Power (i.e., the temptation of a Great Deal between the great powers), inclined to share world domination, which also means multipolarism, spheres of influence and privileged interests. The world will be divided, and deals will be made between great powers, behind closed doors, regarding the future of these spheres of influence. Unfortunately, this is a scenario likely to materialize.

Best case scenario remains the arrangement of the world based on multilateralism, the rule of law, a world based on rules, consensualism in decisions (technically EU values extrapolated globally). While it is promising, the probability of heading in this direction is very low given today's world and the political leaders we have inherited, which must guide the current world through the Coronavirus crisis.

The most likely scenario oscillates between two variants, and these are on the good–bad scale: the G0 world, anarchic, without leadership, with rivalries between great powers and possible wars, with inadequate ambitions and leaders at the forefront, and with the abandonment/marginalization of professionalism and meritocracy, but leaning towards partitocracy and the closure of democratic systems; or Transatlantic Leadership – if it manages to overcome the poor management of the crisis, populism, the temptation to change unfavorable narratives, with a possible change of leadership or options for the main actors, and the need to coordinate global efforts in line with civilized, Western, democratic states.

If the transatlantic rift that many are forcing ends up closing, the United States will no longer be the leader we know; it will need the general support and legitimacy of the contribution of all democratic states in the transatlantic community. It is a simple, well-known, well-beaten and functional road, it has its own common values at its base, it is easy to rebuild, maybe with other leaders, and the will and support of the population can be catalyzed because the need is obvious (France24, 2020).

Unfortunately, the possible catalyst for such a scenario, both likely and close to a best-case, includes the use of the common enemy trope to catalyze all the support, namely the designation of China as a common enemy! American documents and many of those from European and EU Member States are beginning to contain converging elements in such a direction.

The pandemic has created a series of dichotomies in international relations along different lines; as authoritarianism versus democracy during the crisis and winners and losers (Griffiths, 2020). On the other hand, we cannot deny two immediate activities that were determined by the appearance of the virus and the low supply of PPE and medical ventilators, and also by propaganda and information warfare triggered by the temptation of profit from the crisis. First, was an immediate collapse of solidarity and the emergence of excesses of individualism, exceptionalism and crisis selfishness. It is not only about the national tendencies, but also about the local ones, exaggerated formulas of localisms, manifested by keeping room free in the hospitals and the materials necessary for the treatments for "my people" (those from the locality, relatives, acquaintances) that should be treated preferentially, before others (newcomers, foreigners, others). A deeply human behavior, to save peers first is reprehensible, unethical, immoral but deeply human (Mshvidobadze, 2020).

The second process is that of rivalries and the tendency to use the crisis for image gains, political and geopolitical purposes (I put aside the war enriched and the electoral trampolines of the future, which we have already mentioned), through propaganda and information warfare, to show that one system is better than another, that authoritarianism and nationalism solve the problem and liberal democracy and open society amplify it, that the strong-handed leader is the solution, not the democratically taken decision, through debate or consensus (Mshvidobadze, 2020). With the export of soft power through unnecessary or defective pseudo-aid, but with ambitions and desires to achieve political goals – for example, Italy's break with Europe, the pseudo-aid from China and Russia when "the EU does nothing", respectively the tearful anti-European campaign of Salvini and Lega with the famous clip of Macron and Merkel's abandonment of Italy. Here one sees the tendency of absolute nationalism in politics, with the pre-eminence of the providential national leader who is right no matter what they do, and with the closing of borders as the preferred solution.

As absurd as it may be, these declarative, snorting and aggressive approaches are effective, as demonstrated in the early stages of the crisis, in particular through the effect of multiplying propaganda during it and by capitalizing on people's emotions and needs, but also their amplified time on social media due to isolation and an appetite for the exceptional and morbid. But also, for providential saviors – whether they are self-declared nationalist patriots, or authoritarian regimes with ambitions of global hegemony. However, the solutions still come from professional, trained people, from constant investment in research and medicine, and from the effort to find a treatment and a vaccine.

Hence responsible, secure, and lasting solutions – not burdensome and without political conditioning – are also found in modern, rational, and civilized liberal-democratic states, not in the flurry of meaningless words and the bragging of populist leaders or the providential solutions of the authoritarian leaders of the world.

The media tend to see – perhaps too often – options, rifts, wars, and dichotomous choices between good and evil. It is about ease of thinking and communication. And the crisis generated by the Coronavirus brought the same vision to the forefront when the "match," the war or the battle takes place between authoritarianism and democracy. No matter, this real battle of the worlds was also fuelled by the virulent and aggressive campaign, in the space of the information warfare of Russia and China, with predilection, to emphasize how effective the autocracies are. The result is not, however, at all to the liking of those who used propaganda excessively and humanitarian aid for political gains.

The battle in public space should be one between authoritarianism and democracy. Especially when the theme was Who is better? Who claimed victory? Who is more efficient in the design of the political model? Because here the result was not at all advantageous to all the parties that fuelled the dispute, no matter how attractive it seemed. The themes were publicly packaged and sold in the simplest, most raw and direct way possible. Has authoritarianism won? Is it more efficient, has it managed the crisis better? Can China, Russia, or Iran claim victory? And are Italy and the US defeated? What about Turkey and Great Britain, on which side do they stand, winners or losers?

If we look at the result, through effects and consequences, China and Russia, as well as the US and Italy lost equally. And victory, efficiency, the best crisis management capacity goes to Taiwan, New Zealand, Australia, South Korea, Germany, even Singapore, Japan and Hong Kong, who are all at the top as well. All the winners are all democracies! What then separates these states from those "defeated" in the fight against the Coronavirus crisis? What separates at the European level the Nordic states from Italy, Spain, and France?

Many elements come into play. First the peoples, their customs, and characteristic behavior, and especially their conformity to rules, laws, and obedience to authority during a crisis; then the efficiency of the government, the capacity of the political establishment when dealing with both the actions of crisis management, a crisis with multiple human victims, and the support, credibility, and trust that it can attract; especially the ability to associate with credible professionals in the field of infectious diseases and crisis management.

To the contrary, other factors can be noted: populism; the division of society; action rather in words and image gain; a political and ideological approach; not dealing in concrete facts; visibility, perceptible by the population, to solve the crisis or diminish the effects; economic problems existing outside/before the crisis; and pre-existing ruptures in society. These are the ingredients that rather define the failure of Coronavirus crisis management.

That is why China, Russia, the US, and Italy (the Cinque Stelle/Lega parties) fall more into the category of those defeated during the crisis. Not coincidentally, the hardest-hit area in Italy is the rich basin of the Northern League (Lombardy, Padania), while Donald Trump's public statements, such as recommending disinfectant injections as a treatment, the in-fighting amongst states for necessary medical ventilators, or encouraging anti-social distancing demonstrations, were quickly capitalized on by anti-democratic propaganda (Collinson, 2020; Wolf, 2020).

The pursuit of public relations (PR) was the most condemned globally. Not only within democratic populations, but by the whole world. Aspects such as the excessive and ostentatious use of propaganda during the crisis, the temptation to gain political advantages in the midst of the crisis, not to mention those who exploited it for financial gain at the expense of citizens who faced major obstacles – their fears and emotions, the loss of loved ones, job security, and loss of income – and having to deal with the general insecurity and unpredictability of the crisis.

There were failures that destroyed populists all over the world, both authoritarian and democratic alike; principally, excessive pursuit of political gain, alongside too many sweeping statements, not covered by facts. The lessons learned from this are: it does not matter so much how you gained power – although issues of legitimacy and credibility during a crisis are crucial; and it does not matter how balanced and controlled the executive power is – here too democracy gives some guarantees to the population and avoids abuses. Hence, there is a clear separation between authoritarians and democrats, the latter being accountable and acting appropriately.

What matters most in the relationship between victory and defeat in a crisis is what you do with the power gained, how you use it, and who benefits from your actions. How do you make your actions benefit the general population facing the crisis, how does it feel you are helping, how do you increase confidence, and how do you listen, referring to recognized and credible professionals? How do you manage to open the public space, allow the entry of credible figures during the crisis to join government and with the power to manage the crisis? How do governments work in times of crisis? And do the politicians know when to take a step back, leaving room for professionals during the crisis, but without running away from their responsibilities to ensure access to authority and resources to solve the crisis.

This is what distinguishes the general victory of professional democracies, inclined towards solutions, from empty autocracies; this is what distinguishes the uneducated populists from the statesmen of established democracies. Crisis can be seen simultaneously as a threat and an opportunity, such as China as an enemy and the engagement of Putin's Russia (Griffiths, 2020). On the other hand, the idealistic perspective of global management by consensus among all actors, in consensual multilateralist formulas, operates at the limits of acceptability, existing policies, and the growing need to penalize China, to push China to admit its responsibility in the Coronavirus crisis. In this context, the refusal of an international commission to investigate the source and evolution of Coronavirus in Wuhan is not auspicious and creates increasingly strong grounds for turning China into an enemy.

On the other hand, the model of globalization will be fundamentally changed, also to the detriment of China, the current quasi-monopolist. What follows, therefore, is the decoupling and exclusion of China from major global trade chains, the relocation of Chinese industries, and the takeover of the production chains by democratic states. It is underlined by the US National Security Strategy and the US position, which is increasingly mentioned in the European Union, from both Paris and Berlin; it is also confirmed by Romania, through the latest statements of the President of Romania, Klaus Iohannis.

The Chinese model of authoritarian communism and ostentatious and venal information warfare, propaganda and populism unrelated to reality, covering up the responsibilities of leaders and authoritarian states during a crisis, as well as the model of concentrating cheap global manufacturing output, is no longer acceptable. China made mistakes by politicizing aid and imposing conditions on the delivery of much-needed heath supplies, as well as the negative impact due to the quality of cheap Chinese goods that entered the global market once stocks and special sources for export were exhausted.

Obviously, the episode of the Coronavirus crisis also revealed the Chinese pawns in the West, amongst states, leaders, and political parties. It also revealed the motivation of leaders or parties behind the support for, or embracing, the Chinese model; beyond propaganda, it was about bringing in Chinese money, through corruption and the conditioning of authoritarian regimes for private enrichment. At the same time, it led, as we have seen, to the public exposure of hollow populism and undeniably marked the failure of baseless discourse, of chasing a positive image unsupported by facts.

Vladimir Putin's situation is not to be envied today (Higgins, 2020). There were successive mistakes in anticipating the impact of the crisis caused by the Coronavirus, which "did not hit Russia" (Foy & Seddon, 2020), as he declared a month prior to the explosion of the number of cases, its impact causing hysteria amongst the population. The disappearance of the president from public view and the passing of responsibility to mayors and governors of regions, but without federal funds, was frowned upon by the population. His support is drastically declining (Ruvinsky & Aptekar, 2020). Additionally, the oil war with Saudi Arabia to control the market and prices ended very badly for Russia, which was brought to its knees by the agreement with OPEC and the prospect of cutting 10 million barrels of oil production per day, 18 per cent of Russia's production. And that is not all, the estimates are that to stabilize prices requires cuts three times the size (Ambrose, 2020; Egan, 2020; Reed, 2020; Theron, 2020; Thompson & Defterios, 2020).

In this context, Donald Trump's "China's main enemy" policy may bring opportunities, but it forces, as in the case of the European Union, the realignment with the US policy of decoupling China from the global economy. With a clear condition, it must choose between the US or China! *Tertium non datur!*

Russia and Putin have experienced overlapping crises and applied alternative remedial approaches (Ruvinsky & Aptekar, 2020). Vladimir Putin announced confidently, about a month ago, that Russia has no problems with the Coronavirus and that the pandemic and the explosion of the number of cases in China, its neighbor, did not affect it, because it closed the state border (Foy & Seddon, 2020; Pifer, 2020). In fact, the Kremlin relied on the fact that the number of flights between Chinese cities and Moscow is not relevant, and the presence of patients scattered through Siberia – the contact area with China and the Chinese business area – is irrelevant because there the density of the Russian population is so low that a real epidemic cannot "explode" there. And the lack of interest in human rights, the right to life and fair medical treatment, could be added, we should say, to the reason for the misjudgment by the Kremlin, which exposed Putin to a different reality (Kirillova, 2020).

However, the pandemic overwhelmed major population centers in Russia, mainly Moscow and St. Petersburg, the capital of Russia and Putin's hometown respectively, a significant blow to his image. Russia had, from its latest report on 21 April 2020, 5,642 new cases – a steady increase – and 51 new deaths, marking the fact that it is today, along with Britain, on the path of exponential growth and is still far from the peak number of cases (Worldometers).

If we add the fact that the Russian president suddenly disappeared from public view – it was even rumored that he was in self-isolation, after having contact with a doctor who tested positive for Coronavirus later – Vladimir Putin yielded all responsibility and left the battle with Coronavirus to mayors and heads of regions (Foy & Seddon, 2020; Aron, 2020). But what the federal government has forgotten is to offer the basic resources needed to fight the Coronavirus, namely funds for the purchase of protective equipment, support for front-line doctors, and medicine. The fifteen-hour queues of ambulances at the entrances to the hospitals showed the size of the impact of the pandemic in Moscow (Harvard Kennedy School, Belfer Center for Science and International Affairs, 2020).

Vladimir Putin later returned to the forefront, after facing public pressure and at the advice of image consultants, finally acknowledged the following critical period and the fact that even more people would lose their lives – even with the risk of contradicting himself and of associating his image with the bad news. The absence and reluctance of the president in terms of responsibility – still left to mayors and heads of regions – did not help at the level of public opinion (Aron 2020; Financial Times, 2020), as growing economic difficulties exacerbated internal problems and threatened to cause the regime to collapse by implosion (Cordell, 2020; Oil Price.com, 2020; Pismennaya et al., 2020; Ruvinsky & Aptekar, 2020; The Moscow Times, 2020b).

Issues in the management of the current Coronavirus crisis in Russia have overlapped with domestic policy issues and Putin's interests, respectively his forceful attempt to amend the Constitution in a referendum and approve Vladimir Putin's rule for life. The referendum, which was due to take place on April 22, was postponed due to the outbreak, but even without the mistakes made during the crisis management, support for Putin to stay in power after 2024, according to official data, was below 50 per cent, with a low probability for the referendum to meet the required majority (Balmforth, 2020; Foy & Seddon, 2020). Today that probability is even lower.

Indeed, it proves once again that populist leaders have problems on all fronts with this crisis; it is not enough to give speeches, wind up conspiracy theories, and score image points, possibly capitalizing on the charisma and public support for the image of a strong and macho man. First, Vladimir Putin is already sixty-seven years old and does not arouse the same emotion as before. In real crises, we need serious and professional people, who commit themselves to guide the country through the crisis, who give up on or have never had political ambitions, never fought battles of this type, or had electoral interests or ego as a driving factor. Here, what matters are the facts and the concrete, effective ways of managing the crisis.

Usually, the crisis brings with it the principle of rallying around the flag and supporting the leader who is currently in charge of the state. This is usually marked by a significant increase in support in the polls for leaders, both in democratic and authoritarian states (Ruvinsky & Aptekar, 2021). It is not normally a serious and long-term growth; immediately after the end of the crisis, leaders and decision-makers during it become fully liable and support decreases dramatically. Especially when we talk about crises with multiple deaths and victims, and people facing the prospects of both long-term illness and job losses, or collapsing under the effects of the ensuing economic crisis.

This time, Vladimir Putin proved uncompetitive and dropped in the polls; in all polls, conducted by all polling firms (Доверие политикам; RFE/RL, 2020; The Moscow Times, 2020a; Vedomosti, 2020; Zlobin, 2020). This must be a result of a substantial rejection at the grassroots level, a major accumulated emotion, and a punishment for poor crisis management. There are no happy managers during a crisis and in the case of crises with multiple victims, with loss of life, things are even worse (Chifu, 2020b). That is, they usually lose the next election, whether they lose or win the war, whether they manage the crisis well or badly – see Churchill after World War II or the Liberals after the Falklands War (Ruvinsky & Aptekar, 2020).

The only counterexample is the uniqueness of the American position – the starting of a new war, which brings victory to the incumbent president. Today things are much more sophisticated; there is a lot of information, which is why Donald Trump did not benefit – amidst the late and poor management of the crisis, the unfortunate statements about injecting disinfectants to fight the virus, and support for conspiracy theories – from this rally under the flag.

On the other hand, in the case of prolonged wars, the gain goes to those who plead for their end and to bringing the military home – the Vietnam War, or even Trump in 2016 – and losses go to those who perpetuate these long conflicts. But even during crises with multiple victims, with a major impact, there are special crisis managers, good professionals accompanied by leading experts in the field, excellent communicators, who gain trust. The mayor of New York – not directly involved in the fight against terrorism that led to 9/11, but an excellent manager of the crisis that followed, capturing the public eye – ran for president and received the title of 'Mayor of the United States', while Andrew Como, governor of New York, the state most affected by the Coronavirus in the US, gained the most support and highest marks during the crisis, despite the occasional war of words with President Donald Trump, and also by avoiding unnecessary confrontations with him.

Vladimir Putin has today the lowest rating in the last fourteen years (RFE/RL, 2020; The Moscow Times, 2020a; Zlobin, 2020). An official assessment by the state institute, made between April 13–19, 2020 shows that only 28.3 per cent of Russians spontaneously call Putin a politician they trust, in response to an open-ended question, the lowest point since January 2006. There is also a steady decline in confidence in Putin when he is named on a list, reaching 69.8 per cent (Доверие политикам; Zlobin 2020; RFE/RL, 2020).

The Levada Center too, the most prestigious public opinion polling institution, has similar data; when the leader of Russia, in power for twenty-one years, is mentioned in a direct question, with a list being presented, recorded results are 63 per cent, the lowest point since November 2013, six and a half years ago (Vedomosti, 2020). If we add the issue of support regarding the extension of the term of the President after 2024, by amending the Constitution, only 48 per cent of the public supports the idea versus the 47 per cent in opposition to these changes. This was the case in March 2020, and the decrease is probably even more drastic (Balmforth, 2020) today. When compared with the required minimum of 60 per cent support for the public confidence needed to start a new two-term rule, the current poll data already shows the degree of erosion of the Russian president (Foxall, 2020; Harvard Kennedy School, Belfer Center for Science and International Affairs, 2020; Pifer, 2020); therefore, the start of a public disinformation and propaganda campaign with the narrative "if Putin leaves it will get worse! Someone less democratic will take his place," designed to save Putin and his regime, is a worrying prospect.

Lessons Learnt from Delta to Omicron

The aim of this chapter is not to cover all components of information warfare, neither the whole period of the COVID-19 pandemic (that had not ended at the time of writing). Side effects of lockdowns and excessive blockages (China) affected the supply chains and economic development, as well as individual freedoms. The literature is very vast and developed, and our intention is not to cover whole subject areas, but to extract the trends enforced by the combination of pandemic and informational warfare as an intention to create channeled troubles inside a democratic society.

Factually, we must refer to some stages of the pandemic:

(1) **Vaccine and vaccine rollouts –** yes, vaccines have been achieved sooner rather than later. The need for vaccines must always be balanced with the need to jump over some parts of the studies, specifically those with an impact in the mid to long term. The anti-vax movements have energized themselves and reacted rather violently. Together with the conspiracy theories (see below) and the genuine need to know and understand the process, combined with the lack of confidence in the medical/pharmaceutical establishment (due to anti-globalization movements) and in governments, the reaction was harsh and numerous people either refused to take the vaccine or protested against any type of limitation of free movement by introducing obligatory vaccination of any kind. The excessive and boring campaigns of

vaccination, as well as unfriendly and disputable and aggressive campaigns at the national level for vaccination have added the openness to embrace informational warfare against vaccination.

(2) **European certificates and rules of international travel** – it has been an important development in order to maintain the freedom of movement and international travel, specifically by air. But this prompted debates and even justice trials concerning the restrictions imposed on unvaccinated persons. Once again, fear of becoming ill and even dying have been offset by the full range of reactions under the human rights brand, and this legitimate debate opened the way for informational warfare challenging, first and foremost, the governments; specifically, progressive and democratic governments from the point of view of populist, nationalist and far right parties that gained, at some point, an impulse for putting down some governments and changing the political spectrum in the elections.

(3) **The Swedish case** – Sweden became a singular case, using the theory of natural immunization versus the combination of vaccines, lockdowns, and travel bans. After close to three years since the emergence of the pandemic, it seems that Sweden has managed to deal with the COVID-19 virus in a similar way to those vaccinated countries with strict rules for preventing the explosive multiplication of the number of cases which put pressure on the medical systems in some other countries.

(4) **The Chinese case** – a combination of guilt, posturing the strength and need to prove that the fight against COVID-19 is being taken very seriously and with a high degree of responsibility made the Chinese Communist party adopt the so-called Zero-COVID policy. This meant lockdowns for every minor or marginal case identified, and extensive testing and travel bans in China. The combination of repressive policies and Zero COVID-19 has led to economic problems, energy blackouts and especially huge troubles in the supply chains, including the international ones involving China.

(5) **Counting COVID-19 cases** – the way the COVID-19 cases were counted, as well as deaths involving people who tested *positive* for COVID-19, raised another space for informational warfare and challenges against governments and medical establishments, and boosted the spread of conspiracy theories.

(6) **Protests** – as we have seen previously, protests were organized around a large number of themes connected to COVID-19, the most important ones related to the right to choose if a person is vaccinated or not, against any limitation of movement and travel for unvaccinated individuals, and especially against any mandatory vaccination

law and rule. Preventing protests due to the possible spread of COVID-19 has been judged and exacerbated as a breach of basic human rights, another space and niche for the series of disinformation campaigns and targeted informational warfare.

(7) **Delta and Omicron cases: from highly deadly to explosive spread of the virus** – the Delta variant of COVID-19 in the autumn/winter of 2021/2022 had a high mortality rate, and with important consequences. By contrast, in spring 2022, when the Omicron variant appeared, it proved to be highly infectious, spreading with high speed, but with a low level of impact on health. Hospitals were not suffocated, ICUs were only moderately used and numbers of special COVID sections of hospitals needed to be closed down and removed for other uses.

(8) **Responsibilities of the Governments and reviews of informational warfare** – governments, especially the democratic ones, bare the responsibilities for all acts taken during the crisis. Such is the case with the COVID-19 pandemic. And it comes with the explanation for the acquisition of extreme numbers of vaccines, side effects of vaccination, and high rates of death. In the case of the death toll, there were three causes that impacted those figures: first, the COVID-19 virus per se; however, this proved to not always be the most significant reason for excessive death rates. The second cause – and the most important – is the many cases of chronic illnesses that were not diagnosed in time or were not treated correctly during the pandemic due to the huge influx of COVID-19 cases and the blockages within the healthcare system, including low numbers of doctors and insufficient hospital spaces for those with severe symptoms. We could add the fear of some individuals to go to the doctor in order to avoid any possibility of being infected with COVID. And the third cause is the number of sudden death cases appearing in younger adults, young people, and sporty and physically fit people, all of whom were vaccinated.

In all those cases, the informational warfare plays an important role in pressurizing governments, asking for officials to take responsibility for some victims of certain categories, and for the quality and side effects of the vaccines. The companies that made vaccines did not assess the risks for mid- to long-term side effects, since they did not have the time to study those effects, and the full responsibility to move to mass vaccination was passed to the governments. Those realities created new avenues to exert pressure and raise informational attacks to discredit policies, vaccines, and governments alike.

A rich literature covered the multiple aspects of the interaction and impact of the pandemic with informational warfare. The lockdowns, as seen before, have amplified the context in which the impact of the informational warfare on the human being, society, and family relations increased dramatically. Anxieties were raised among individuals and societies everywhere as a result of the health system and economic difficulties brought on by the SARS-CoV-2 virus, a brand-new and unstoppable threat. Many individuals turn to millennial/millenarian and apocalyptic explanatory models for security and purpose as a result of scientific uncertainties, harrowing hospital images, major changes to our daily lives, and an abundance of partially contradicting information from experts and politicians (Sturm & Albrecht, 2021).

Information is undoubtedly important. It matters for the psychology of family interaction, public health, and public safety. The US president himself was subjected to a pattern of widespread misinformation during COVID-19, and in April 2020 he presented his own eccentric course of action: "And then I see the disinfectant where it knocks it out in a minute. One minute. And is there a way we can do something like that, by injection inside or almost a cleaning?" (Connock, 2020).

In the year 2020, many apocalyptic dreamers in right-wing and conservative cultural milieus began to hold the view that they must protect the American people and their individual bodies from a worldwide conspiracy centered on COVID-19. Western far-right conspiracy theorists have, for instance, claimed that the current crisis will cause the collapse of the capitalist world system, which will be followed by the establishment of what is believed to be the true catastrophe: the emergence of communism and globalism, shortly after its onset in the United States and Europe. In an Infowars episode in March 2020, Alex Jones said that the Corona crisis is a "*globalist plot*" to establish a communist One-World-Government. Jones spreads millennial prophetic discourse about the purported impending breakdown of American culture, which is allegedly "planned by a clandestine cabal of Satan-worshiping socialists." Jones claims that the virus is real but that the crisis has been fabricated in order to sow fear and persuade people to adopt a totalitarian New World Order as the answer to the problem. Jones makes an effort to reaffirm images of the evil that have become ingrained in American conservative and far-right environments by blaming the United Nations and China for the catastrophe.

It is hardly surprising that the far right became a significant actor in the COVID-19 pandemic. Extremist subcultures did seem to be well situated in Western liberal democracies at the nexus of the convergent fluxes of fear, rage, and disenfranchisement to profit from the disruption of regular political rhythms (Chapelan, 2021). You cannot avoid COVID-19 by consuming bleach or by destroying mobile towers. One of the most absurd conspiracy theories that governments in Europe and other parts of the world have had to deal with in their efforts to stop the flood of misinformation is the unfounded assertion that 5G mobile networks accelerate the spread of the Coronavirus (Baker, 2020).

Whether it is absurd or not, the online rumour regarding the contribution of 5G to the epidemic has had real-world repercussions throughout Europe. Moreover, sixty-seven arson assaults on 5G towers had been reported as of May 2, 2020 in the UK, twenty-two in the Netherlands, seventeen in France, three in Ireland, two in Cyprus, one each in Belgium, Italy, and Sweden, and one suspected arson incident in Finland, according to the GSMA.

Narratives and strategies for Russian and Chinese COVID-19 information operations kept changing as the rate of vaccination and the emergence of Coronavirus variants differed around the world. Along with the development of information operations in 2021, CEPA's (Centre for European Policy Analysis) data largely confirmed earlier findings that the Chinese Communist Party (CCP) used more corrosive and conspiratorial narratives than in its previous forays into the information space, while Russia recycled previous narratives and did not significantly alter its strategy from previous crises (Dubow et al., 2021). China's information efforts around COVID-19 during the pandemic were significantly more assertive than in earlier campaigns. Contrary to popular assumption, China's most important statements remained largely encouraging and centered on its reaction to the virus. However, during the pandemic, China's strategy changed. The fourth most common narrative in the data set from Chinese sources focused on the Western response to the virus and Western blaming of China, especially from the United States, while still keeping China at the centre of its narratives. China's negative and critical narratives about other nations were a shift from its pre-COVID-19 toolkit and a step closer to the Russian playbook, even though it still mainly avoided deception.

Certain regions were given priority in Chinese COVID-19 storylines that emphasized China's aid to other nations, demonstrating China's use of the pandemic for geopolitical objectives. The pandemic was no exception to China's growing desire to take the lead in the Global South under Xi. It was no accident that South Africa and Ethiopia were given priority as keywords while promoting the narrative that China was assisting the international community in responding to COVID-19.

When it comes to the Russian Federation, the amount of time that Russian media outlets spent reporting on COVID-19 case statistics peaked in spring 2020 and then began to decline as the number of cases in the US initially decreased. Russia started Phase I testing for its Sputnik V vaccine on June 18, 2020, and on September 4 it released preliminary findings showing the vaccine was secure and efficient. Russian state media and authorities launched an offensive after an open letter in the medical journal *The Lancet* questioned the veracity of the country's findings. In an opinion piece that adopted left-wing criticism, Russia Today interviewed the COVID-19 vaccine's creators and posed the question, "*Are Western attacks on the Russian COVID-19 vaccine a corporate cold war against humanity?*".

The European External Action Service (EEAS) noted various forms of health-related misinformation in a report released on April 24, 2020, including "attempts to downplay the pandemic and indicate that it is a hoax, for example by suggesting that the fatality rate is inflated" (Baker, 2021). China was mentioned in the study as the country from which the pandemic first spread. It stated: "Reports indicate that there are ongoing efforts to shift responsibility for the pandemic outbreak, employing both overt and covert techniques."

Two tactics have been used internationally to combat gullibility and a lack of discernment in the community brought on by COVID-19 fears: the dissemination of scientifically based health information, such as that provided above from the World Health Organization; and the use of a variety of consumer protection laws (developed since the time of the Spanish Flu) to stop the gullible and the vulnerable from being taken advantage of by advertisements for goods or devices (Freckelton, 2020). In order to tackle the disinformation campaigns, the creation of a communication strategy is of paramount importance. The communication strategy must consider who should communicate with whom and to whom, as well as the how, when, and audience considerations in addition to having a coherent story. The audience question is directly tied to the nation in which this audience is present as well as how – through which medium – this audience may be specifically targeted (Bodenheimer & Leidenberger, 2020).

It is helpful to look at crisis communication when developing a plan to communicate the story because it originates from the corporate context but may also serve as a beneficial starting point in this case. Truthful and circumspect (factual, open, and devoid of conjecture), understandable (short, snappy messages), quick (active and early), consistent (uniform, coordinated, and continuous), and explanatory, are the characteristics of crisis communication (committed to gaining knowledge), addressing various target audiences through proper means. When communication in times of crisis possesses these qualities, it can satisfy the public's desire for reliable information and greatly spur interest in professional expertise.

Conclusion

The Coronavirus offers a good framework for the study of informational warfare. The main themes involved were the source of the virus, the best regime to deal with crisis – autocracy versus democracy – and the disinformation linked with the crisis of Coronavirus – wrongdoings by the authorities, capacity to obtain the medical supplies and antibiotics, mandatory vaccination procedures, certificates that conditioned travel and entrance to shops in crowded spaces, as well as aircraft. The rivalries and positioning for the post-Covid-19 world made the clash of narratives very vivid and harsh, with high tones involving the roles of China in the crisis, its responsibilities as well as its behaviour once the crisis emerged.

China emerged as a very assertive actor during this crisis in terms of Infodemics, using in large measure Russia's practices and the cooperation with Russia's approach at the European and American level, and with Iran's in the Arab world. The first concern was to create a blurred and indefinite context with multiple alternative narratives, in order to hide any type of responsibility. Propaganda also promoted the so-called masked diplomacy, assisting with medical products for European and Western countries – hiding the fact that a large number were paid for and that the quality was not always of the expected standard.

The informational warfare surged between the two categories of regimes – autocratic versus democratic – and was also instrumentalized via informational warfare. The fact is that if autocracies reacted quicker, the democracies reacted in a more structured manner, and solidarity emerged in the EU after the first episodes of competition for the same products, where China, as a principal supplier played political games. The result was a fragile positioning in the future world, all dependent upon the second wave of the pandemic as well as on the vaccine. The split in the West and episodes of fierce competition to obtain medical supplies revealed that the primary concern of governments was to fulfil the needed supply for their own nations and put aside partnerships and alliances in times of high crisis. That came from the fact that all countries had to buy the same products in high quantities, at the same time, from the same supplier, China.

The entire lessons learnt turned on the possibilities to obtain alternative sources for manufactured products, with shorter supply chains and closer to home, simultaneously avoiding the monopoly and the competition. At the same time, at the level of the EU, it offered an opportunity for the Commission to set a new objective for the supply of medical and pharmaceutical equipment, a new area of integration. It also proved that the economic part of the crisis is better approached and mitigated at the EU level rather than by each member state, and it was a new opportunity for further integration by the initial ties assumed at the European level.

Informational warfare proved to be insidiously linked to events like crises and to high levels of emotion of entire populations of a country or the world, as in the case of COVID-19. The beneficiaries of the anti-vax groups and of anarchic challenges of authority proved to be far-right groups, populists, and nationalists, each claiming that democratic government abused their position and the crisis in order to extend their attributes and force people to do things against their own will – especially vaccination. A thorough revision of the policies during the COVID-19 crisis should be made following the pandemic, including those related to creating possibilities for informational warfare to obtain extensive public reach. On another point, what needs to be said about the pandemic crisis is that it was a new crisis, uncharted, where officials relied on experts and virologists at the global level, the finest knowledge available, but nevertheless, the full responsibilities for each and every action was assumed by the authorities in place, which had to suffer the consequences of each and every unpopular but necessary restriction in order to save lives. The most important strand of the informational warfare – undeserved and inaccurate it may be for the authorities, in the context of the pandemic crisis, and a situation requiring clear answers – is that promoted in terms of costs, pain, and death as side effects of the policies adopted in order to counter the virus that were, supposedly, if not more important, at least comparable with those created by the impact of the pandemic itself. That needs clarification with figures, such as concrete data and scientific medical proof.

References

Ambrose, J. (2020), OPEC Poised to Slash Oil Output as Coronavirus Cuts Demand, The Guardian, 5 March 2020, www.theguardian.com/business/2020/mar/05/opec-poised-to-slash-oil-output-as-coronavirus-cuts-demand.

Aron, L. (24 April 2020), The Coronavirus Could Imperil Putin's Presidency, in Wall Street Journal, at www.wsj.com/articles/the-coronavirus-could-imperil-putins-presidency-11587682524.

Baker, J. (2020), Europe's Disinformation Epidemic: Who's Checking the Facts? https://eu.boell.org/en/2020/05/06/disinfo-reaches-epidemic-proportions-and-needs-serious-eu-response.

Balmforth, T. (26 March 2020), Poll Finds Russians Split Over Allowing Putin to Extend Rule, REUTERS, www.reuters.com/article/us-russia-putin-poll/poll-finds-russians-split-over-allowing-putin-to-extend-rule-idUSKBN21E1T2.

Bodenheimer, M. & Leidenberger, J. (2020), COVID-19 as a Window of Opportunity for Sustainability Transitions? Narratives and Communication Strategies Beyond the Pandemic, *Sustainability: Science, Practice and Policy*, 16(1), pp. 61–66, DOI: 10.1080/15487733.2020.1766318.

Chapelan, A. (2021), "Swallowing the Red Pill": The Coronavirus Pandemic and the Political Imaginary of Stigmatized Knowledge in the Discourse of the Far-right, *Journal of Transatlantic Studies*, 19, pp. 282–312. https://doi.org/10.1057/s42738-021-00073-2.

Chifu, I. (2020a), Sfârşitul coşmarului: scenariile prospective pentru criza de coronavirus – pauza de 3 luni, îngheţ un an sau ameninţare eternă a om enirii, Adevărul, 3 May, at https://adevarul.ro/international/in-lume/sfarsitul-cosmarului-scenariile-prospective-criza-coronavirus-pauza-3-luni-inghet-an-amenintare-eterna-omenirii-1_5e86bdb55163ec427167438b/index.html.

(2020b), Coronavirusul. Lecţii pentru decidenţi în vremuri de criză, March 13, Adevărul, at https://adevarul.ro/news/eveniment/coronavirusul-lectii-pentrudecidenti-in-vremuri-criza-1_5e6b5fd25163ec4271e89459/index.html.

Chifu, I. & Şaranuţa, D. (2020), *Coronavirus Pandemic and the Changes of the World*, Bucharest: Publishing House of the Institute of Political Sciences and International Relations "Ion I. C. Brătianu" of the Romanian Academy.

Collinson, S. (2020), Trump Predicts "This Is Going to Be Bad" but Vows to Reopen America, CNN, 24 March, https://edition.cnn.com/2020/03/24/politics/donald-trump-coronavirus-strategy/index.html.

Connock, A. (2020), The Coronavirus Infowar: What Went Wrong?, https://ora.ox.ac.uk/objects/uuid:55e22eea-047d-4a8c-a7a5-653e69a23399.

Cordell, J. (2020), Explainer: What Happened to Oil Prices and What Does It Mean for Russia?, The Moscow Times, 21 April, www.themoscowtimes.com/2020/04/21/explainer-what-happened-to-wti-oil-prices-and-what-does-it-mean-for-russia-a70055.

DiGeorgia, J. (2020), Russian President Vladimir Putin Will End Up Assassinated Because of the Oil Market Crash, Wall Street Rebel, 29 April, https://wallstreetrebel.com/wsr/articles/russian_president_vladimir_putin_will_end_up_assassinated_because_of_the_oil_mar/2020-04-29-18-13-54.html.

Dubow, B., Lucas, E. & Morris, J. (2021), Jabbed in the Back: Mapping Russian and Chinese Information Operations During the COVID-19 Pandemic, CEPA, 2 December, https://cepa.org/comprehensive-reports/jabbed-in-the-back-mapping-russian-and-chinese-information-operations-during-the-covid-19-pandemic/.

Egan, M. (2020), Oil Prices Keep Tumbling, With No Sign of Stopping, CNN Business, 12 March, https://edition.cnn.com/2020/03/12/investing/oil-prices-trump-europe-travel-ban-coronavirus/index.html.

Financial Times (2020), Russia's Economic Woes Will Clip Vladimir Putin's Wings, April 27, www.ft.com/content/d000617c-8620-11ea-b872-8db45d5f6714.

Freckelton, I. (2020), COVID-19: Fear, Quackery, False Representations and the Law, *International Journal of Law Psychiatry*, September–October, 72:101611. DOI: 10.1016/j.ijlp.2020.101611.

France24 (2020), Trump and Macron Urge Increased Cooperation on Coronavirus at the UN, 4 April, www.france24.com/en/20200404-trump-and-macron-urge-for-increased-cooperation-on-coronavirus-from-the-un.

Foxall, A. (29 April 2020), COVID-19 Is Testing Putin's Regime, The Spectator, at www.spectator.co.uk/article/COVID-19-is-testing-putin-s-regime.

Foy, H. & Seddon, M. (23 April 2020), Economic Woes Undermine Putin's Pledge Pandemic Is "Under Control,'" Financial Times, at www.ft.com/content/63ea4134-ce5c-48dc-a0e0-0fde43948d86.

Fukuyama, F. (2019), *Identity: Contemporary Identity Politics and the Struggle for Recognition*, London: Profile Books.

Griffiths, J. (2020), China's Model of Control Has Been Blamed for the Coronavirus Crisis, but for Some It's Looking Increasingly Attractive, CNN, 29 April, https://edition.cnn.com/2020/04/28/asia/china-coronavirus-model-democracy-intl-hnk/index.html.

Harvard Kennedy School, Belfer Center for Science and International Affairs, (2020), Russia Analytical Report, 20–27 April, at www.russiamatters.org/news/russia-analytical-report/russia-analytical-report-april-20-27-2020.

Higgins, A. (30 April 2020), Putin, Russia's Man of Action, Is Passive, Even Bored, in the Coronavirus Era, New York Times, at www.nytimes.com/2020/04/30/world/europe/russia-putin-coronavirus.html.

Jones, A. (2020), The Great Reset Conspiracy Flourishes Amid Continued Pandemic, www.adl.org/resources/blog/great-reset-conspiracy-flourishes-amid-continued-pandemic.

Kirillova, K. (2020), Russia Tries to Capitalize on the Coronavirus Pandemic, Eurasia Daily Monitor, 17 (49), 13 April, https://jamestown.org/program/russia-tries-to-capitalize-on-the-coronavirus-pandemic/.

Mshvidobadze, M. (2020), Trio Pandemic Propaganda: How China, Russia and Iran Are Targeting the West, GFSIS, 23 April, www.gfsis.org/blog/view/1067.

Oil Price.com (2020), Oil Crisis Challenges Putin's Power, 18 April, https://oilprice.com/Geopolitics/International/Oil-Crisis-Challenges-Putins-Power.html.

Pifer, S. (2020), Putin's Not-So-Excellent Spring, Brookings Institution/Stanford CISAC, 23 April, www.brookings.edu/blog/order-from-chaos/2020/04/23/putins-not-so-excellent-spring/.

Pismennaya, E., Arkhipov, I. & Meyer, H. (2020), Russia Paid a Heavy Price to End the Oil Price War, World Oil, 13 April, www.worldoil.com/news/2020/4/13/russia-paid-a-heavy-price-to-end-the-oil-price-war.

RFE/RL (28 April 2020), Poll Finds Russians' Trust In Putin Dips To Lowest Point In 14 Years, at www.rferl.org/a/poll-finds-russians-trust-in-putin-dips-to-lowest-point-in-14-years/30581727.html.

Roger, J. (January 2019), Audit of Geopolitical Capability. An Assessment of 20 Major Powers, Henry Jackson Society, ISBN: 978-1-909035-50-8, 64 pp, at https://henryjacksonsociety.org/wp-content/uploads/2019/01/HJS-2019-Audit-of-Geopolitical-Capability-Report-web.pdf.

Reed, S. (2020), Oil Prices Nose-Dive as OPEC and Russia Fail to Reach a Deal, The New York Times, 6 March, at www.nytimes.com/2020/03/06/business/opec-oil-prices-russia.html.

Ruvinsky, V. & Aptekar, P. (2021), The Coronavirus Gave Europe's Leaders a Popularity Boost. Putin Hasn't Fared As Well, The Moscow Times, 1 May at www.themoscowtimes.com/2020/04/30/the-coronavirus-gave-europes-leaders-a-popularity-boost-putin-hasnt-fared-as-well-a70157.

Sturm, T. & Albrecht, T. (2021), Constituent COVID-19 Apocalypses: Contagious Conspiracism, 5G, and Viral Vaccinations, *Anthropology & Medicine*, 28(1), pp. 122–139, DOI: 10.1080/13648470.2020.1833684.

The Moscow Times (2020a), Russians' Trust in Putin Hits 14-Year Low – State Poll, Updated: April 28, www.themoscowtimes.com/2020/04/28/russians-trust-in-putin-hits-14-year-low-state-poll-a70128.

(2020b), Coronavirus Hits Russian Markets | May 1, The Moscow Times, 1 May, www.themoscowtimes.com/2020/05/01/russia-coronavirus-ruble-stock-market-reaction-live-updates-a69640.

Theron, M. (2020), Oil Plunges 11%, the Most Since 2014, After Russia Refuses Saudi Arabia's Request to Cut Output, Business Insider, 6 March, at https://markets.businessinsider.com/news/stocks/oil-price-plunges-coronavirus-fears-opec-russia-clash-cuts-2020-3-1028971722.

Thompson, M. & Defterios, J. (2020), Oil Just Had Its Worst Day in 11 Years As OPEC and Russia Fall Out Over the Coronavirus Crisis, CNN Business, 6 March, at https://edition.cnn.com/2020/03/06/business/oil-prices-opec-disarray/index.html.

Vedomosti (2020), Levada Opinion Poll: Trust Level and Rating of Vladimir Putin, at www.vedomosti.ru/society/articles/2020/03/26/826370-konstitutsionnaya-popravka.

Wolf, Z. B. (2020), *Dollars vs. Deaths Is the Sickening Choice Created by Coronavirus*, CNN, 23 March, at https://edition.cnn.com/2020/03/23/politics/what-matters-march-23/index.html.

Worldometers, www.worldometers.info/coronavirus/.

Zlobin, A. (27 April 2020), Уровень доверия к Путину упал до минимума за 14 лет, Forbes Russia, at www.forbes.ru/newsroom/obshchestvo/399199-uroven-doveriya-k-putinu-upal-do-minimuma-za-14-let.

Доверие политикам, https://wciom.ru/news/ratings/doverie_politikam/.

12 Conclusion: Summing Up the Knowledge and Answering the Questions

Greg Simons and Iulian Chifu

It was the intention from the outset to write a book that challenges the current orthodoxy of knowledge. This challenge is undertaken by presenting different aspects of the current complex global picture of the politics of contemporary warfare from a multi-disciplinary and multi-perspective, critical approach; this does not avoid the essential academic, political and practitioner problems of the day or try to artificially compartmentalise these aspects in isolation from each other. The aspects and roles of academic, policymaker and practitioner are related and interact with each other; lessons can be drawn from across these roles in order to change this orthodoxy of knowledge and practice in order to make it more and better informed about the whole chain of actions and consequences.

It was the intention of this book to engage in the interaction and relationship between the elements of politics, information and communication on contemporary warfare. This was achieved by presenting a mixture of theory, policy and practice across the chapters. Each of the chapters gives a particular insight into a specific theoretical, empirical, political or practical problem. The resulting influences on why and how wars are fought at the beginning of the twenty-first century become apparent. But the picture is a complex one, and more nuanced than may be assumed at first glance. In addition to the usual 'suspects', such as politicians, media and the military, there are the influences of powerful intangible mechanisms such as language and culture. These powerful intangible factors are not universal either but influenced by the individual and specific circumstances of the nation/civilisation and/or the operational environment in question. A result of this is to add a further layer of complexity to an already complicated environment, where some factors may not be considered, or they are misunderstood. The net result is a dire need to rethink the current toxic global foreign and security policy trajectory before the prospect of another iconic scenario such as the Cuban Missile Crisis reoccurs and either forces us to learn or, at worst, destroys humanity as we currently know and understand it.

Even though the ten chapters contain a great deal of varied subject matter and resulting qualitative analysis and assessments, some common 'red thread' can be observed that unites the sum of knowledge that is made up from informational pieces. The current state of global instability and the increasing tensions and conflict that are alarmingly being observed are both a cause and a consequence of the transforming global order from a unipolar into a multipolar configuration.

In the Introduction to this book, three research questions were posed. These were intended to enable the common connections and relations between the different political, informational and communicational elements to be visualised more readily and easily. In the following sections of the Conclusion, these three research questions will be answered systematically in turn.

Is Warfare in the Twenty-First Century Qualitatively Different from Earlier Periods in Human History?

Warfare has been traditionally delineated between varied types of conflict: regular (standing armies, a defined and open battlefield, for example World Wars One and Two); irregular (where at least one side uses a non-standing army, may be associated with insurgency and terrorism, has an ill-defined area of operations, and where it is difficult to tell combatants from non-combatants, for example the current and on-going Global War on Terrorism (GWOT)); high-intensity (often associated with conventional conflicts), and low-intensity (often associated with irregular wars or terrorist campaigns) conflicts. There are different relatively new terms or concepts assigned to categorise various kinds of warfare, such as that of the fluid concept of 'hybrid warfare' or 'vicarious warfare'. One may also speak of different generations of warfare, where the ways in which wars are waged is influenced by innovations in technology, economics and politics. However, this informational representation of the physical reality is an oversimplified one of a single category existence, whereas an actual armed conflict can display more than one of these labels. For example, the Vietnam War (1945–1975) exhibited at varying points in its history regular and irregular, high-intensity and low-intensity conflict, and third and fourth generation warfare.

Betz (2015), Hageback and Hedblom (2022) and Whyte, Thrall and Mazanec (2021) have noted that warfare, in terms of why and how it is waged, is evolving with time and circumstance and therefore differs from earlier periods in human history. However, they note that there are also some similarities in terms of the nature of the deployment of armed forces in conventional conflicts as well as the use of various types of specific weaponry. It is also necessary to consider the intangible elements in addition to the already-stated tangible elements, as these should be considered in tandem and not separately. This is evident in Chapter 4 on the political and intangible elements of modern wars. There is the distinct evidence of the effect of historical thinking (such as Sun Tzu or von Clausewitz) on how contemporary warfare is envisaged. It impacts upon the various cultures of wars and warfare, which is demonstrated in Chapter 5 on the culture and language of armed conflict. These aspects influence the cognitive realm of various stakeholders, which binds ethos and logos to how warfare is imagined and conceived, and consequently the relationships between those audiences and warfare. However, it should be kept firmly in mind that these imaginations and conceptions of warfare do not exist as a universal global singularity as the specific circumstances of nations and civilisations form over time through experience and lessons offered by events through history.

In Chapter 8 the discussion on the impact of COVID-19 on politics and international relations demonstrates the role of opportunity when it coincides with an actor's motivation to vigorously pursue a goal by various means that include force or coercion. This fits with the notion that extraordinary circumstances (crisis situations) present more ruthless and daring actors with extraordinary opportunities that are not bound by the conventions and rules that guide behaviour and expectations in ordinary times. Therefore, crises should be seen as not only a threat, but also an opportunity for others. In this regard, there is some continuity as crises can precipitate warfare as desperate times can create desperate measures to defend a position of power and privilege or to attack someone else's with the likelihood of reduced ethical or moral costs for the aggressor.

Chapter 9 on Western hybrid warfare confirms the above-noted role of the conceptual and operational realisation of a crisis situation. Historically, within the topic of waging warfare, this could have included the initiation of wars on the basis of vanity, opportunity or emotions (see Chapter 6), which would be more difficult to prosecute in the current social and political climate. Language and symbolism were vital in creating the necessary cognitive disposition that would permit actions that often violated international law and the rules of war, which will be discussed in more detail in the following section. Therefore, the favoured and promoted concepts of 'humanitarian warfare' and R2P are used as enablers of military operations that would otherwise not be feasible or sustainable owing to the likely high political costs incurred by the aggressor.

Warfare is increasingly being used as a foreign policy goal in the twenty-first century; for instance, by leading Western countries to set an example to those who dare defy the influence of their global hegemony and to defend their position as global hegemon. It is also used by those countries that are in the challenger (of the global hegemony) category in the international relations market of actor positioning. For example, US-led attempts at regime change in Syria prompted Russia to defend its client state militarily, but the Syrian conflict could also have been used to weaken the global influence of the US and potentially marginalise its position in the wider MENA region. It is the equivalent to the law of physics applied to the politics of warfare; for every military action there is an equal but opposite action. This is the longer-term perspective of actions and consequences, rather than being policy and practitioner blinkered on having short-term military hard power dominance projected over a limited geographical terrain. This is also seen in the folly of the 2003 US-led invasion of Iraq and the longer-term consequences of the rapid US 'victory' in the conventional high-intensity phase of the conflict.

Technology and tactics can also have an impact upon the battlefield as well as the hearts and minds of audiences. As noted by Waldman (2021) and others, the US has been innovative in trying to circumvent the military reaction threshold of states that it targets and is far from being alone in this strategy. Sometimes technology is used to achieve this; for example, the use of drones in targeting people in other countries as seen in the use of US drones against civilian targets (often referred to as 'Taliban suspects') in Northern Pakistan. As observed in the 2020 conflict in Nagorno-Karabakh in the South Caucasus, drones are now not only a feature of irregular conflicts, but regular ones too. Cyber-attacks were also another technological means of trying to avert a military response, such as the Stuxnet attack on the Iranian centrifuges, although the US is also now regularly subjected to such attacks. Another more recent non-technological tactic, is the use of assassination of key figures in another country's leadership, which violates the orthodoxy of ethical rules and norms owing to the rather risky and unpredictable results and outcomes. The assassination of Iranian General Soleimani in Iraq was intended as a new and innovative approach in the contemporary age (though used historically, such as attempts to kill Cuba's Fidel Castro) that would not generate a military response. This was a serious miscalculation as Iran launched a series of missiles against a US military base in Iraq as a retaliatory and somewhat symbolic countermeasure.

There are differences and similarities observed in earlier versions of how warfare was waged historically compared to the currently evolving iteration of the twenty-first century. This has been illustrated above in terms of the constraints and possibilities through of the influence of technology, history and experience of warfare. Perhaps one of the more troubling trends is the gradual breakdown of the 'old' rules and agreed-upon constraints of international law, rules of war and a lack of strategic thinking and actions in a global environment of a transforming global order. These rules and systems of managing competition in international relations between different powers were introduced after the events of the Cuban Missile Crisis in 1962, together with the last-minute realisation of the potential risks and costs of sleepwalking into a conflict with the destructive power of the weapons of the time. This increasing lack of consideration of this important historical lesson, and undoing the orthodoxy of knowledge and practice to try and gain a competitive advantage in global politics and geopolitics, needs to be urgently and seriously reconsidered and rethought, moving from unstable short-term tactical goals to more stable long-term strategic visions.

In the next section of this Conclusion, the topic turns to the issue of the influence and effect of information and politics in the prosecution of contemporary warfare. This concerns a focus on the processes and strategies of the informational realm to affect the cognitive realm's perception and understanding of the 'realities' of the physical realm.

How Do Politics and Information 'Inform' and Influence Twenty-First Century Warfare?

Payne (2005), Western (2005) and DiMaggio (2009) have characterised mass media as an instrument of modern warfare, which is an essential element given the need and nature of warfare in the twenty-first century to create and engineer the façade of public consensus (at a minimum non-resistance) on the policy course as a basis of political legitimacy and political capital. This refers to the historical dicta of Sun Tzu and von Clausewitz that war is politics by another means (see Chapter 4). Another point noted by Payne (2005) is that the individual battles are decided by military action; the acumen of the sides as well as chance can decide the outcome. At the level of wars (a culmination of a series of battles in a particular theatre of military operations) the outcome is decided by politics; and politics is guided and decided by intangible elements, and not solely by tangible ones. The role and influence of information and politics is nothing new to warfare and has been witnessed since the arrival of organised warfare throughout periods of classical history (Taylor, 2003). Several cogent lessons and knowledge emerge from the chapters of this book.

Chapters 2, 4 and 7 combine the lessons of the interaction and effect of politics and information within the context of military and political conflicts. This concerns the political or military objectives of damaging an opponent or competitor's public persona, brand and reputation as a means of limiting their operational choices and increasing the aggressor's possibilities to exploit a target that has been weakened by an operation using intangible means (through the informational realm that is directed at the cognitive realm of the various stakeholders). This creates an information environment where politics creates the demand and technological development creates the opportunity for engaging in mass persuasive communication campaigns, including the use of information warfare.

A finding of Chapter 3 is that intangible elements of warfare (found in the informational and cognitive realms) have a definite and sometimes profound effect on the tangible elements (found in the physical realm of warfare). This has been understood and realised in early human history, which gives rise to such conceptual tools of warfare as psychological operations and military deception. As such, intangible elements can increase the effectiveness of an actor's performance in warfare by enhancing the efficiency of its tangible elements as it can decrease the effectiveness of an opponent's performance in warfare by decreasing the effectiveness of its tangible elements. The historical case of the US involvement in the Vietnam War is used as an example to illustrate these points, with a more contemporary point being the direction and trajectory of the GWOT and the Forever Wars in which the US-led West is currently bogged down.

As Chapters 5 and 6 illustrate, the role of communicating information on intended or current armed conflicts is a precise and difficult political task. There is a multitude of environmental factors that can affect the feasibility and effectiveness of attempting to communicate to various stakeholders the 'necessity' for war, especially as there are increasingly wary public evaluations that current thinking is to wage warfare as a foreign policy choice rather than out of necessity and a lack of viable alternatives. This includes the difficult balancing act of communicating the appropriate moral and ethical tone, with its resonating language and symbolism, via the most suitable medium of communication, at the right time and to the right audience. The art of successful communication is a situation-specific phenomenon that is tied to the specific sets of physical, informational and psychological circumstances experienced by the target audience, both historically as well as contemporarily. For example, the perceived excesses of R2P as applied to regime change in Libya by NATO in 2011 has served as an effective restraint against future regime change operations that seek to use R2P for the sense of legitimacy, which has also been seen in the Syrian War. This has the logical significance and meaning that not all forms of information and communication are equal and can be divided into different categories falling between truthful and deceitful (white to black).

In Chapters 8 and 11, the role of crisis is explored and analysed in different contexts. Chapter 8 engages with the notion of crisis and the creation of the extraordinary and specific circumstances created at the social, political and international relations level by the Coronavirus pandemic. Chapter 9, however, investigates and analyses the purpose and role of invoking the notion of the 'crisis' concept with the purpose of subversion and the eventual regime change of a targeted country and its government. As noted in Chapter 8, the Coronavirus pandemic has had the effect of amplifying existing tensions in global politics and geopolitics as the different actors and interests seek to position themselves to gain the best advantage or incur the least damage from opponents and competitors as possible. Although this is communicated in terms other than specific and exclusive vested self-interest, this does not prime and mobilise publics to support their cause. Instead, this is framed and narrated as serving the best interests of a particular audience, a specific country or specific sets of benevolent 'universal' values and norms. A similar pattern is detected and analysed in Chapter 9 concerning Western hybrid warfare, where highly subjective and selective foreign policy is used to subvert and change an 'undesirable' foreign government. However, the politics of the policy of overt regime change are not communicated in this manner but in terms of sets of projected binary value and norm-based scenarios: a good future with a lifestyle that is allegedly enjoyed in the West through successful regime change; or a bad future of tyranny and lack of freedom in the event of unsuccessful or no regime change. Crisis is used as an operational concept to remove the obstacles in place during 'ordinary' circumstances to enable greater freedom of choice in the mechanisms of subversion and regime change. As noted by Zollmann (2017) and others (DiMaggio, 2009; Carruthers, 2011; De Franco, 2012), there is a tendency for 'independent' mainstream media outlets to support planned and executed military operations. This fits in with the observation that mass media indeed tend to serve as politicised instruments of war.

Chapters 9 and 10 contain a more empirical focus in terms of the dynamics of policymaking and practice, in their influence and effects upon the armed conflicts in Ukraine and Syria. At first glance these conflicts represent a geopolitical mirror of each other; the US and its allies (especially the EU) supported the regime change of the then President Yanukovych on value- and normative-based logic and installed a Euro-Atlantic friendly central government. Since this time, with the declaration of an anti-terrorist operation in Donbas they have been attempting to assert control of all Ukrainian territory that has been stymied in the conflict that has emerged and is supported by Russia. Syria was an unsuccessful attempt by the US and its allies to regime change the government of President Al Assad, however the fortunes of war swung more in favour of the Syrian government when Russia became directly involved in the conflict. Currently, the US attempts at a destabilising role in the conflict to keep the war simmering are a mirror to the Ukrainian situation. The early stages of the conflict narrative and framing are important in shaping public opinion and perception of the event, which translates into the presence or absence of political legitimacy and political capital. As Bernays (1947) noted in the context of engineering public consent for political possibilities and outcomes in policy and governance, there is no coincidence in the timing or the nature of the news. Today, this news and information comes from a variety of sources and opinions on any issue, if one has the time and requisite knowledge of how and where to look for it, either in mainstream media, or other niche media sources and social media. This makes the communication management of the conflict much more difficult to achieve, let alone try to control.

The old method of seeking information dominance in the informational realm, which military and political actors still try to achieve in Western countries, was historically based on a one-to-many form of passive communication to mass audiences, such as the basis for the Nazi or Soviet system of mass propaganda, in order to engineer the population towards certain state goals and modes of behaviour/thought. It was also an era when the state and mass media retained a monopoly on the mass communication of information. The current information society age, enabled through the development of powerful tools of information communication technologies, is characterised by a many-to-many form of active communication (Cunningham, 2010). It is also an environment where audiences of information can also be producers (prosumers), where the state and mass media have had their monopoly and gatekeeper roles eroded. Therefore, the focus for domain dominance may be shifting from the informational to the cognitive realm owing to the new dynamics of the information communication environment, which means it is a more difficult task to control and dominate the flows of information and conversation owing to the active nature of communications. A 'natural' operational evolution looks to create an environment of cognitive dominance, which requires far less centralised control and greater creativity and innovation in managing information and communicating messages to target audiences.

When reflecting on the lessons and knowledge of the individual chapters, the different roles and approach of information operations in relation to combat operations becomes clear. In the case of state-based actors, information operations play a supporting role to combat operations (both offensively and defensively), as the intention is to contribute to the effectiveness of military operations in the physical realm by providing the necessary supporting intangible elements; this is because warfare tends to be waged on a value and norm basis. However, for non-state actors (such as insurgents or terrorists) the roles are reversed, and military operations play a supporting role to information operations as this actor is usually facing a much stronger and better-equipped adversary. Therefore, events and actions in the physical realm are intended to shape the cognitive realm of friend, foe and neutral alike. The basis of waging warfare is different too, where this is found in the binary ideas and projections of justice and injustice (social, economic, political, etc.).

What Are the Possible Trends and Transformations of Twenty-First Century Warfare?

As the global order gradually transforms from a US-led Western-centric unipolar order towards a Non-Western multipolar order there is likely to be a continued sense of lack of order and increased chaos and upheaval in the conceptual approaches (academic and political) and practice (the implementation of those concepts by practitioners, such as military forces and foreign ministries) in international relations and international affairs. This is a 'natural' product of different powers and actors seeking to position themselves to prosper from or suffer less losses as a result of a reconfiguration of the global order. This is seen politically in the Asia-Pacific region where the Philippines under President Duterte has moved closer to China and further from US interests and influences, and where Singapore announced at a regional forum that the time has come for countries of the region to choose between the US or China.

Various observers of international relations and global geopolitics have noted the gradual and relative decline of US global geopolitical hegemony and the political hegemony of global liberalism (Simons & Glaser, 2019; Cooley & Nexon, 2020). However, the US is currently paying for the folly of its global foreign and security policy in the wake of the bipolar world of the Cold War that saw the US as the single largest, uncontested military and economic power. This created a system of a lack of checks and balances, which during the Cold War saw a more balanced and nuanced policy of compromise in addition to competition between the US and the Soviet Union in order to create a more stable and predictable system to avoid or lessen the possibility of inadvertent war occurring.

This led to the era of the Humanitarian Wars (Hehir & Murray, 2013), such as the 1991 Gulf War and the 1999 Kosovo War (Willcox, 2005). The global stability situation has been further eroded in the wake of the declaration of the GWOT after the 9/11 attacks on the US mainland. A result has been to begin numerous politically motivated small-scale wars of different types since 2001, where not one single conflict has been successfully ended. The financial and human costs of these wars is enormous,[1] even if not widely known or publicised.

These wars have imposed a large additional cost in terms of political legitimacy and political capital among establishment politics, serving as a restraint and a constraint for future warfare. This has prompted a more widespread use of deception and psychological operations to go to war (Whaley, 1982; Linebarger, 2009), but also attempts to find more covert and less accountable means of waging warfare (Waldman, 2021). This is being done at a time of relative economic decline and a gradual loss of global influence. The breaking down of the old system of regulation of global hazards and risks by the main global actors, such as the dwindling number of nuclear arms control treaties and the emergence of what some have labelled as the New Cold War (Sakwa, 2008), create not only threats, but potential opportunities for some actors that seek to leverage their strengths against their opponent's weaknesses. This can come in the form of extraordinary circumstances creating opportunities (and threats) that can generate the context for further pushing the boundaries of the international rules and norms remaining to pursue competition. The current international environment of the worldwide Coronavirus pandemic has managed to increase the existing tensions of the global order by making the competition and conflict more overt (Simons, 2020). There is also the issue of the impact and influence of technological development on twenty-first century warfare.

The technological development, not only in weapon systems, but also information communications technology in an increasingly fragmented and divided world is having a profound effect on how warfare is conceived and waged beyond the kinetic military operations to include the cognitive elements and where the state has lost its monopoly on waging global kinetic and non-kinetic warfare (Bazarkina, Pashentsev & Simons, 2020). Thus, technology can be directed at the cognitive realm of the target audience to influence their thinking and actions. Technology can also be deployed virtually against tangible targets, such as computer network attacks to disable, disrupt or destroy an opponent's critical

[1] Please see the Watson Institute at Brown University's Cost of War project for a detailed breakdown of these costs – Costs of War (brown.edu).

infrastructure in order to gain an advantage or simply to disadvantage the target (Simons, Danyk & Maliarchuk, 2020). There are perceived advantages to this kind of warfare in terms of being below the threshold of a military response, although this seems to be in the process of changing. Technology has also led to the widespread use of remote killing, not only on defined battlefields, but also private homes, with the use of drone warfare.

However, the short-term shortcuts and benefits, along with the political expediency of using these transforming and evolving conceptualisations and practices of warfare, also create a great deal of risk in the potential for inflaming and intensifying an already delicate and volatile balance of the global order. The calculations and predictions of Western decline and the eventual loss of its political, economic and military hegemony is being increasingly visualised by more open and assertive attempts by challengers (China, Iran, North Korea and Russia for example) to test the strength and resolve of the US-led West. By creating an environment of decreased predictability at a time of increased risk and change, the threshold of the different actors is less known, and the system intended to regulate the actions and predictability of the international system increasingly runs the risk of inadvertent war. This is in no one's interests; instead, a more stable and long-term considered and effective approach to managing the transformation of the global order is needed to reduce the level of hazard and risk. Poor or inappropriate policymaking and its practical implementation can have the effect of hastening the process of the already transforming global order.

References

Bazarkina, D., Pashentsev, E. & Simons, G. (Eds.) (2020), *Terrorism and Advanced Technologies in Psychological Warfare: New Risks, New Opportunities to Counter the Terrorist Threat*, New York: Nova Science.

Bernays, E. (1947), The Engineering of Consent, *The Annals of the American Academy of Political and Social Science*, 250(1), pp. 113–120.

Betz, D. (2015), *Carnage and Connectivity: Landmarks in the Decline of Conventional Military Power*, Oxford: Oxford University Press.

Carruthers, S. L. (2011), *The Media at War*, 2nd Edition, Basingstoke: Palgrave Macmillan.

Cooley, A. & Nexon, D. (2020), *Exit from Hegemony: The Unravelling of the American Global Order*, New York: Oxford University Press.

Cunningham, T. (2010), Strategic Communication in the New Media Sphere, *JFQ* Issue 59, 4th Quarter, pp. 110–114.

De Franco, C. (2012), *Media Power and the Transformation of War*, Basingstoke: Palgrave Macmillan.

DiMaggio, A. R. (2009), *Mass Media, Mass Propaganda: Examining American News in the 'War on Terror'*, Lanham: Lexington Books.

Hageback, N. & Hedblom, D. (2022), *AI: For Digital Warfare*, Boca Raton: CRC Press.

Hehir, A. & Murray, R. (Eds.) (2013), *Libya: The Responsibility to Protect and the Future of Humanitarian Intervention*, London: Palgrave Macmillan.

Linebarger, P. M. A. (2009), *Psychological Warfare*, Nevada City: Gateways Books and Tapes.

Payne, K. (2005), The Media as an Instrument of War, *Parameters*, Spring, pp. 81–93.

Sakwa, R. (2008), 'New Cold War' or Twenty Years' Crisis? Russia and International Politics, *International Affairs*, 84(2), pp. 241–267.

Simons G. (2020), The Corona Virus Pandemic and Global Transformations: Making or Breaking International Orders? *Outlines of Global Transformations: Politics, Economics, Law*, 13(5), pp. 20–37.

Simons, G., Danyk, Y. & Maliarchuk, T. (2020), Hybrid War and Cyber-attacks: Creating Legal and Operational Dilemmas, *Global Change, Peace & Security*, 32(3), pp. 337–342.

Simons, G. & Glaser, M. (2019), New Cold War and the Crisis of the Liberal Global Order, *Outlines of Global Transformations: Politics, Economics, Law*, 12 (3), pp. 61–77.

Taylor, P. M. (2003), *Munitions of the Mind: A History of Propaganda from the Ancient World to the Present Day*, 3rd Edition, Manchester: Manchester University Press.

Waldman, T. (2021), *Vicarious Warfare: American Strategy and the Illusion of War on the Cheap*, Bristol: Bristol University Press.

Western, J. (2005), *Selling Intervention and War: The Presidency, the Media, and the American Public*, Baltimore: Johns Hopkins University Press.

Whaley, B. (1982), Toward a General Theory of Deception, *Journal of Strategic Studies*, 5(1), pp. 178–192.

Whyte, C., Thrall, A. T. & Mazanec, B. M. (Eds.) (2021), *Information Warfare in the Age of Cyber Conflict*, New York: Routledge.

Willcox, D. R. (2005), *Propaganda, the Press and Conflict: The Gulf War and Kosovo*, London: Routledge.

Zollmann, F. (2017), *Media, Propaganda and the Politics of Intervention*, New York: Peter Lang.

Index

Printed in the USA
CPSIA information can be obtained
at www.ICGtesting.com
LVHW010955070124
768340LV00006B/245